"This collection of essays gathers together important strands in the current studies of ecosomatics. It includes many 'practice pages' that open doors to the feelings that have generated the commitment of the writers to creating common grounds for deep conversation about the way people live in the ecologies of the world. The combination of affective strength, so difficult to articulate, with practical exercises—such as the many approaches to breathing as a form of ecoproprioception—will draw readers into places/geographies where artmaking and philosophy join together and suggest new languages for thinking and talking about engaging with this Earth."

Lynette Hunter, *Professor of Theatre and Dance,*
University of California, Davis

"This seminal collection of essays maps the contours of an emerging field: ecosomatics. At the intersection of dance studies, movement studies, philosophy, and ecology, ecosomatics encourages ways of thinking and doing that cultivate a human's sensory awareness of their bodily enmeshment in enabling places and worlds—nexuses of material relationships which call for respect, reciprocity, and responsibility. In essays written by an international cast of contributors, ecosomatics demonstrates its fierce commitment to social and environmental justice; a ready embrace of Indigenous knowledges, histories, and rights; thoughtful engagement with established fields of phenomenology, eco-philosophy, and dance studies; a lived, dialectical production of theory and practice, and an overriding mission to participate as consciously as possible in generating worldviews and bodily practices that sensitize humans to the ongoing health and wellbeing of the Earth in us and around us."

Kimerer L. LaMothe, *PhD, author of* Why We Dance:
A Philosophy of Bodily Becoming

"*Geographies of Us* provides an exciting snapshot of a diversifying field: of the different methods, playful encounters, bodymind approaches, and land politics that make up the contemporary ecosomatic inquiry, with plenty of invitations to join in the dance. At its heart, this collection is about local and grounded connection, about reaching out—in intergenerational liveliness and critterly entanglement, in touch and in movement, in human and more-than-human worlds."

Petra Kuppers, *author of* Eco Soma: Pain and Joy in Speculative
Performance Encounters; *Anita Gonzalez Collegiate Professor of*
Performance Studies and Disability Culture, University of Michigan

"I consider this the most important work to emerge in interdisciplinary dance/performance studies this century. The depth and quality of engagement available to the reader in these pages has the potential to widely transform thought, practice, institutions, environments, and the lived relations between."

Karen Bond, *Chair of Dance, Temple University*

GEOGRAPHIES OF US

Geographies of Us: Ecosomatic Essays and Practice Pages is the first edited collection in the field of ecosomatics.

With a combination of essays and practice pages that provide a variety of scholarly, creative, and experience-based approaches for readers, the book brings together both established and emergent scholars and artists from many diverse backgrounds and covers work rooted in a dozen countries. The essays engage an array of crucial methodologies and critical/theoretical perspectives, including practice-based research in the arts, especially in performance and dance studies, critical theory, ecocriticism, Indigenous knowledges, material feminist critique, quantum field theory, and new phenomenologies. Practice pages are shorter chapters that provide readers a chance to engage creatively with the ideas presented across the collection. This book offers a multidisciplinary perspective that brings together work in performance as research, phenomenology, and dance/movement; this is one of its significant contributions to the area of ecosomatics.

The book will be of interest to anyone curious about matters of embodiment, ecology, and the environment, especially artists and students of dance, performance, and somatic movement education who want to learn about ecosomatics and environmental activists who want to learn more about integrating creativity, the arts, and movement into their work.

Sondra Fraleigh is Professor Emeritus, Department of Dance, State University of New York, Brockport, U.S.A.

Shannon Rose Riley is Professor of Humanities & Creative Arts, San José State University, U.S.A.

Routledge Studies in Theatre, Ecology, and Performance

Studies in Theatre, Ecology, and Performance (STEP) explores intersections between theatre, ecology, and performance studies in a global context. Amidst the current pandemic, the threat of environmental degradation—in particular climate change, challenges to clean water and air, food insecurity, and species extinction—seems to lurk just off stage. Yet, ecology and the environment remain overriding concerns of the Anthropocene and the pandemic is incriminated in received interspecies practice. Theatre and performance studies can add unique contributions to current conversations around future human behaviours. STEP seeks proposals from those working in intersecting fields of theatre, performance studies, and ecology across the globe. Ecology can be defined as the study of organisms and their interactive environment or "home."

The Stage Lives of Animals
Zooesis and Performance
Una Chaudhuri

Earth Matters on Stage
Ecology and Environment in American Theater
Theresa J. May

Choreographing Dirt
Movement, Performance, and Ecology in the Anthropocene
Angenette Spalink

For more information on the series please visit the series page: www.routledge.com/Routledge-Studies-in-Theatre-Ecology-and-Performance/book-series/STEP#

GEOGRAPHIES OF US

Ecosomatic Essays and Practice Pages

Edited by Sondra Fraleigh and Shannon Rose Riley

Routledge
Taylor & Francis Group

LONDON AND NEW YORK

Designed cover image: Daniel Ìgbín'bí Coleman

First published 2024
by Routledge
4 Park Square, Milton Park, Abingdon, Oxon OX14 4RN

and by Routledge
605 Third Avenue, New York, NY 10158

Routledge is an imprint of the Taylor & Francis Group, an informa business

British Library Cataloguing-in-Publication Data
A catalogue record for this book is available from the British Library

Library of Congress Cataloging-in-Publication Data
Names: Fraleigh, Sondra Horton, 1939– editor. | Riley, Shannon Rose, editor.
Title: Geographies of us : ecosomatic essays and practice pages /
[edited by] Sondra Fraleigh & Shannon Rose Riley.
Description: Abingdon, Oxon ; New York, NY : Routledge, 2024. |
Series: Routledge studies in theatre, ecology, and performance |
Includes bibliographical references and index. |
Identifiers: LCCN 2023044004 (print) | LCCN 2023044005 (ebook) |
ISBN 9781032479996 (hardback) | ISBN 9781032488271 (paperback) |
ISBN 9781003390985 (ebook)
Subjects: LCSH: Somesthesia. | Mind and body. | Human beings–Effect of environment on. | Human ecology.
Classification: LCC QP448 .G46 2024 (print) | LCC QP448 (ebook) |
DDC 128/.2–dc23/eng/20231220
LC record available at https://lccn.loc.gov/2023044004
LC ebook record available at https://lccn.loc.gov/2023044005

ISBN: 9781032479996 (hbk)
ISBN: 9781032488271 (pbk)
ISBN: 9781003390985 (ebk)

DOI: 10.4324/9781003390985

Typeset in Sabon
by Newgen Publishing UK

LAND ACKNOWLEDGMENTS

Sondra writes from Saint George, Utah, the traditional home and unceded territories of the Ute, Goshute, Paiute, Navajo (Diné), and Shoshone peoples.

Shannon writes from Fremont, California, the traditional home and unceded territories of the Muwekma Ohlone peoples—the place where six nations were enslaved at Mission San Jose—the Lisjan (Ohlone), Karkin (Ohlone), Bay Miwok, Plains Miwok, Delta Yokut, and Napian (Patwin).

We acknowledge that we are settlers on these lands and that we have gratitude and love for the places we occupy.

CONTENTS

CONTRIBUTORS

Adesola Akinleye is Assistant Professor of Dance at Texas Woman's University in Denton, Texas. Akinleye creates live performance, installation, and film and is the author of *Dance, Architecture, and Engineering* (2021), and the editor of *Narratives in Black British Dance: Embodied Practices* (2018) and *(Re:) Claiming Ballet* (2021).

Annette Arlander is Visiting Researcher in the Academy of Fine Arts at the University of the Arts, Helsinki, Finland. She is a member of the editorial board of *Journal for Artistic Research*; a member of the executive committee of International Federation for Theatre Research; and co-convener of the Artistic Research Working Group of Performance Studies International.

Karen Barbour is Associate Professor and Head of Te Kura Toi School of Arts at the University of Waikato in New Zealand. Barbour is the author of *Dancing across the Page: Narrative and Embodied Ways of Knowing*. She publishes widely in journals and is co-editor of several volumes.

Christine Bellerose is an artist-researcher, a somatics professional, and Adjunct Faculty in the Department of Theatre, University of Ottawa. Bellerose is the author of *On the Lived, Imagined Body: A Phenomenological Praxis of a Somatic Architecture* and co-founder of the Somatic Engagement Working Group with the Canadian Association for Theatre Research.

Tria Blu Wakpa is Assistant Professor of Dance Studies in the Department of World Arts and Cultures/Dance at the University of California, Los Angeles. Her research and teaching center community-engaged, decolonizing, and dance studies methodologies. She is a scholar, poet, and practitioner of Indigenous dance, North American Hand Talk, martial arts, and yoga.

Edward S. Casey is Distinguished Professor of Philosophy at SUNY Stony Brook University in New York. Casey is the author of 11 books, most of them in the general area of phenomenology. His most recent is titled *Turning Emotion Inside Out: Affective Life Beyond the Subject* (2022).

Daniel Ígbín'bí Coleman is Assistant Professor of Women's, Gender, and Sexuality Studies at Georgia State University, Atlanta. Coleman (he/they) is the author of, *Refusals and Reinventions: Engendering New Indigenous and Black Life Across the Americas*. Other publications appear in *QED: A Journal of GLBTQ World Making* and *Transgender Studies Quarterly*. They have performed internationally with La Pocha Nostra.

Alison (Ali) East is former Chair of Dance Studies at The University of Otago, New Zealand, and is Adjunct Professor in Dance Ethnography at Tezpur University, Assam. She has created numerous eco-political works and taught and published on intuitive ecosomatic practice. Her book *Teaching Dance as if the World Matters* was published in 2011.

Sondra Fraleigh is Professor Emeritus of Dance and Somatic Studies at State University of New York, Brockport, and the Founder and Director of Eastwest Somatics and Shin Somatic Methods. She is the author and editor of numerous books, including *Back to the Dance Itself: Phenomenologies of the Body in Performance* and *Moving Consciously: Somatic Transformations through Dance, Yoga, and Touch*.

Nathalie Guillaume is a doctor of acupuncture and oriental medicine, a somatics professional, and a licensed acupuncturist practicing in New York City and Hawaii. She is certified with the Mark Morris Dance Company where she has provided specialized rehabilitation classes for Parkinson's Disease patients as a dance therapist at the Juilliard School.

Debra Lacey is a dancer, choreographer, and photographer. Her long career in dance includes extensive training in the Graham Modern and the Vaganova Ballet techniques. She was the Senior Associate Director for LVCDT and the Founder of RUPa art Studio. Her film shorts and photography have won awards in festivals and photography venues.

Glen A. Mazis is Distinguished Professor of Philosophy and Humanities Emeritus at Penn State, Harrisburg. A philosopher and poet, he is the author of numerous works, including *Humans, Animals, Machines: Blurring Boundaries* (2008), *Merleau-Ponty and the Face of the World* (2017), *Bodies of Time and Space* (2022), and *The River Bends in Time: Poems* (2015).

Missy Pfohl Smith is Director of the Program of Dance and Movement and the Institute for the Performing Arts at the University of Rochester, New York. Pfohl Smith is also the Founder and Director of BIODANCE, a collaborative dance company in Rochester. Her choreographic and performance practice is influenced by the environment and time spent in nature.

George Quasha is an artist who explores a performative "principle of axiality" across mediums, including language, sound, and bodywork. A 2006 Guggenheim Fellow in Video Art, his poetry includes *Somapoetics* (1973), *Ainu Dreams* (1999), and *Waking from Myself (preverbs)* (2022). Co-editor of *America a Prophecy* (1973/2012), his collaborations with artist and co-publisher Susan Quasha include Station Hill Press.

Shannon Rose Riley is Professor of Creative Arts and Humanities at San José State University. She is the author of *Performing Race and Erasure: Cuba, Haiti, & US Culture, 1898–1940* and co-editor of *Mapping Landscapes for Performance as Research*. Her creative works have been staged and exhibited internationally; she is a member of ONO.

Fei Shi (石飛) is Professor of Arts and Humanities at Quest University, Squamish, British Columbia, Canada. Shi's main academic interests are Chinese theater and film, women writers, and gender and queer theory. He teaches courses in film studies, theater, literature, and Chinese language and has directed and produced numerous plays.

Stephen Smith is Professor in the Faculty of Education at Simon Fraser University in Canada. His scholarly work focuses on curricular and instructional matters of health and physical activity promotion, and on eco-pedagogical processes of active wellness and interactive well-becoming. An ongoing line of scholarship concerns interspecies relations.

Lani Weissbach is a teacher, choreographer, and performer in the field of contemporary dance, somatics, and yoga, and her work has been presented at numerous venues nationally. She is currently the Director of Embodied Learning for the Indianapolis Movement Arts Collective and offers workshops in somatic movement and dance throughout the region.

FIGURES

EAR AND HEART TO EARTH WITH GRATITUDE (ACKNOWLEDGEMENTS)

Sondra Fraleigh and Shannon Rose Riley both attended the 2003 International Federation for Theatre Research Conference in Kraków, Poland. Unfortunately, they presented at the same time on different panels, but Warren Fraleigh, Sondra's husband, came to see Shannon's paper on Authentic Movement and Butoh and then introduced them to one another afterward (Shannon, who had quoted Sondra at least twice in her paper, remains ever grateful to Warren for this introduction!). For a couple of years afterward, Shannon pursued the first few levels of training in Sondra's Eastwest Somatics Institute. But then life intervened. Indeed, it wasn't until the COVID-19 pandemic brought Sondra's workshops to Zoom that Shannon continued with that training in June of 2021!

In October of that same year, Eastwest Somatics Institute held a conference in Elmira, New York, on the theme of Somatic Practices in Nature. Sondra and Shannon had already been discussing the need for an edited collection in the area of ecosomatics but it was clear from the various presentations at the conference that they had struck upon a very good idea. Sondra and Shannon would like to acknowledge all of the folks at the last several Eastwest conferences for being a part of the conversation on this topic. They are especially grateful to conference-goers who wrote new works for this collection!

Indeed, the editors are incredibly blessed by the contributions of all of the book's authors (in order of appearance): Tria Blu Wakpa, Daniel Ìgbín'bí Coleman, Lani Weissbach, Karen Barbour, Stephen Smith, Fei Shi, Alison (Ali) East, Annette Arlander, Adesola Akinleye, Christine Bellerose, Nathalie Guillaume, Glen A. Mazis, Debra Lacey, George Quasha, Missy

Pfohl Smith, and Edward S. Casey! Without every one of you, this collection would not exist. As editors, Sondra and Shannon are also grateful for one another—for camaraderie, for lots of reading and editing, and for each other's chapters, which help make up the whole. The editors are incredibly delighted to be working with the very fine folks at Routledge Studies in Theatre, Ecology, and Performance series; thank you for your belief in this project! The editors also thank Cathy Hannabach and the folks at Ideas on Fire for the wonderful index.

The editors are also thankful to those who have given us permission to reproduce their works in these pages (also in order of appearance): ArcGIS Online hosted by ESRI (Esri, USGS | Esri, FAO, NOAA, USGS), George Blue Bird (Oglala Lakota), Daniel Ìgbín'bí Coleman, Charles Borowicz, Rodrigo Hill, Michele Black, Carson Au, Michele Black and Stephen Smith, Ilya Noé, Amanda Ribas Tugwell, Shannon Rose Riley, Fei Shi, Alix Melchy, Ali East, Annette Arlander, Cheniece Warner, Christine Bellerose, Becky Yee, Debra Lacey, Ella S. Smith, Sondra Fraleigh, Tom Gallo, Alycia Bright Holland, and Mark Howe.

The exquisite cover image is used with permission of the artist and author, Daniel Ìgbín'bí Coleman.

We dedicate this book to the memory of Warren Fraleigh (1925–2022), who brought the first two of "us" in this collection together.

INTRODUCTION

Locating Geographies of Us

Sondra Fraleigh and Shannon Rose Riley
St. George, Utah, U.S.A: 37.0941° N, 113.5749° W
Fremont, California, U.S.A.: 37.5483° N, 121.9886° W

Locating "Us"

Geographies of Us: Ecosomatic Essays and Practice Pages is the first edited collection in the multidisciplinary field of ecosomatics. With a combination of essays and practice pages that provide a variety of scholarly, creative, and experience-based approaches for readers, the book brings together established scholars and artists like Adesola Akinleye, Annette Arlander, Karen Barbour, Edward S. Casey, Alison (Ali) East, Sondra Fraleigh, Debra Lacey, Glen A. Mazis, George Quasha, Shannon Rose Riley, Fei Shi, Missy Pfohl Smith, Stephen Smith, and Lani Weissbach with more emergent voices like Christine Bellerose, Tria Blu Wakpa, Daniel Ígbín'bí Coleman, and Nathalie Guillaume. International in scope, *Geographies of Us* brings together work being done in, or with deep roots in, Canada, China, Finland, Germany, Haiti, Japan, México, New Zealand, South Africa, Sweden, the United Kingdom, and the United States. Each chapter, whether essay or practice pages, offers one or more sets of GPS coordinates below the title in order to index the locations of ecosomatic research and practice or otherwise locate "us" where we write, move, or perform.[1] From these coordinates, we have created a world map that constitutes one of the layers of "geographies of us" that this book explores; a map that in some way marks the book's places, people, plants, assemblages, and critters (*see* Figure 0.1 and the detail, Figure 0.2).

David Wood, a founding member of the early 1970s Oxford Group of philosophers who promoted animal rights, speaks of the need to expand "our sense of who 'we' are so as to break down the apparent

DOI: 10.4324/9781003390985-1

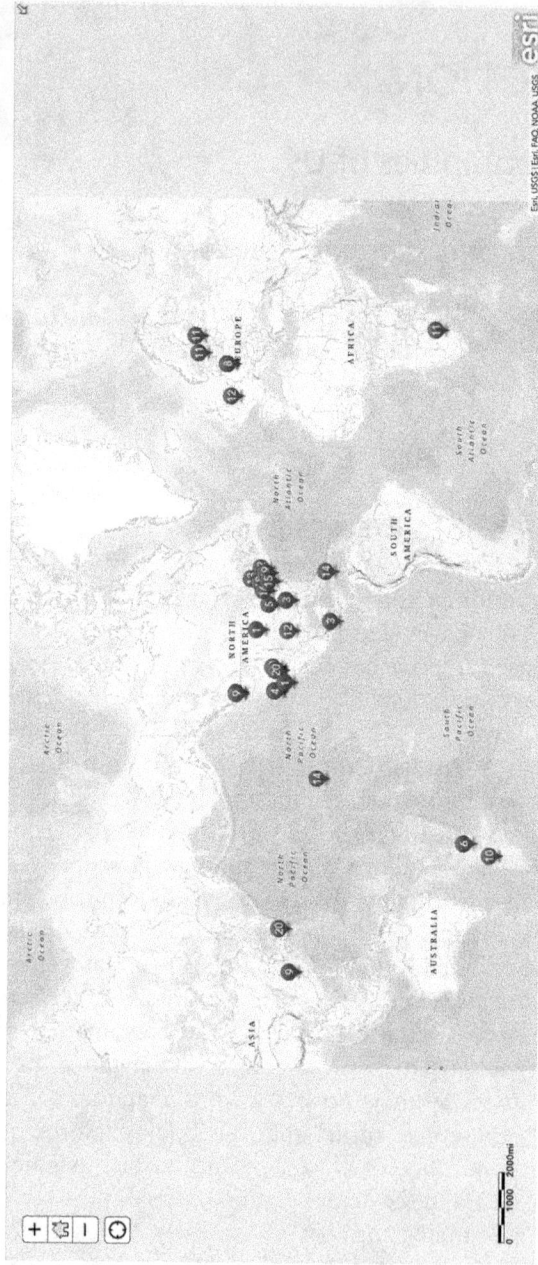

FIGURE 0.1 *Geographies of Us*, map of GPS coordinates by chapter number, made by Shannon Rose Riley with ArcGIS Online hosted by ESRI (Esri, USGS I Esri, FAO, NOAA, USGS), 2023.

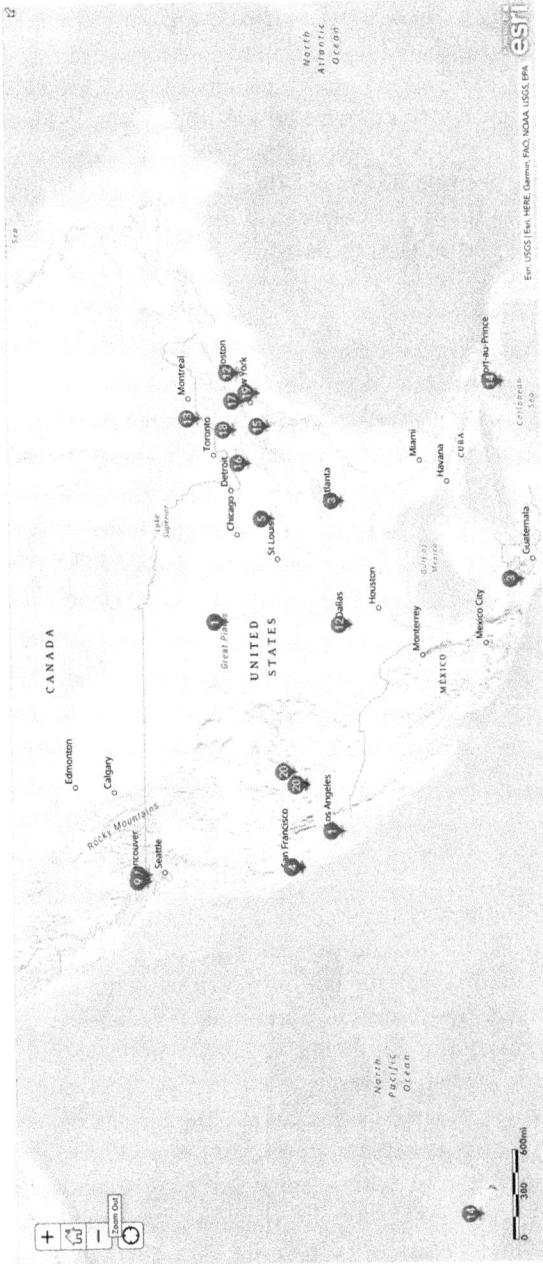

FIGURE 0.2   *Geographies of Us*, detail map of GPS coordinates by Chapter number, made by Shannon Rose Riley with ArcGIS Online hosted by ESRI (Esri, USGS | Esri, FAO, NOAA, USGS), 2023.

opposition between altruism and self-interest in coming to see the fates of humans and other species as linked."[2] It is from this perspective that this book reimagines a "geographies of us" and, in particular, how the term, "us," must be resituated in relation to each other, to the more-than-human world, and to all of materiality, if any of "us" are to survive, or thrive, in the geologic epoch known as the Anthropocene.[3] The main title, *Geographies of Us*, intends to remap that pronoun—"us"—by upending its anthropocentric position. It extends the notion of "us" to become more inclusive of the more-than-human world. With the term "geographies," we hope to foreground the impacts of human activity on Earth in the context of the Anthropocene and to map the extended community of authors that comprise this collection. We span many locations, multiple decades, and many races and ethnicities; we are trans and straight and gay and genderqueer; we are artists and scholars and somatics practitioners; our bodies have differing capacities; and we are all drawn to this work and to stretching the idea of who or what counts as "us." Several of us confront this "us-ness" directly in our chapters.

While we are proud that our map includes many locations around the world, it is apparent that there is much to be done to gather the ecosomatic approaches going on in other geographies. Some of the limitations have to do with language and translation; others are more structural, regarding the limits of a single edited collection, but we want to acknowledge that the need to gather more is clear. To be sure, geographies of us must remain ever emergent or run the risk of becoming reified and exclusive.

Locating Ecosomatics

Ecosomatics is a multidisciplinary area of practice and research that connects somatic practices in dance, movement, performance studies, and disability cultures with ecological attunement or awareness. The ecological turn in philosophy and the arts is deeply entangled and wildly rhizomatic. Its structures extend far into various fields and clusters of knowledge. To be sure, awareness of the connections between the human, the more-than-human, and the living environment is deeply rooted in global Indigenous knowledges even as these bodies of wisdom have been long ignored or worse, erased. Ecological awareness was brought to bear on academic and creative disciplines in the late twentieth century as many thinkers and artists began to acknowledge and grapple with the negative impacts of human activity on planet Earth. Work in eco-phenomenology, eco-philosophy, and feminist science studies, in particular, have made powerful contributions, including inverting the anthropocentrism of traditional philosophy and bringing environmental concerns to the center of philosophical discourse. These scholars are essential to work being done in ecosomatics today and include David Abram, Karen Barad, Jane Bennett, Edward S. Casey, Jacques Derrida, Donna Haraway, Bruno Latour, Michael Marder, and David Wood, to name just a few.[4] Indeed, the careful reader will notice that these thinkers are cited many times throughout this collection. The eco-turn is equally indebted to Nordic eco-philosophy and the articulation of "deep ecology" by Arne Næss, to eco-psychology as first articulated by Theodore Roszak, and to various forms of feminist and queer critical theory, including ecocriticism, ecofeminism, and queer ecologies. Ecosomatics also has deep roots in the somatic movement and dance work of Sondra Fraleigh and others. Her book, *Land to Water Yoga* (2009), and the collection, *Moving Consciously: Somatic Transformations through Dance, Yoga, and Touch* (2015), have been particularly influential on a new generation of scholars and artists in somatic movement and ecosomatics.[5]

The ecosomatic paradigm is the primary theoretical framework for this book. Each of the authors is engaged in analysis, argument, or creative practice that challenges the notion of a human-centric "us," by moving forward into other kinds of affinities and naturecultures; each examines what an ecosomatic ethics might afford at a time when our planet and many species are in crisis and resilience is needed. As suggested above, several of the essays and practice pages share theoretical stances that emerge from feminist science studies and feminist materialism, eco-philosophy, eco-phenomenology, and eco-psychology, and practice-based research in the arts, especially dance, movement, and performance as research.

The term, "ecosomatic" or "ecosomatics," has developed in several directions over the last 15 or so years. Rebecca Enghauser first used the term in a 2007 scholarly article in the *Journal of Dance Education* titled "The Quest for an Ecosomatic Approach to Dance Pedagogy."[6] Enghauser uses an ecological model to reframe essential somatic concepts; she argues that ecosomatics focuses on sensitization of the body's relationship not only with self and others but also with the environment and the natural world. Susan Bauer uses the term in an essay in *Conscious Dancer* (2008). For Bauer, the term "opens an inspiring dialogue about the dynamic relationship between the earth and the living beings inhabiting it."[7] In 2010, the Walker Art Museum published an interview with the interdisciplinary artist, Olive Bieringa, on what she calls the "ecosomatics classroom." For Bieringa, "[e]cosomatics is an emerging interdisciplinary field which connects embodiment practices such as dance and the healing arts with ecological consciousness."[8] Matthew Nelson's essay, "Embodied Ecology: The Ecosomatics of Permaculture," was published in *Choreographic Practices* (2018) and draws parallels between somatic and permaculture practices.[9]

In 2013, Matthew Cella used the concept to frame up a literary criticism that brings together disability studies and ecocriticism; his book, *Disability and the Environment in American Literature: Toward an Ecosomatic Paradigm* (2017), is the first monograph to use the term in its title.[10] In the last few years, there have been several themed journals on the topic, primarily from a performance studies perspective: a 2018 special edition of *Choreographic Practices* titled "Performing Ecologies in a World in Crisis," co-edited by Sondra Fraleigh and Robert Bingham; a special issue of *CSPA Quarterly 31*, a publication of the Centre for Sustainable Arts, on the theme of "EcoSomatics," edited by Petra Kuppers (January, 2021); and a themed issue titled "Performing (in) Place: Moving on/with the Land" in a leading journal in performance studies, *Performance Matters*, co-edited by Jenn Cole and Melissa Poll (December, 2021).[11] In 2021, some of the first MFA and PhD theses were completed in the area, Elise Nuding wrote the first situated critique of ecosomatics in the *Journal of Dance and Somatic Practices* and Petra Kuppers' book, *Eco Soma: Pain and Joy in Speculative Performance Encounters*, was published.[12] Kuppers' is the first monograph that contributes to the emerging discourse on ecosomatics from a performance studies perspective; it documents and theorizes her innovative "eco soma" method and her practice within dance, speculative performance, and poetry. Like Cella, Kuppers shares a focus on disability cultures. Most recently, and from a less scholarly perspective, Cheryl Pallant's book, *Ecosomatics: Embodiment Practices for a World in Search of Healing*, was published by a trade publisher specializing in spirituality in 2023.[13]

Geographies of Us: Ecosomatic Essays and Practice Pages contributes to this growing body of literature even as it is forward-looking. It does not aim to take stock of where ecosomatics has been as much as it aims to chart where it is going. The essays engage an array of crucial methodologies and critical/theoretical perspectives, including practice-based research in the arts, especially in performance and dance studies; critical theory; ecocriticism; material feminist critique; Barad's quantum field theory; and new phenomenologies. Practice pages are shorter chapters that provide the reader a chance to engage creatively with the ideas presented in and across the collection. Through practice-based research, the methodologies of these short pieces are guided by somatic designs in which movement matters and ecology is central. Practices pages develop somatic performances that curry corporeal mingling with material nature and ethics as lived and reiterated through networks of naturecultures. These short chapters draw upon the aesthetic and performance backgrounds of their authors and often include methods of autoethnography. In particular, this book offers a multidisciplinary perspective that brings together work in performance as research, phenomenology, and dance/movement; this is one of its significant contributions to the area of ecosomatics.

Structure of the Book

Geographies of Us is comprised of 20 chapters—10 scholarly essays and 10 sets of practice pages—divided into 4 sections:

Part I: Enworlding, Rewilding, Decentering, Transing/ Pluraling,
  Performing, Attending to, Dancing
Part II: Horse, Lion, Queer Animal, Skin
Part III: Forest, Tree, Carbon, Stone
Part IV: Place, Plasma, Pluriverse, Potato

The section themes constitute a theoretical poetics that shapes the book and which emerged from the chapters themselves. The theme of the first section, "Enworlding, Rewilding, Decentering, Transing/ Pluraling, Performing, Attending to, Dancing," offers key verbs from the chapters and practice pages in the collection. These verbs are understood as ways of being and doing that constitute the ecosomatic in a performative sense. We conceive of these eight verbs as a poem, perhaps even a chant, in order to foreground both poetics and performance as key methodologies of ecosomatics and the ecosomatic paradigm. The remaining three thematic sections outline a poetic taxonomy that aims to complicate and subvert the classical anthropocentric structures of man/animal/plant/mineral. Each section theme speaks to the chapters they contain but also creates fruitful poetic frictions across the sections. The collection offers more essays than practice pages in Parts I and II and then shifts to more practice pages and fewer essays in Parts III and IV. This shift is intentional as ecosomatics is inclined toward action. The inclusion of practice pages across all sections makes clear that creativity and the arts are central to multidisciplinary ecosomatics.

Part I consists of four essays and two sets of practice pages; all six chapters foreground concepts, actions, and processes that constitute aspects of the ecosomatic turn. Among the essays in Part I, Shannon Rose Riley's "A Critical Ecosomatics: Cultivating Awareness and Imagination" outlines several key areas for developing awareness: of one's position in relation to the land; of one's intra-relation with the more-than-human world and vibrant materiality; of one's place within the Anthropocene. In "What Native American Dance Does and the Stakes of Ecosomatics," Tria Blu Wakpa makes clear that "a critical and socially-just conception of ecosomatics must acknowledge and reckon with ongoing colonization and its oppression of Indigenous peoples." In "Material/Material: Thousandfold

Somas and Poetry of Emergence," Sondra Fraleigh considers enworlding, dancing, and ecosomatics from a phenomenological perspective and in "Shaky Islands and Rising Seas: Dancing Entanglements in the Global South," Karen Barbour uses Karen Barad and others to reflect on environmentally centered somatic work and the impacts of ecosomatics in education. The two sets of practice pages include Daniel Ígbín'bí Coleman's "Ecosomatic Performance Research for the Pluriverse,"—an exploration of his ecosomatic performance practice in Chiapas, México, that includes suggestions for the reader—and Lani Weissbach's "Decentering the Human through Butoh," which offers an embodied performance map that the reader can use to explore decentering by playing with edges and being off balance in heightened relation to one's environment.

The two essays and two sets of practice pages in Part II are primarily concerned with reimagining the human animal in relation to the more-than-human world, and contemplating the limits of perceived boundaries, whether through the periphenomenology of affect, intra-species attunement, or via the porous interface of skin. Stephen Smith's essay on "rewilding" our horse senses calls for cultivating what he calls "*critter*cal somaticity" in relation to playing and working with horses, while Riley's practice pages, "Moving with Cats," document her performance of *Lie-in (Lion)* at Month of Performance Art-Berlin in 2013 and practicing Authentic Movement with a housecat in a private multispecies dwelling; she provides the reader with tips on how to incorporate moving with cats into one's ecosomatic practice. In "Embodying Islands: Ecosomatics and the Transnational Queer," Fei Shi shares a deeply personal narrative of his relationship with two islands—one in Canada and one in China—in order to reimagine possibilities for ecosomatics. The final chapter in Part II is a set of practice pages, "Skinbody and Skin of the Earth" by New Zealand dance artist and scholar, Ali East, in which she offers a beautifully theorized and rich ecosomatic performance map for the reader.

The three sets of practice pages and one essay in Part III poetically reimagine the categories of plant, mineral, and the man-made in terms of global Indigenous knowledges, quantum field theory, and the more-than-human. Annette Arlander explores standing in tree pose with trees in various locations in Finland, Sweden, and South Africa; she invites the reader to develop their own practice of balancing and building relationships with local trees. Adesola Akinleye's essay reflects on their dance choreographies at the waters' edge in London and Boston through which they develop a particularly moving articulation of self as "river-me" in light of the concept of "fearless belonging." The section concludes with two practice pages: Christine Bellerose's "How to Apprentice with Land in Enchanted Kinship," which documents her recent ecoperformance work at the junction of the Ottawa and Gatineau Rivers on unceded Indigenous land in Rockcliffe Park, Ottawa, and offers the reader several short, accessible activities toward developing what she calls "enchanted kinship" with place. In "Feel the Carbon Under your Footprint–Indigenous Approaches to Grounding," Nathalie Guillaume, a Doctor of Acupuncture and Oriental Medicine (D.A.O.M.), offers suggestions that draw from Chinese medicine and Indigenous practices for the reader to connect to the ground.

In the final section of the book, three sets of practice pages and three essays invite readers to consider their place in relation to the unseen, the very small, the incredibly vast, and the everyday capacities of ecosomatics. The alliterative set of terms that make up this final section theme, "Place, Plasma, Pluriverse, Potato," playfully frames the ecological/biological materiality of plasma and the temporal-spatial fluidity of the pluriverse alongside the particularities of place and the simple potato. There is a wonderful quantum poetry in this vast spiraling out of plasma and pluriverse and sudden return to potato's muddy earth. And with pluriverse, we harken back to one of the first chapters in the collection. The section's essays are some of the more philosophical in tone and each is followed by a set of practice pages that further complicates and opens up possible intra-connections and meanings. In "My Place is a Chiasmatic Dance," philosopher Glen A. Mazis offers personal and phenomenological reflection on what constitutes home. Debra Lacey's practice pages, "Cosmic Plasma Echoing in (Our) Place," offer the reader the opportunity to further reflect on place and to move in relation to the fourth state of matter. In his experimental essay, artist and author, George Quasha, coins and develops the useful term, "ecoproprioception," which he presents as self-perception intrinsically interwoven with various registers of environmental awareness and articulated poiesis. Missy Pfohl Smith's practice pages document three ecosomatic dances in a forgotten wooded landscape in the Finger Lakes region of western New York and offer reflection upon their sudden destruction just weeks after the performance. Pfohl Smith offers the reader a chance to reflect on loss and growing anxiety and to engage in ecosomatic practice by noticing and yielding through embodied responses. The last philosophical essay in the collection belongs to well-known phenomenologist, Edward S. Casey; titled "Awe and Empathy," it speaks with great clarity and beauty and perhaps could only be written from a certain vantage point of experience. The final chapter is Sondra Fraleigh's "Enworlding Place Dances and Potatoes," a deeply poetic autoethnographic reflection on her performance of place in Snow Canyon, Utah; Yokohama, Japan; and her original home, Circleville, Utah. Fraleigh's practice pages invite the reader to also reflect upon their own roots of belonging and their own Place Dances.

Thoughts on How to Use This Book

Begin anywhere. You can start at the front and work to the back of the book or you can open serendipitously to a random page or section. You can focus on a particular thematic section and read all of the essays and practice pages contained therein. You might begin with all of the practice pages—or you can begin by reading the essays. There is no proper way to use this book. That said, we do recommend that you explore the practice pages with friends and other readers or that you pre-record yourself reading the practice section of those chapters aloud so that you can follow along with the instructions as you move.

This book invites you to think deeply and also to move—in whatever way movement is accessible to you. All bodies and all abilities are not only welcome but are absolutely necessary to this work. Above all, we invite you to experience your own ecosomatic exploration through the pages of this book and to enjoy your own process of learning and moving in newly found intra-relation.

Notes

1 The GPS coordinates in this book are not meant to publish any private address. In cases where coordinates would mark a private home, we use instead the author's university, a city building, or a more generic local point.
2 David Wood, *Thinking Plant Animal Human: Encounters with Communities of Difference* (Minneapolis: University of Minnesota Press, 2020), 151.
3 The term, "more-than-human world" was coined by David Abram in *The Spell of the Sensuous: Perception and Language in a More-than-Human World*, 2017 ed. (New York: Vintage Books, 1996), ix.
4 Abram, *Spell of the Sensuous*; Bruno Latour, *Pandora's Hope: Essays on the Reality of Science Studies* (Cambridge, MA: Harvard University Press, 1999); Jacques Derrida, "The Animal That Therefore I Am (More to Follow)," *Critical Inquiry* 28, no. 2 (Winter, 2002): 369–418; Donna J. Haraway, *When Species Meet* (Minneapolis: University of Minnesota Press, 2007); Karen Michelle Barad, *Meeting the Universe Halfway: Quantum Physics and the Entanglement of Matter and Meaning* (Durham: Duke University Press, 2007); Michael Marder, *Plant-Thinking: A Philosophy of Vegetal Life* (New York: Columbia University Press, 2013); Donna J. Haraway, *Staying with the Trouble: Making Kin in the Chthulucene* (Durham: Duke University Press, 2016); Wood, *Thinking Plant Animal Human*; Edward S. Casey, *Turning Emotion Inside Out: Affective Life beyond the Subject* (Evanston: Northwestern University Press, 2022).
5 Sondra Fraleigh, *Land to Water Yoga: Shin Somatics Moving Way* (New York and Bloomington: iUniverse, Inc., 2009); Sondra Fraleigh, *Moving Consciously: Somatic Transformations through Dance, Yoga, and Touch* (Urbana: University of Illinois Press, 2015).

6 Rebecca Enghauser, "The Quest for an Ecosomatic Approach to Dance Pedagogy," *Journal of Dance Education* 7, no. 3 (2007). https://doi.org/10.1080/15290824.2007.10387342.

7 Susan Bauer, "Body and Earth as One: Strengthening Our Connection to the Natural Source with Ecosomatics," *Conscious Dancer* 2 (Spring, 2008): 8.

8 Ashley Duffalo, "Olive Bieringa on the Ecosomatics Classroom," Walkerart.org, 2010, accessed September 12, 2021. www.walkerart.org/magazine/olive-bieringa-on-the-ecosomatics-classroom.

9 Matthew Nelson, "Embodied Ecology: The Ecosomatics of Permaculture," *Choreographic Practices* 9, no. 1 (April 1, 2018). https://doi.org/10.1386/chor.9.1.17_1. *See also* Nala Walla, "Ecosomatics at Work and Play: In the Landscape," *playGROUND* 1.0 (2009). http://www.bcollective.org/ESSAYS/playGROUND.v1.0.pdf

10 Matthew J. C. Cella, "The Ecosomatic Paradigm in Literature: Merging Disability Studies and Ecocriticism," *ISLE: Interdisciplinary Studies in Literature and Environment* 20, no. 3 (2013), https://doi.org/10.1093/isle/ist053; Matthew J. C. Cella, *Disability and the Environment in American Literature: Toward an Ecosomatic Paradigm* (Lanham: Lexington Books, 2017).

11 Sondra Fraleigh and Robert Bingham, eds. "Performing Ecologies in a World in Crisis," Special Issue, *Choreographic Practices* 9, no. 1 (2018); Petra Kuppers, ed. "EcoSomatics," Special Issue, *CSPA Quarterly* 31 (2021); Jenn Cole and Melissa Poll, eds. "Performing (in) Place: Moving on/with the Land," Special Issue, *Performance Matters* 7, no. 1–2 (2021).

12 Kimberly "Bela" Watson, "Ecosomatics: The Embodiment of Nature and Other Worlds" (MFA in Dance Saint Mary's College of California, 2021), www.proquest.com/dissertations-theses/ecosomatics-embodiment-nature-other-worlds/docview/2562843919/se-2?accountid=201395; Christine Bellerose, "I Dance Land: An Apprenticeship with Wind and Water: Depatterning Somatic Amnesia, Repatterning Ecosomatic Senses" (Ph.D. York University, 2021), https://yorkspace.library.yorku.ca/xmlui/handle/10315/38698; Elise Nuding, "Approaching Eco-Somatics: A Consideration of Potential Pitfalls and their Implications," *Journal of Dance & Somatic Practices* 13, no. 1–2 (2021), https://doi.org/10.1386/jdsp_00034_1; Petra Kuppers, *Eco Soma: Pain and Joy in Speculative Performance Encounters* (Minneapolis: University of Minnesota Press, 2021).

13 Cheryl Pallant, *Ecosomatics: Embodiment Practices for a World in Search of Healing* (Rochester: Bear & Company, 2023).

Bibliography

Abram, David. *The Spell of the Sensuous: Perception and Language in a More-Than-Human World*. 2017 ed. New York: Vintage Books, 1996.

Barad, Karen Michelle. *Meeting the Universe Halfway: Quantum Physics and the Entanglement of Matter and Meaning*. Durham: Duke University Press, 2007.

Bauer, Susan. "Body and Earth as One: Strengthening Our Connection to the Natural Source with Ecosomatics." *Conscious Dancer* 2 (Spring, 2008): 8–9.

Bellerose, Christine. "I Dance Land: An Apprenticeship with Wind and Water: Depatterning Somatic Amnesia, Repatterning Ecosomatic Senses." Ph.D., York University, 2021. https://yorkspace.library.yorku.ca/xmlui/handle/10315/38698.

Casey, Edward S. *Turning Emotion Inside Out: Affective Life beyond the Subject.* Evanston: Northwestern University Press, 2022.

Cella, Matthew J. C. "The Ecosomatic Paradigm in Literature: Merging Disability Studies and Ecocriticism." *ISLE: Interdisciplinary Studies in Literature and Environment* 20, no. 3 (2013): 574–596. https://doi.org/10.1093/isle/ist053.

———. *Disability and the Environment in American Literature: Toward an Ecosomatic Paradigm.* Lanham: Lexington Books, 2017.

Cole, Jenn, and Melissa Poll, eds.. "Performing (in) Place: Moving on/with the Land." Special Issue. *Performance Matters* 7, no. 1–2 (2021).

Derrida, Jacques. "The Animal That Therefore I Am (More to Follow)." *Critical Inquiry* 28, no. 2 (Winter, 2002): 369–418.

Duffalo, Ashley. "Olive Bieringa on the Ecosomatics Classroom." Walkerart.org, 2010, accessed September 12, 2021. www.walkerart.org/magazine/olive-bieringa-on-the-ecosomatics-classroom

Enghauser, Rebecca. "The Quest for an Ecosomatic Approach to Dance Pedagogy." *Journal of Dance Education* 7, no. 3 (2007): 80–90. https://doi.org/10.1080/15290824.2007.10387342.

Fraleigh, Sondra. *Land to Water Yoga: Shin Somatics Moving Way.* New York; Bloomington: iUniverse, Inc., 2009.

———. *Moving Consciously: Somatic Transformations through Dance, Yoga, and Touch.* Urbana: University of Illinois Press, 2015.

Fraleigh, Sondra, and Robert Bingham. eds. "Performing Ecologies in a World in Crisis." Special Issue. *Choreographic Practices* 9, no. 1 (2018).

Haraway, Donna J., *When Species Meet.* Minneapolis: University of Minnesota Press, 2007.

———. *Staying with the Trouble: Making Kin in the Chthulucene.* Durham: Duke University Press, 2016.

Kuppers, Petra. *Eco Soma: Pain and Joy in Speculative Performance Encounters.* Minneapolis: University of Minnesota Press, 2021.

———. ed. "Ecosomatics." Special Issue. *CSPA Quarterly* 31 (2021).

Latour, Bruno. *Pandora's Hope: Essays on the Reality of Science Studies.* Cambridge, MA: Harvard University Press, 1999.

Marder, Michael. *Plant-Thinking: A Philosophy of Vegetal Life.* New York: Columbia University Press, 2013.

Nelson, Matthew. "Embodied Ecology: The Ecosomatics of Permaculture." *Choreographic Practices* 9, no. 1 (April 1, 2018). https://doi.org/10.1386/chor.9.1.17_1.

Nuding, Elise. "Approaching Eco-Somatics: A Consideration of Potential Pitfalls and Their Implications." *Journal of Dance & Somatic Practices* 13, no. 1–2 (2021): 29–39. https://doi.org/10.1386/jdsp_00034_1.

Pallant, Cheryl. *Ecosomatics: Embodiment Practices for a World in Search of Healing.* Rochester: Bear & Company, 2023.

Walla, Nala. "Ecosomatics at Work and Play: In the Landscape." *playGROUND* 1.0 (2009). www.bcollective.org/ESSAYS/playGROUND.v1.0.pdf.

Watson, Kimberly "Bela". "Ecosomatics: The Embodiment of Nature and Other Worlds." MFA in Dance, Saint Mary's College of California, 2021. www.proquest.com/dissertations-theses/ecosomatics-embodiment-nature-other-worlds/docview/2562843919/se-2?accountid=201395.

Wood, David. *Thinking Plant Animal Human: Encounters with Communities of Difference.* Minneapolis: University of Minnesota Press, 2020.

PART I

Enworlding, Rewilding, Decentering, Transing/ Pluraling, Performing, Attending to, Dancing

1

A CRITICAL ECOSOMATICS

Cultivating Awareness and Imagination

Shannon Rose Riley

Grau Pond, Fremont, California, U.S.A.: 37.5735° N, 121.9847° W

Somatic Awareness, Extended

If somatics invites us to increase awareness of our breath, our embodiment, ourselves, then ecosomatics invites us to increase awareness of ourselves in entangled relation with and as "the more-than-human world" and all of "vibrant matter." The phrase, "more-than-human world," was coined by performer and phenomenologist David Abram as a way to decenter the category of the human from its imagined apex and to reawaken human perception to the magic of what he calls "the many-voiced landscape."[1] In the provocative work, *Vibrant Matter: A Political Ecology of Things*, political theorist Jane Bennett argues further that even inanimate things should be reconsidered in terms of their relation to the human.[2] Her vital materialism is a challenge to "the life-matter binary and its correlate, the machine model of nature."[3] She proposes instead that we begin "encountering the world as a swarm of vibrant materials entering and leaving agentic assemblages."[4]

The ecosomatics I am sketching here comprises philosophies, critical-theoretical positions, methods, practices, and worldviews which rethink the notion of a discrete human subject and instead locate it as constituted within and through various kinds of relations and entanglements. It is indebted to theories and practices in somatic movement but especially to phenomenology and eco-phenomenology, quantum field theory, the material feminisms of Bennett and others, and to various Indigenous knowledges. All of these challenge, in one way or another, Cartesian dualisms in regard to overly simplistic human-animal-plant-matter divides

DOI: 10.4324/9781003390985-3

and notions of a self-possessed human subject with claims to superiority and a singular agency. In ecosomatics, we cultivate awareness of entanglement and intra-connection.

At the level of perception and cognition, we are always already ecosomatic. As I have argued elsewhere, philosopher and neurologist Antonio Damasio "postulates that what we mistakenly call 'the mind' as if it were an object located in the brain, is more accurately an interactive relational process between brain, body, and environment." Mind is a process constituted by what he calls "multiple, parallel, converging streams," which are coded as images that flow throughout the body and brain in response to the environment. For Damasio, mind is "embodied, in the full sense of the term, not just embrained."[5] Mind is also environed— emerging entangled and intra-active with/in place.

Feminist science studies scholar and theoretical particle physicist, Karen Barad, develops a theoretical framework of "agential realism" that reconceives of materiality as both "agentive and intra-active."[6] For Barad, "[b]odies do not simply take their places in the world. They are not simply situated in, or located in, particular environments. Rather, 'environments' and 'bodies' are intra-actively co-constituted."[7] Barad argues that to be entangled is "not simply to be intertwined" as separate entities but rather "to lack an independent, self-contained existence."[8] Individuals do not preexist interaction but instead emerge "through and as part of their entangled intra-relating."[9] This is a performative understanding of being-as-relation.

Also in science studies, philosopher Bruno Latour describes an ecology of human and nonhuman elements and prefers the notion of the "collective" as opposed to the categories of "nature" and "culture."[10] This parallels traditional and contemporary Indigenous knowledges in many ways.

> For many Indigenous peoples, collectives are not anthropocentric [...] they do not exclude animals, plants, and ecosystems as members with the responsibilities of active agents in the world. In many cases, plants, animals, and ecosystems are agents bound up in moral relationships of reciprocal responsibilities with humans and other nonhumans.[11]

The Lakota language represents a similar concept of intra-connection and intra-action with the saying, *Mitákuye Oyás'iŋ*, or, "we are all related."[12] Perhaps it is in this vein that Abram argues, "we are human only in contact, and conviviality, with what is not human."[13]

Bennett raises questions that I consider to be vital to ecosomatics, including whether a "discursive shift from environmentalism to vital materialism might enhance the prospects for a more sustainably-oriented public."[14] She notes that

> if environmentalists are selves who live *on* earth, vital materialists are selves who live *as* earth [...] if environmentalism leads to the call for the protection and wise management of *an ecosystem that surrounds us*, a vital materialism suggests that the task is to engage more strategically with *a trenchant materiality that is us* as it vies with us in agentic assemblages.[15]

Here too, I hear parallels with Indigenous perspectives; as the U.S. Poet Laureate, Joy Harjo (Mvskoke Creek), puts it: "These lands aren't our lands. These lands aren't your lands. We are this land."[16] I hear Sondra Fraleigh's ecosomatics, too, when she states, "we are not separate from nature's elements; rather, they dwell within, fluid as water and murky. Nature is not simply in surroundings. *Nature is in us.*"[17] It should be clear that I take ecosomatics to be a vital materialist position and practice.

Following Fraleigh's approach to exploring somatic movement, we can begin to develop ecosomatic awareness without judgment, from wherever we are, by further extending our awareness slowly, over time.[18] But to ex-tend (to stretch out)—from the Latin, *extendĕre*, < *ex-* out + *tendĕre* to stretch—is not the full extent of the stretch. Ecosomatic awareness requires that we also at-tend (stretch toward) and in-tend (stretch inward) in different forms of intra-relation.[19] Ecosomatics invites a critical shift in perspective about how we experience our already intra-connected selves in relation, how we engage our response-ability with and to the more-than-human world, and how we cultivate an awareness of and respect for the vibrancy of all matter.

There are many ways we can become more aware of our entangled intra-relation. Starting "here and now" as Fraleigh suggests may consist of increasing your awareness of small things; it did for me. In a personal essay titled, "Loving a Hornworm in End Times: Connecting to the Small in Nature," I chronicle paying close attention to a tobacco hornworm over a period of a few weeks in August 2019 and reflect on how that relationship opened my eyes to the incredible beauty and struggle of a creature I would have once dismissed as unworthy of attention or even care. That relationship sparked an entirely new awareness of my garden as an agentic assemblage and my growing awareness of the many critters around my mobile home.[20] As we do in somatics, we can start small, begin wherever we are, by extending our awareness simply to what is here, now.

Critical Ecosomatics: Acknowledging and Foregrounding Ecosomatic Inheritances

There are several key steps in developing an ecosomatic awareness that stretches toward critical perception of its blind spots and inheritances. As we just discussed, the first is simply to bring one's attention and awareness to the more-than-human on a daily basis—seek to encounter the ignored, the overlooked, and the very small. This can become a practice of attending, or stretching toward, the other.

The second step is to develop critical awareness of the land one occupies. Many people, including myself, are settlers on Indigenous lands; I write this chapter from the traditional home and unceded territories of the Muwekma Ohlone peoples—from a place where six Nations were enslaved at Mission San José. One of the first things to do in developing this awareness is to learn whose land you are on. Indeed, many scholars, artists, and activists have defined "settler colonial domination" precisely as "violence that disrupts human relationships with the environment."[21] Indigenous philosopher and environmental justice scholar, Kyle Whyte, further argues that "settler colonialism commits environmental injustice through strategically undermining Indigenous collective continuance."[22]

As an example of this kind of violence, a report by the international nongovernmental organization, Global Witness, documents that between 2012 and 2022, more than 1700 people were killed globally for their environmental activism.[23] "More than 40 percent of the murders were of Indigenous people who represent only five percent of the world's population."[24] Most of these deaths are related to, or more accurately, result from, mining industry projects.[25]

To be sure, not everyone reading this chapter is on someone else's land; perhaps you are of European descent and you still live in Europe or you are Japanese and still live in Japan, and so on. But this is too simplistic; think of the unsettled borders and histories of genocide in Europe alone. If my ancestors can be said to be Indigenous to anywhere, it is to Northern Europe, perhaps to Ireland; and many say Ireland is not free from colonial occupation today. Nonetheless, settler colonial and settler-capitalist ideologies of nature (as resource for development and extraction) are among our inheritances globally given the worldwide flows of capital.[26] As we benefit from the mining that kills Indigenous activists, we enter an assemblage of settler-capitalism. As a result, in addition to problematizing any clear or clean response to the question of whose land one is on, we must also increase our ecosomatic awareness by learning about the often-erased and normalized violent histories of extraction of the land. All of this is to locate ourselves historically, conceptually, ethically—and ecosomatically.

As consumers within global capitalism, we all must come to grips to some degree or another with the intra-relations of our desires, choices, other humans, the more-than-human world, and various agentic assemblages, like the "Great Pacific Garbage Patch," a concentration of marine debris made mostly of microplastics in the North Pacific Ocean.[27] In what ways do we participate in—enter into and, perhaps more importantly, exit— assemblages of ecocide and empire? I suggest that deep reflection on these matters is very much a part of the work of ecosomatics.

Ah, plastic... afloat, adrift, settling in mounds on the ocean floor.... It is worth noting here that "trash" in the gutter inspired Bennett's first example of an assemblage in which "*objects* appeared as *things*, that is, as vivid entities not entirely reducible to the contexts in which (human) subjects set them, never entirely exhausted by their semiotics."[28] She concludes that a "vital materiality can never really be thrown 'away,' for it continues its activities even as a discarded or unwanted commodity."[29] There can be no doubt that the marine debris in the North Pacific Ocean is an agentic assemblage continuing its activities and will for some time: its plastic continues to break down to smaller and smaller microplastics and it impacts—acts upon—various marine species with consequences that move up the food chain. As NOAA warns, "debris found in any region of the ocean can easily be ingested by marine species causing choking, starvation, and other impairments."[30]

Indeed, ecosomatics offers the opportunity to begin to develop awareness of one's position within settler colonialism—it offers an invitation to acknowledge settler-capitalism in historical and current forms and to begin to perceive one's participation in its various and nefarious assemblages. And with increased perception of the impacts of settler-capitalism on the land and water—from the extraction of minerals and heavy metals to the running of oil pipelines and the accumulation of plastic waste—we have already shifted our attention to the third key area in developing ecosomatic awareness, which is to consider our place in the current geological epoch. As such, we ask what it might be like to firmly situate ecosomatics in the context of the Anthropocene. For one, they have emerged alongside— ecosomatics and the Anthropocene—entangled and intra-active from the start.

Most simply, the Anthropocene is "the epoch of geological time during which human activity is considered to be the dominant influence on the environment, climate, and ecology of the earth."[31] Atmospheric chemist Paul Crutzen and biologist Eugene Stoermer first used the term in 2000.[32] In 2019, the Anthropocene Working Group published the results of two binding votes. The first acknowledged the presence of the Anthropocene by virtue of determining a "golden spike" that represents a significant and permanent change in the geological epoch. The second vote confirmed that the "golden spike" that marks the start of the Anthropocene is located with the release of atomic radionuclides in the mid-twentieth century.[33]

Increased perception of the Anthropocene and our intra-relation with its assemblages may invite us to become more curious about the foods we eat or whether they are produced in sustainable ways. An ecosomatic view encourages us to consider some of the deep conflicts that embed in capitalist consumption. Indeed, there have been situated critiques of the term "Anthropocene," including the important work of Donna Haraway, who proposes that *capitalocene* or *plantationocene* would have been more accurate terms. Haraway's concern is that the term "Anthropocene" naturalizes the problem as a "species act" rather than placing it properly with the effects of capitalism.[34] Ecosomatics invites us to look at these intersections of power in our own lives—and as Fraleigh reminds us, we "practice non-judgment" and "stay present to improvement."[35] In this process, I learned, for example, that while I financially support a non-profit that protects orangutan habitat, I was entangled in the destruction of those same habitats in various ways from the chips I used to eat to the deodorant I used to use, all heavy in unsustainably produced palm oil or its derivatives. I stay present to improvement.

To be clear, this is not about purity politics—far from it; it is about cultivating the increased perceptual awareness of self and other that somatics is already known for. When you become aware of the intra-action that constitutes you, other humans, the more-than-human world, and all materiality, you may simply begin to reconsider where your money, time, and energy move in relation. Yet remember that for Barad, "intra-action constitutes a radical reworking of the traditional notion of causality" and thus forces us to rethink our involvement in any simple terms.[36] This is not about placing blame or judgment, it is about paying attention to the entanglements within global capitalism and how we perpetuate or continue, through acts of consumption, the very things that we might like to see changed in the world. We must bear witness to intra-action—and we must also act—perhaps as activists but even more so as ecosomatic philosophers and practitioners because a change in worldview seems to be necessary not only for our ongoing thriving and survival but also to imagine alternative pasts, presents, and futures.

Breathing with Trees

I begin to ask, how can I develop an ecosomatic performance practice that engages critically in the three steps outlined above: developing awareness of our intra-relation with the more-than-human and all of vibrant materiality;

learning the histories of the land one is engaging with; and situating the work firmly within the context of the Anthropocene. What might this look like when so much ecosomatic performance seems to focus on the beauty and awe of the natural world? What happens if we engage something more akin to the terror-inducing sublime?[37] One possibility is an ecosomatics that takes seriously Anna Tsing's provocation to "tell terrible stories,"

> We have to tell stories where we're not winning, where there's just terrible things happening and we might not win [… and] that we "stay in the trouble" as Donna Haraway puts it, that we get involved, so that's our challenge. […] the challenge of our time is: "how do we tell terrible stories beautifully."[38]

What is an ecosomatics that can "stay in the trouble?" Tsing's digital work, *Feral Atlas: The More-Than-Human Anthropocene*, may offer some possibilities. The highly creative work explores "the ecological worlds created when nonhuman entities become tangled up with human infrastructure projects." What she calls "feral ecologies" are "encouraged by human-built infrastructures, but […] have developed and spread beyond human control." Tsing argues that these infrastructural effects "are the Anthropocene."[39]

As a performance artist and ecosomatic practitioner, I wonder what might happen if we were to embody and apply Tsing's theoretical approach. What might happen if we were to practice ecosomatic performance or touch at such sites? Or at sites impacted by the effects of global warming? What does an ecosomatic practice or performance look like in the fire zone? On land that is parched and cracked from drought?

In 2021, with these provocations in mind, I considered traveling to one of our recent fire sites in Northern California or to one of our low reservoirs given California's recent history of severe, extreme, and exceptional drought.[40] Of course, watersheds are generally off-limits to humans and driving out to a former fire site seemed wrong. I did not want to participate in a kind of ghoulish ecotourism of devastated sites, so I returned again to perception. Like all somatic work, this project begins with perception. When I directed my attention closer to home, I realized that a majority of the trees in my town were struggling due to the drought. I started walking by them with intention; then I asked quietly for permission to touch them. I began lightly touching the trees that I passed regularly, offering them some words of encouragement: "hang in there." On one walk along the Alameda Creek, I saw a tree with an open tear in its trunk and I laid my hands on either side of the wound. These were my points of entry and some trees seemed to be willing collaborators.

Philosopher Michael Marder's project in *Plant-Thinking: A Philosophy of Vegetal Life* is useful here. In it, his goal is "to reduce, minimize, put under erasure, bracket, or parenthesize the real and ideal barriers humans have erected between themselves and plants," to acknowledge plants as beings (what he calls the pursuit of "vegetal democracy"), and to learn to "let beings be."[41] Marder asks, "How is it possible for us to encounter plants? And how can we maintain and nurture, without fetishizing it, their otherness in the course of this encounter?"[42]

I extended my research on the agentic capacities of plants to determine a possible way forward—to try to find a way to stretch toward the trees, to intra-act with them in some kind of temporary ecosomatic assemblage. In the 2017 art exhibition catalog, *Moving Plants*, Line Marie Thorsen argues that "Plants move, they move other things, they move people, and they are themselves being moved around."[43] Scientific research supports the notion that plants and fungi are actors and that they act upon and with us—in *Entangled Life: How Fungi Make Our Worlds, Change Our Minds & Shape Our Future*, biologist Merlin Sheldrake makes our ecosomatic entanglements clear when he notes that fungi are "inside you and around you. They sustain you and all that you depend on. As you read these words, fungi are changing the way life happens."[44] He describes the truffle's compelling power over humans and more-than-human alike when he notes that "truffles create nested layers of attraction around themselves: Humans train dogs to find truffles because pigs are so attracted to them that they devour" them on the spot.[45] In acknowledging that the very oxygen we breathe requires trees and their process of photosynthesis for its production, I knew I had uncovered an approach: breathing with trees. Entangled, enmeshed, ecosomatic, indeed.

With renewed awareness of my intra-relation with trees, I went for a walk with some friends and their dogs one Sunday morning in September of 2021. We were in a lovely local park, surrounded by Alameda Creek, a couple of ponds, and numerous squirrels, humans, other dogs, birds, insects, plants, fungi—and of course, trees. One of my friends paused to point out an area of the pond that had dried up completely in recent years due to the drought. We stepped off the sidewalk and down into the space that once was under 3 or 4 feet of water. The ground was cracked and dusty; gnarled roots were exposed along the bottoms of the large trees living there. I felt drawn to the trees and the place itself.

My first impression—like my friends'—was that the pond had only recently dried up due to drought. Here, I pause. Phenomenology teaches us how to bracket the worldviews that we inherit or that we create for ourselves. Marder recommends it above. With an ecosomatic approach,

I thus bracketed that first assumption and asked whose land I was on and what human-built infrastructures might be at play on the site. When I returned home, I confirmed that my parched little site—Grau Pond in Niles Community Park in the City of Fremont—was usually filled with water in non-draught times, but only recently. The Muwekma Ohlone people, whose land this is, fished and hunted along the Alameda Creek and called the nearby town where they lived, Oroysom—but there was no quarry, and probably no Grau Pond. Fremont is also the site of the original Mission San José, established by the Spanish in 1797 as a tool of settler colonial domination. Around 1836, portions of the mission were sold off to Mexican Rancheros, who developed the land for animal production.

It should be no surprise that the impacts of human infrastructure projects are all over the Niles Community Park and nearby Fremont Quarry Lakes. When the western leg of the transcontinental railroad was built in the mid-nineteenth century, the area around Alameda Creek was quarried for its gravel, an essential component to laying the tracks. The quarries were filled with water long after construction operations were terminated creating the Quarry Lakes. In 1992, the East Bay Regional Parks and the Alameda Water District purchased the land and "restored" the lake in order to transform the area into "usable land for recreation."[46] Indeed, the site is a palimpsest of erased Indigenous presence, settler colonialism, and human infrastructure projects with claims to have been "rewilded" to some degree. But even the bald cypress trees were imported from elsewhere because they can grow in very wet conditions as well as in draught. Grau Pond, which seems to have been formed by the run-off when the quarry was filled, and indeed, all of Quarry Lakes constitutes a historically layered feral ecology.

On October 3, 2021, I returned to the Pond with three other humans and a special needs doggy named Ginger to practice breathing with the trees. My husband found a shady grove of bald cypress that would protect us from the hot sun and we all picked a spot to begin sitting or standing near one of them. We simply focused on trading breath with our hosts for about 20 minutes—visualizing that we breathe in what they exhale and they do the same in return. We directed our breath toward their branches as an offering of gratitude as we gathered under their shade. Marder describes the life of plants in terms of "being-with," which is a "mode of being in relation to all the others, being *qua* being-with." As such, we simply practiced "being-with" the trees and as Marder further suggests, we practiced "letting-be" by letting the trees simply be as they were.[47]

I remain interested in trading breath with trees and in developing humanarboreal relationships. As mentioned above, the work at Grau Pond was initially conceived as a temporary ecosomatic assemblage. But the recent heavy rains have returned the site to its pre-draught conditions and I am intrigued with continuing to work there in some way. As I consider these possibilities, I will reach out to Muwekma Ohlone Tribe of the San Francisco Bay Area to invite them to consider collaborating, participating, or providing feedback on this work and process.

Ecosomatic Possibilities

Ecosomatics includes moving and dancing our intra-relation with all of vibrant materiality; it moves us to imaginative action. The final component in developing ecosomatic awareness, as I see it, consists of enacting ecosomatic practices toward healing and imagining alternative pasts, presents, and futures.

Intra-Related Healing

Clearly, our need for healing is great and ecosomatic practices—even just time spent in reflection and intra-relation with the more-than-human world—offer healing opportunities. Indeed, many humans experience being "in nature" as a powerfully healing resource.[48] Ecosomatic practices may also offer ways to process our grief over the current environmental crisis and teach us how to bear witness to suffering. There can be no doubt that human and more-than-human alike suffer—that our suffering is intra-related and intra-active. As I have written elsewhere, "ecoanxiety" is

> described in a 2017 report by the American Psychological Association as the stress and worry associated with the ecological changes that mark our time. Symptoms include emotions of fear, anger, powerlessness, and the exhaustion that come with both gradual and sudden changes in climate, the "unrelenting day-by-day despair" of living through events like drought, fear of extreme weather caused by living through wildfires or hurricanes, or simply from watching slow impacts unfold—the melting of icecaps, the suffering of species.[49]

Solastalgia, too, is a kind of sadness specific to nostalgia for a place that no longer exists due to environmental impacts; areas burned, shorelines reconfigured due to erosion and rising seas. If we need new words to describe how we feel, imagine, for a moment, how displaced wildlife might feel.

An ecosomatic view reveals our deep intra-connection with the more-than-human world and suggests that the human and more-than-human must heal simultaneously—and much contemporary legal environmental theory suggests that we will only truly achieve rights in terms of one another; even our rights are intra-related. Some of these questions include, can trees have legal standing? Might wildlife have claims to property? What are environmental human rights?[50] The Carr Foundation's project at the Gorongosa National Park, Mozambique, is of interest because it articulates "an integrated multi-partner approach to conservation and to people-centred [sic] development,"—the conservation efforts and related community work go hand-in-hand.[51] What are the stakes of intra-related healing in the Anthropocene? As Robin Wall Kimmerer puts it, "In a world of scarcity, interconnection and mutual aid become critical for survival. So say the lichens."[52]

Undoing the Future

Bennett argues, and I agree, that "to begin to *experience* the relationship between persons and other materialities more horizontally, is to take a step toward a more ecological sensibility."[53] Indeed, ecosomatic scholars, artists, practitioners, and philosophers must develop practices, workshops, and visualizations that support this work—in short, to create *experiences* that allow people to shift the "mechanistic worldview" we have all inherited to some degree and, instead, learn to see "more horizontally."[54]

Taking up ecosomatics as a philosophical stance, as critical theory, as method, or as practice offers an invitation to shift one's worldview to something more horizontal, to something more entangled and intra-active—and this may be its most important work. Perhaps one of the greatest challenges and opportunities for ecosomatics is to help to cultivate new ways of imagining and being in the world: how can ecosomatics help us to encounter the world as "a swarm of vibrant materials" or as a "many-voiced landscape"?[55] How can we real-ize—make real—the re-membering of alternative pasts, presents, and futures... nothing less than the re-making of worlds? Here, Barad makes clear that as our image of something changes (in their example, the atom), "our practices of imaging and imagining and intra-acting with them have changed, and so have we."[56] This is another example of co-constitution through intra-action. And it speaks to the power of how we are changed when we change how we imagine something. In conversation with EL Putnam, Barad speaks of "trying to trouble the nature of nature" in terms of quantum field theory. Barad invites us to observe the mysteries of "quantum temporalities" that in turn present radical political imaginaries for cohabiting this planet more justly, by "undoing the future."[57] From this quantum perspective, imagining other possibilities is what makes other possibilities possible.

This is some of the work of ecosomatics, as I see it: to create situations and practices in which participants can experience embodied change and make these perceptual shifts.

A Terrible Little Story

I want to end this essay with a terrible little story—as Tsing suggests. I took a break from writing this chapter to get some bagels with the same folks who went to Grau Pond to trade breath with the cypress trees. We drove the back way from Fremont up to Berkeley, winding through the San Leandro Hills along Redwood Road and then down into Oakland. As we descended into town, we entered a suburban area with houses lined along both sides of the road and the entire Bay stretched out in front of us, all sparkling and light. On the left side of the street, a scraggly and thin coyote caught my eyes. It was pacing and trying to dart into the nearest lane—mange and open wounds flecked its partly hairless body. Its ears were flattened out to the side, which I took as a sign of distress. We drove slowly past, surprised to see the animal in a fully developed human neighborhood and knowing we didn't know enough to get out and help. I haven't been able to get this image out of my mind; worse, I don't know if I've been able to decipher what the coyote was trying to say. But I know that the more-than-human among us, close to home, are speaking—and I am listening. I am listening and stretching toward you.

Notes

1 David Abram, *The Spell of the Sensuous: Perception and Language in a More-than-Human World*, 2017 ed. (New York: Vintage Books, 1996), ix.
2 Jane Bennett, *Vibrant Matter: A Political Ecology of Things* (Durham: Duke University Press, 2010).
3 Ibid., 92.
4 Ibid., 107.
5 Shannon Rose Riley, "Embodied Perceptual Practices: Towards an Embrained and Embodied Model of Mind for Use in Actor Training and Rehearsal," *Theatre Topics* 14, no. 2 (2004): 451.
6 Karen Michelle Barad, *Meeting the Universe Halfway: Quantum Physics and the Entanglement of Matter and Meaning* (Durham: Duke University Press, 2007), 170. *See* their chapter, "Agential Realisms: How Material-Discursive Practices Matter," 132–185.
7 Ibid.
8 Ibid., ix.
9 Ibid.
10 Latour, qtd. in Bennett, *Vibrant Matter*, 103. *See also* Bruno Latour, *Pandora's Hope: Essays on the Reality of Science Studies* (Cambridge, MA: Harvard University Press, 1999), 198.

11 Chris Caldwell, Marie Schaefer, and Kyle Whyte, "Indigenous Lessons about Sustainability Are Not Just for 'All Humanity'," in *Sustainability: Approaches to Environmental Justice and Social Power*, ed. Julie Sze (New York: New York University Press, 2018), 155.

12 Daniel P. Modaff, "Mitakuye Oyasin (We Are All Related): Connecting Communication and Culture of the Lakota," *Great Plains Quarterly* 39, no. 4 (2019): 346, https://doi.org/10.1353/gpq.2019.0055. *See also* the work of Carl Mika (Māori of the Tuhourangi iwi): "Indigenous Notions of Interconnection and Formation by the World," *Oxford Research Encyclopedia of Education* (February 23, 2021). https://doi.org/10.1093/acrefore/978019 0264093.013.1558.

13 Abram, *Spell of the Sensuous*, ix. To be sure, Abram is strongly influenced by various Indigenous knowledges.

14 Bennett, *Vibrant Matter*, 111.

15 Italics added. Ibid.

16 Joy Harjo, *An American Sunrise: Poems* (New York: W. W. Norton & Company, 2019), 108. Harjo served three terms as the 23rd Poet Laureate of the United States of America.

17 Sondra Fraleigh, "Canyon Consciousness," in *Dance and the Quality of Life*, ed. Karen Bond (Cham: Springer, 2019), 30.

18 Fraleigh's approach to somatic movement is to "start here and now," to "practice non-judgment," and to "stay present to improvement." *See* Sondra Fraleigh, "Somatic Movement Arts," in *Moving Consciously: Somatic Transformations through Dance, Yoga, and Touch*, ed. Sondra Fraleigh (Urbana: University of Illinois Press, 2015), 45. More recently, Fraleigh reflects, "I start where I am in ability to move, learn from there and progress as I do." Sondra Fraleigh, "A Future Worth Having: Somatic Ethics of Flow and Curiosity," in *Somatics in Dance, Ecology, and Ethics: The Flowing Live Present*, ed. Sondra Fraleigh (Chicago: Intellect Ltd.; University of Chicago Press, 2023), 68.

19 "Extend, v.". *OED Online*. Oxford University Press (accessed June 14, 2023).

20 Shannon Rose Riley, "Loving a Hornworm in End-Times: Connecting to the Small in Nature," *Catamaran Literary Journal* 8, no. 1 (Spring, 2020). https://static1.squarespace.com/static/50a3e7b0e4b0216a96954b82/t/60297d656bde4077e06461b0/1613331822698/Loving+a+Hornworm+in+End+Times+-.pdf.

21 Kyle Whyte, "Settler Colonialism, Ecology, and Environmental Injustice," *Environment and Society* 9, no. 1 (2018): 125. https://doi.org/10.3167/ares.2018.090109.

22 Ibid., 126. Whyte uses the term "collective continuance" to refer to "an Indigenous conception of social resilience and self-determination" that he develops from Anishinaabe (Neshnabé) intellectual traditions. *See* ibid., 125–126.

23 Global Witness, "Decade of Defiance: Ten Years of Reporting Land and Environmental Activism Worldwide," *globalwitness.org* (2022). www.globalwitness.org/en/campaigns/environmental-activists/decade-defiance/#decade-killings-globally.

24 "More than 1,700 Land Activists Murdered in the Past Decade," *AlJazeera* (September 29, 2022). www.aljazeera.com/news/2022/9/29/over-1700-environmental-activists-killed-in-past-decade-report.

25 Ibid.

26 On "settler-capitalism," see Shannon Speed, *Incarcerated Stories: Indigenous Women Migrants and Violence in the Settler-Capitalist State* (Chapel Hill: University of North Carolina Press, 2019).

27 National Ocean Service and National Oceanic and Atmospheric Administration, "What is the Great Pacific Garbage Patch?", *National Ocean Service* (2023). https://oceanservice.noaa.gov/facts/garbagepatch.html.

28 Italics in original. Bennett, *Vibrant Matter*, 5.

29 Ibid., 6.

30 National Ocean Service and National Oceanic and Atmospheric Administration, "What is the Great Pacific Garbage Patch?".

31 "Anthropocene, n." *OED Online.* Oxford University Press (accessed January 12, 2022).

32 Paul J. Crutzen and Eugene F. Stoermer, "The 'Anthropocene'," *Global Change Newsletter* 41 (May, 2000). www.igbp.net/download/18.316f183213234701 77580001401/1376383088452/NL41.pdf#page=17.

33 Anthropocene Working Group, *Working Group on the "Anthropocene": Results of Binding Vote by AWG*, quaternary.stratigraphy.org (Subcommission on Quaternary Stratigraphy, 2019). http://quaternary.stratigraphy.org/working-groups/anthropocene/.

34 Donna Haraway, *Anthropocene, Capitalocene, Chthulucene: Making String Figures with Biologies, Arts, Activisms* (YouTube, 2016); Donna Haraway and Martha Kenney, "Anthropocene, Capitalocene, Chthulucene," in *Art in the Anthropocene: Encounters Among Aesthetics, Politics, Environments, and Epistemologies*, eds. Heather Davis and Etienne Turpin (London: Open Humanities Press, 2015), 259.

35 Fraleigh, "Somatic Movement Arts," 45.

36 Barad, *Meeting the Universe Halfway*, 33.

37 Immanuel Kant, *Observations on the Feeling of the Beautiful and Sublime* (Berkeley: University of California Press, 1960).

38 Anna L. Tsing, *Episode 32: Anna Tsing*, podcast audio, Conversations in Anthropology 2020. https://soundcloud.com/anthro-convo/ep32-tsing.

39 Anna L. Tsing, *Feral Atlas: The More-than-Human Anthropocene* (Stanford: Stanford University Press, 2021). http://feralatlas.org/.

40 As of May 2023, the State of California is approximately 92% drought free thanks to months of rain and snow. *See* Brianna Taylor and Michael McGough, "California Will Be Cool and Wet in May. Here's What More Rain Means for Drought Conditions," *The Sacramento Bee* (May 10, 2023). www.sacbee.com/news/california/water-and-drought/article275055516.html.

41 Michael Marder, *Plant-Thinking: A Philosophy of Vegetal Life* (New York: Columbia University Press, 2013), 5.

42 Ibid., 3.

43 Line Marie Thorson, *Moving Plants* (Denmark: Rønnebaeksholm, 2017), 11.

44 Merlin Sheldrake, *Entangled Life: How Fungi Make Our Worlds, Change Our Minds and Shape Our Futures* (New York: Random House, 2020), 3. For additional information on plant ethics, the posthuman and more-than-human, *see also* Marder, *Plant-Thinking* and David Wood, *Thinking Plant Animal*

Human: Encounters with Communities of Difference (Minneapolis: University of Minnesota Press, 2020).

45 Sheldrake, *Entangled Life*, 33.

46 Be prepared to work hard to uncover your local histories. *See* Museum of Local History, "The Ohlones and the Mission San José," *Rancho Higuera Historical Park*. https://museumoflocalhistory.org/wordpress2/wp-content/uploads/2021/02/Rancho-Higuera5-Final_201310011520454745.pdf; Mission San José, "The Story of Mission San José," 2023, accessed July 3, 2023, https://missionsanjose.org/history; Niles Canyon Railway, "About Us," accessed July 3, 2023, www.ncry.org/about/; City of Fremont, "About Fremont," accessed July 3, 2023, www.fremont.gov/about; New City Adventures, "History of Quarry Lakes," accessed July 2, 2023. https://newcityadventures.com/quarry-lakes-regional-park/.

47 Marder, *Plant-Thinking*, 51, 3.

48 There is substantial literature on the healing impacts of being in nature. Theodore Roszak, Mary E. Gomes, and Allen D. Kanner, *Ecopsychology—Restoring the Earth, Healing the Mind* (San Francisco: Sierra Club Books, 1995); Linda Buzzell and Craig Chalquist, *Ecotherapy: Healing with Nature in Mind* (San Francisco, Berkeley: Sierra Club Books, 2009); Rochelle Calvert, *Healing with Nature: Mindfulness and Somatic Practices to Heal from Trauma* (Novato: New World Library, 2021).

49 Riley, "Loving a Hornworm in End-Times: Connecting to the Small in Nature," 77. *See also* Susan Clayton et al., "Mental Health and Our Changing Climate: Impacts, Implications, and Guidance," (March, 2017). www.apa.org/news/press/releases/2017/03/mental-health-climate.pdf; Glenn Albrecht, *Earth Emotions: New Words for a New World* (Ithaca: Cornell University Press, 2019); and Glenn Albrecht et al., "Solastalgia: The Distress Caused by Environmental Change," *Australasian Psychiatry* 15, no. S1 (2007). https://doi.org/10.1080/10398560701701288.

50 On legal approaches to considering human and more-than-human rights as related, *see* Christopher D. Stone, *Should Trees Have Standing: Law, Morality, and the Environment* (Oxford: Oxford University Press, 2010). David R. Boyd, *The Rights of Nature: A Legal Revolution that Could Save the World* (Toronto: ECW Press, 2017); Karen Bradshaw, *Wildlife as Property Owners: A New Conception of Animal Rights* (Chicago: University of Chicago Press, 2020); Anthony R. Zelle et al., *Earth Law: Emerging Ecocentric Law: A Practitioner's Guide* (New York: Wolters Kluwer, 2021); Daniel P. Corrigan and Markku Oksanen, *Rights of Nature: A Re-Examination* (Abingdon; New York: Routledge, 2021); Natalia Kobylarz and Evadne Grant, eds. *Human Rights and the Planet: The Future of Environmental Human Rights in the European Court of Human Rights* (Cheltenham; Northampton: Edward Elgar Publishing, 2022).

51 The Gorongosa Project. "The Gorongosa Project," https://gorongosa.org, 2020, accessed July 2, 2023 https://gorongosa.org/our-mission-2/.

52 Robin Wall Kimmerer, *Braiding Sweetgrass: Indigenous Wisdom, Scientific Knowledge and the Teachings of Plants* (Minneapolis: Milkweed Editions, 2013), 272.

53 Bennett, *Vibrant Matter*, 10.
54 On the "mechanistic worldview," *see* Barad, *Meeting the Universe Halfway*, 354.
55 Bennett, *Vibrant Matter*, 107; Abram, *Spell of the Sensuous*, ix.
56 Barad, *Meeting the Universe Halfway*, 354.
57 Mobius Artists Group. "EL Putnam Chairs Karen Barad: Remembering Time/s for the Time Being, May 12, 2022," 2022, accessed December 27, 2022. www.mobius.org/events/mobius-artist-event-el-putnam-re-member ing-times-for-the-time-being.

Bibliography

Abram, David. *The Spell of the Sensuous: Perception and Language in a More-than-Human World*. 2017 ed. New York: Vintage Books, 1996.

Albrecht, Glenn. *Earth Emotions: New Words for a New World*. Ithaca: Cornell University Press, 2019.

Albrecht, Glenn, Gina-Maree Sartore, Linda Connor, Nick Higginbotham, Sonia Freeman, Brian Kelly, Helen Stain, Anne Tonna, and Georgia Pollard. "Solastalgia: The Distress Caused by Environmental Change." *Australasian Psychiatry* 15, no. S1 (2007): S95–S98. https://doi.org/10.1080/1039856070 1701288.

Anthropocene Working Group. Working Group on the 'Anthropocene': Results of Binding Vote by AWG. quaternary.stratigraphy.org (Subcommission on Quaternary Stratigraphy: 2019). http://quaternary.stratigraphy.org/working-groups/anthropocene/.

Barad, Karen Michelle. *Meeting the Universe Halfway: Quantum Physics and the Entanglement of Matter and Meaning*. Durham: Duke University Press, 2007.

Bennett, Jane. *Vibrant Matter: A Political Ecology of Things*. Durham: Duke University Press, 2010.

Boyd, David R. *The Rights of Nature: A Legal Revolution that Could Save the World*. Toronto: ECW Press, 2017.

Bradshaw, Karen. *Wildlife as Property Owners: A New Conception of Animal Rights*. Chicago: University of Chicago Press, 2020.

Buzzell, Linda, and Craig Chalquist. *Ecotherapy: Healing with Nature in Mind*. San Francisco; Berkeley: Sierra Club Books, 2009.

Caldwell, Chris, Marie Schaefer, and Kyle Whyte. "Indigenous Lessons about Sustainability Are Not Just for 'All Humanity'." In *Sustainability: Approaches to Environmental Justice and Social Power*, edited by Julie Sze, 149–179. New York: New York University Press, 2018.

Calvert, Rochelle. *Healing with Nature: Mindfulness and Somatic Practices to Heal from Trauma*. Novato: New World Library, 2021.

City of Fremont. n.d. "About Fremont." Accessed July 3, 2023. www.fremont.gov/about.

Clayton, Susan, Christie Manning, Kirra Krygsman, and Meighen Speiser. "Mental Health and Our Changing Climate: Impacts, Implications, and Guidance." (March 2017). Accessed August 31, 2019. www.apa.org/news/press/releases/2017/03/mental-health-climate.pdf.

Corrigan, Daniel P., and Markku Oksanen. *Rights of Nature: A Re-Examination*. Abingdon; New York: Routledge, 2021.

Crutzen, Paul J., and Eugene F. Stoermer. "The 'Anthropocene'." *Global Change Newsletter* 41 (May 2000): 17–18. www.igbp.net/download/18.316f18321323 470177580001401/1376383088452/NL41.pdf#page=17

Fraleigh, Sondra. "Somatic Movement Arts." In *Moving Consciously: Somatic Transformations through Dance, Yoga, and Touch*, edited by Sondra Fraleigh, 24–49. Urbana: University of Illinois Press, 2015.

———. "Canyon Consciousness." In *Dance and the Quality of Life*, edited by Karen Bond, 23–44. Cham: Springer, 2019.

———. "A Future Worth Having: Somatic Ethics of Flow and Curiosity." In *Somatics in Dance, Ecology, and Ethics: The Flowing Live Present*, edited by Sondra Fraleigh, 59–84. Chicago: Intellect Ltd.; University of Chicago Press, 2023.

Global Witness. "Decade of Defiance: Ten Years of Reporting Land and Environmental Activism Worldwide." *globalwitness.org* (2022). www.global witness.org/en/campaigns/environmental-activists/decade-defiance/#decade-killings-globally.

The Gorongosa Project. "The Gorongosa Project." https://gorongosa.org, 2020. Accessed July 2, 2023. https://gorongosa.org/our-mission-2/.

Haraway, Donna. *Anthropocene, Capitalocene, Chthulucene: Making String Figures with Biologies, Arts, Activisms*. YouTube, 2016.

Haraway, Donna, and Martha Kenney. "Anthropocene, Capitalocene, Chthulucene." In *Art in the Anthropocene: Encounters among Aesthetics, Politics, Environments, and Epistemologies*, edited by Heather Davis and Etienne Turpin, 255–270. London: Open Humanities Press, 2015.

Harjo, Joy. *An American Sunrise: Poems*. New York: W. W. Norton & Company, 2019.

Kant, Immanuel. *Observations on the Feeling of the Beautiful and Sublime*. Berkeley: University of California Press, 1960.

Kimmerer, Robin Wall. *Braiding Sweetgrass: Indigenous Wisdom, Scientific Knowledge and the Teachings of Plants*. Minneapolis: Milkweed Editions, 2013.

Kobylarz, Natalia and Evadne Grant, eds. *Human Rights and the Planet: The Future of Environmental Human Rights in the European Court of Human Rights*. Cheltenham and Northampton: Edward Elgar Publishing, 2022.

Latour, Bruno. *Pandora's Hope: Essays on the Reality of Science Studies*. Cambridge, MA: Harvard University Press, 1999.

Marder, Michael. *Plant-Thinking: A Philosophy of Vegetal Life*. New York: Columbia University Press, 2013.

Mika, Carl. "Indigenous Notions of Interconnection and Formation by the World." *Oxford Research Encyclopedia of Education* (February 23, 2021). Accessed June 15, 2023. https://doi.org/10.1093/acrefore/9780190264093.013.1558.

Mission San José. "The Story of Mission San José." 2023. Accessed July 3, 2023. https://missionsanjose.org/history.

Mobius Artists Group. "EL Putnam Chairs Karen Barad: Remembering Time/s for the Time Being, May 12, 2022." 2022. Accessed December 27, 2022. www.mobius.org/events/mobius-artist-event-el-putnam-re-membering-times-for-the-time-being.

Modaff, Daniel P. "Mitakuye Oyasin (We Are All Related): Connecting Communication and Culture of the Lakota." *Great Plains Quarterly* 39, no. 4 (2019): 341–362. https://doi.org/10.1353/gpq.2019.0055.

"More than 1,700 Land Activists Murdered in the Past Decade." *AlJazeera* (September 29, 2022). Accessed July 2, 2023. www.aljazeera.com/news/2022/9/29/over-1700-environmental-activists-killed-in-past-decade-report.

Museum of Local History. n.d. "The Ohlones and the Mission San José." *Rancho Higuera Historical Park*. Accessed July 3, 2023. https://museumoflocalhistory.org/wordpress2/wp-content/uploads/2021/02/Rancho-Higuera5-Final_201310011520454745.pdf.

National Ocean Service, and National Oceanic and Atmospheric Administration. "What Is the Great Pacific Garbage Patch?" *National Ocean Service* (2023). Accessed July 2, 2023. https://oceanservice.noaa.gov/facts/garbagepatch.html.

New City Adventures. n.d. "History of Quarry Lakes." Accessed July 2, 2023. https://newcityadventures.com/quarry-lakes-regional-park/.

Niles Canyon Railway. n.d. "About Us." Accessed July 3, 2023. www.ncry.org/about/.

Riley, Shannon Rose. "Embodied Perceptual Practices: Towards an Embrained and Embodied Model of Mind for Use in Actor Training and Rehearsal." *Theatre Topics* 14, no. 2 (2004): 445–471.

———. "Loving a Hornworm in End-Times: Connecting to the Small in Nature." *Catamaran Literary Journal* 8, no. 1 (Spring, 2020): 75–80. https://static1.squarespace.com/static/50a3e7b0e4b0216a96954b82/t/60297d656bde4077e06461b0/1613331822698/Loving+a+Hornworm+in+End+Times+-.pdf.

Roszak, Theodore, Mary E. Gomes, and Allen D. Kanner. *Ecopsychology—Restoring the Earth, Healing the Mind*. San Francisco: Sierra Club Books, 1995.

Sheldrake, Merlin. *Entangled Life: How Fungi Make Our Worlds, Change Our Minds and Shape Our Futures*. New York: Random House, 2020.

Speed, Shannon. *Incarcerated Stories: Indigenous Women Migrants and Violence in the Settler-Capitalist State*. Chapel Hill: University of North Carolina Press, 2019.

Stone, Christopher D. *Should Trees Have Standing: Law, Morality, and the Environment*. Oxford: Oxford University Press, 2010.

Taylor, Brianna, and Michael McGough. "California Will Be Cool and Wet in May. Here's What More Rain Means for Drought Conditions." *The Sacramento Bee* (May 10, 2023). www.sacbee.com/news/california/water-and-drought/article275055516.html.

Thorson, Line Marie. *Moving Plants*. Denmark: Rønnebaeksholm, 2017.

Tsing, Anna L. Episode 32: Anna Tsing. Podcast audio. Conversations in Anthropology 2020. https://soundcloud.com/anthro-convo/ep32-tsing.

———. Feral Atlas: The More-than-Human Anthropocene. Stanford University Press, 2021. http://feralatlas.org/.

Whyte, Kyle. "Settler Colonialism, Ecology, and Environmental Injustice." *Environment and Society* 9, no. 1 (2018): 125–144. https://doi.org/10.3167/ares.2018.090109.

Wood, David. *Thinking Plant Animal Human: Encounters with Communities of Difference.* Minneapolis: University of Minnesota Press, 2020.

Zelle, Anthony R., Grant Wilson, Rachelle Adam, and Herman F. Greene. *Earth Law: Emerging Ecocentric Law: A Practitioner's Guide.* New York: Wolters Kluwer, 2021.

2

WHAT NATIVE AMERICAN DANCE DOES AND THE STAKES OF ECOSOMATICS

Tria Blu Wakpa

University of California, Los Angeles, California,
U.S.A.: 34.0700° N, 118.4442° W

South Dakota State Penitentiary, Sioux Falls, South Dakota,
U.S.A.: 43.5668° N, 96.7250° W

> "Dancing is [tribal and Indigenous peoples'] way to sanctify the earth and all living things."
>
> George Blue Bird (Oglala Lakota)[1]

> "The meaning of dance comes from an initiative to be balanced in the center of our lives. It is inspired a majority of the time from our animals and the blessings of confidence that exist in their families."
>
> George Blue Bird (Oglala Lakota)[2]

This chapter describes some of the decolonial possibilities of Native American dance—that is, it delineates what Indigenous dance does—considering that settler-capitalist academic fields and discourses have often overlooked and misunderstood this practice. In what is frequently called the United States, "settler-capitalism" is a predominant and ongoing social structure that privileges Whiteness and Eurocentric understandings and norms—often to the detriment of Indigenous peoples, people of color, and more-than-humans as well as their ways of knowing and being.[3] Settler-capitalism emerged in the United States when settlers arrived and began to make their homes on Native lands.[4] The logics of settler-capitalism, which construct land as a "resource," depart from Indigenous understandings, which recognize land as a sacred "relative."[5]

Native dance takes place on- and off-stage and can be incredibly diverse; at the same time, it is possible to draw comparisons among Indigenous dances given commonalities in Native ways of knowing and moving.[6] The

DOI: 10.4324/9781003390985-4

settler-capitalist logics of anthropocentrism and Cartesian dualism attempt to portray humans as separate from and superior to more-than-humans, and the "mind" as separate from and superior to the "body," respectively; yet, Indigenous understandings recognize that humans and more-than-humans are interconnected, knowledgeable, and sacred. Inextricable from Native understandings, Indigenous dances can challenge settler-capitalist logics of anthropocentrism and Cartesian dualism by honoring firstly, more-than-human relatives, such as non-human animals, air, water, land, and the cosmos, and secondly, bodies and bodily movements as vital sources of knowledge.[7] In this way, Native dance can be viewed as a longstanding and dynamic "ecosomatic" and decolonial practice.

Dance scholar Elise Nuding "examine[s] the term 'eco-somatics' and the wider prevalence of ecology and the ecological in somatics discourse, situating both somatics and ecological/environmental discourse within modernity-coloniality."[8] From this, I argue that the anthropocentric and Cartesian dualist logics of settler-capitalism make the term "ecosomatics" useful. That is, part of the very premise of the field of "ecosomatics" is to challenge preexisting anthropocentric and Cartesian dualist structures by illuminating human and more-than-human relationships and bodily insights. By centering on Native dance, this chapter articulates it as an important ecosomatic practice and foregrounds that Indigenous understandings and practices have been, since time immemorial, what we only now call "ecosomatic."

When a new field like ecosomatics emerges in the "modern-colonial" academy, it can problematically further settler-capitalism by obscuring Native peoples, practices, and sovereignties.[9] By articulating what Native American dance does, this chapter illustrates, (1) that despite enduring colonization, Native lifeways and understandings—again, which might be described as "ecosomatic"—endure and innovate through dance, and (2) a critical and socially-just conception of ecosomatics must acknowledge and reckon with ongoing colonization and its oppression of Indigenous peoples. Settler colonial structures have attempted not only to annihilate Native peoples and practices but also to relegate them to a historic past as an attempt to obscure ongoing violence against Indigenous peoples.[10] Since colonization affects humans differently based on interlocking social structures—including race, gender, class, sexuality, nationality, religion/spirituality, ability, and age—I also underscore the importance of applying an intersectional approach to theorizing ecosomatics.

Indeed, Nuding argues that "the prefix 'eco' thus serves [...] to redirect, promote, and insist on the sociopolitical relevance of the ecological and environmental element—or potential—of somatics."[11] The field of somatics "advocates for holistic understandings of human somas as dynamic, living systems-in-process in relational exchange with their environment."[12] This definition aptly describes, at least in part, what Native dance does and is clearly in conversation with the epigraphs from George Blue Bird about human and more-than-human connections that open this chapter.

The impetus for this chapter's research and analysis must be credited to Blue Bird who initially asked me to speak about why Native American dance is important at the 2021 powwow held at the South Dakota State Penitentiary for Indigenous men who were then imprisoned there and their guests.[13] In his own words, Blue Bird is "a father, grandfather, self-taught artist, traditional dancer, ceremony man, activist, singer, songmaker, writer, and pow wow announcer."[14] Blue Bird, who "grew up dancing," is fluent in the Lakota language and serves as the President for the Native American Council of Tribes, a non-profit organization founded and run by Native people who are incarcerated at the South Dakota State Penitentiary.[15] He has been imprisoned at the Penitentiary for nearly four decades and is serving a life sentence. Throughout this chapter, I weave in some of Blue Bird's insights about Native dance from our ongoing correspondences, which began in 2019. Native confinement furthers settler-capitalism by attempting to separate Native people who are incarcerated from their human and more-than-human relatives. Focusing on the insights of a Lakota dancer who is imprisoned illuminates the decolonial possibilities of Native dance and how powerful this practice can be in the face of the heightened settler-capitalist violence that pervades confinement.

A vital aspect of this research is a decolonizing methodology which aims to counter the ways that academic research has often been exploitative and extractive.[16] Decolonizing methodologies build and sustain respectful, reciprocal, collaborative, and accountable relationships with Native individuals, communities, and/or nations who are involved in the research process. My approach to decolonizing methodologies differs depending on the people I am working with and the contexts and constraints of our work and their lives. For this chapter, Blue Bird is my primary collaborator. I have consulted with him throughout my writing process and provided him with honoraria for his time and expertise, including for the images that he created specifically for this chapter (*see* Figures 2.1 and 2.2). Another crucial aspect of decolonizing methodologies is sharing my positionality and physical location, and therefore locating the knowledge this chapter articulates. I am a woman of Filipino, European, and tribally unenrolled Native ancestries and the mother of two Lakota children and have been conducting research on Lakota lands with Lakota people for over a decade.

My children and I currently reside on unceded, Gabrielino-Tongva tribal lands, where I work at the University of California, Los Angeles, a land grant institution.

In the sections that follow, I first define how the term "Native American" is being used and then outline seven interconnected, decolonial possibilities of Native dance. I selected the number seven because it is a well-recognized number in tribally specific and pan-Indigenous understandings, which connotes Native communities, futurities, and more-than-human interconnections.[17] This chapter's conclusion positions returning land to Indigenous peoples—what Native academics and activists frequently refer to as "land back"—as a powerful ecosomatic practice that can counter the violence of settler-capitalism by supporting Indigenous peoples in revitalizing and innovating their dances.

Describing Native American and Native American Dance

Considering the immense variety among Native American dances, it is a nearly impossible task to create a comprehensive definition or description of Native American dancing. These dances are practiced by people from hundreds of tribes and take place in diverse contexts, including concert halls, outdoors, and community settings. Nevertheless, in order to show connections between Native dance and the field of ecosomatics, and to clarify how Native dancing informs ecosomatics, I offer a working definition of "Native American" and "Native American dance."

Although Indigenous peoples and dances exist throughout the world, this chapter focuses on Native American dances on Turtle Island, within the settler colonial boundaries of what is often referred to as the United States.[18] The politics of Indigeneity and what constitutes an Indigenous person are multifaceted, contested, and vary beyond and within U.S. borders.[19] However, in the United States, a primary determinant of Native identity is an individual having citizenship in a federally-recognized or, at times state-recognized, Indigenous nation. That is, Native people are often dual citizens of the United States and their Indigenous nation, which differentiates them from all immigrant groups and demonstrates that Indigenous identity is a political category, rather than a racial one.[20]

When settler state policies and practices conflate Native identity with race, this can obscure Native peoples' unique political identity and have detrimental material consequences. Settler colonial narratives often exclude the fact that Native tribes are Native *nations*—or in some cases, bands of

Native nations—and have a government-to-government relationship with the United States. Although Native nations have "inherent sovereignty," the project of settler colonialism has often denied and restricted Native nations' self-determination.[21] Currently in the United States, there are 574 federally-recognized tribes; however, this number is subject to change as more Native tribes petition for federal recognition or, conversely, are unjustly stripped of their sovereignty.

The frequent focus on federally-recognized tribes as legitimate can problematically obscure the Native nations who are most vulnerable: those who are not federally-recognized and therefore do not receive structural support from the United States.[22] By refusing to recognize these tribes, the U.S. government reinforces settler-capitalism, which frees the federal government from its treaty obligations to tribes.[23] Considering the politics of settler-capitalism, it is no surprise that federally-unrecognized tribes may exist and persist in areas where land is most desirable and expensive, such as on Gabrielino-Tongva and Ramaytush Ohlone lands, or what is frequently referred to as Los Angeles and San Francisco. By having the authority to determine if a tribe is "legitimate," the U.S. government exerts oppressive control over Native land and livelihoods. For this reason, some tribes have elected *not* to petition for federal recognition, despite the benefits that federal recognition could provide.[24]

Within this variety of settings, Native American dancing presents a similar array of characteristics and purposes: Indigenous dances can be tribally specific/intratribal, or pan-tribal/intertribal. Some dances have tribally specific origins but have developed to become pan-tribal.[25] Native people and experts will make a distinction between Native "ceremonial" dances and Native "social" dances, with the implication that "social" dances are secular.[26] Ceremonial dances may occur in intratribal and intertribal contexts. Due to differences between Indigenous and Eurocentric/mainstream understandings of dancing, it is crucial to use a culturally relevant and/or tribally specific lens to analyze Native dances and other practices.

Colonization continues to impact the frameworks used to study Native dancing, seen most vividly in Native logics and gender norms, which can complicate the male/female binary that is prevalent in Eurocentric constructions.[27] In some Native dances, like the "cross dance" or "switch dance," practitioners "switch" to a different gender: women will dance a style that is associated with men or vice versa. Leta Wise Spirit (Hunkpapa) has shared with me that doing a "cross dance" or "switch dance" can be a way of honoring people of that gender.[28] A cross/switch dance can also provide an opportunity for people who are Two Spirit or nonbinary to dance a form consistent with and celebratory of their gender, beyond the

Eurocentric binaries of female and male.[29] Given homophobia in settler society, Two Spirit and nonbinary people performing in mainstream powwows have endured discrimination, and today there are Two Spirit powwows, which strive to center nonbinary people and dancers as normative and thereby combat stigmatization.[30] Some Native dances, such as the Potato Dance, frequently feature couples. Sexuality can also be overt in Native fusion styles that combine powwow dance with pole dancing and burlesque.[31]

Centering Native dance also highlights the importance of attending to how social structures frequently misrepresent Indigenous peoples and practices as deviant, criminal, static, or extinct.[32] A prevalent and false stereotype about "traditional" Native dances is that they cannot be choreographed in the contemporary day.[33] A dance studies approach demonstrates that no dance can be perfectly replicated, because dances change each time they are performed.[34] Native people have indigenized movement modes such as pole dancing, burlesque, hip hop, skateboarding, and yoga. Evidencing how settler colonial structures have mischaracterized Native peoples' practices as deviant and criminal, from the late 1880s to the 1978 American Indian Religious Freedom Act, the U.S. government outlawed some Native dances.[35] One of the reasons why the U.S. government banned Native dancing is because of the fear of dancing's power and transformative capacities. Even during this prohibition, Indigenous people continued to covertly practice in their communities, taking their dances "underground."[36] As Blue Bird highlighted, "We danced underground and kept this important practice alive. No one saw us or knew what we were doing. Dancing kept us secretly unified."[37]

Today, Indigenous artists and practitioners choreograph new "traditional" dances. In recent decades, cell phone cameras with audiovisual capabilities and social media platforms have circulated Native dancing and brought attention to these practices.[38] Indigenous dances have been performed in shopping malls, on the frontlines of social protests, and for healing and solidarity. To show the multifaceted purposes of Indigenous dance, and its relevance to ecosomatics, I outline seven, interconnected possibilities of Native dance, as well as their sociopolitical contexts, practitioners, and contributions. This information is not comprehensive; instead, my aim is to succinctly describe and offer tangible illustrations of Native dance's decolonial potential in order to inform, situate, and contribute to ecosomatic discourses. Here, I specify "discourses" in lieu of "discourses and practices" because I do not condone or encourage non-Native ecosomatic practitioners to include Indigenous practices in their work. This is because of the long and common occurrence of non-Native people appropriating Indigenous practices.

Each of the decolonial possibilities that I delineate below is connected to "ecosomatics" explicitly or implicitly, because Native bodies and spiritualities are inseparable from Native lands and more-than-humans.[39] Because the decolonial possibilities I offer can combat settler-capitalism, anthropocentrism, and Cartesian dualism, they can also be broadly conceived of as "resistance." Yet, I prefer to use the term "decolonial possibilities" rather than the word "resistance," because the former moves beyond a refusal to comply and instead implies how Native dance can bring Indigenous futurities into being.

Decolonial Possibility #1: Spirituality, Worldviews, Narratives, and Dreams

First and foremost, many Indigenous artists and practitioners emphasize Native dance as a spiritual practice, which, again, is inextricably connected to more-than-humans. Blue Bird once instructed me to, "Teach those that walk with you what it means to … [use] dance as a constant expression of spiritual movement."[40] In this way, Indigenous dance can be viewed as countering U.S. policies of assimilation, conversion, and Native land dispossession, such as those apparent in Indian boarding schools.[41] Originating in the late nineteenth century, Indian boarding schools forcibly and coercively removed Indigenous children and young people from their families and communities in order to aid settler-capitalism. Indian boarding schools frequently prohibited Native children from speaking their languages and undertaking their spiritual practices, including Indigenous dance. In the present day, some prisons, such as the South Dakota State Penitentiary where Blue Bird is imprisoned, *do* allow Native people who are incarcerated to dance, which is vital to their survival while incarcerated. During the COVID-19 pandemic when the prison prohibited Native people who were then imprisoned from engaging in their cultural practices in order to help prevent the spread of the infectious disease, Blue Bird wrote, "Any way we can receive spiritual help and support [...] means that our burdens of being locked down away from our ceremonies and dances are tremendously eased."[42]

For this decolonial possibility, I purposefully enact "ethnographic refusal" by not providing an explicit example of Native dance as a spiritual practice, although there are many.[43] As Alex Zahara writes, "Ethnographic refusal is a practice by which researchers and research participants together decide not to make particular information available for use within the academy. Its purpose is not to bury information, but to ensure that communities are able to respond to issues on their own terms."[44] Because many Native people view Indigenous dance as sacred knowledge, they may regard writing about such practices as controversial if not also disrespectful. Additionally, writing about these practices may unwittingly contribute to cultural appropriation since there are long histories of non-Native choreographers, such as Ruth St. Denis, Ted Shawn, and Martha Graham, who appropriated Indigenous dance.[45]

Like Native ceremonies, Indigenous theatrical productions can express Native worldviews. Theatrical works by Daystar (Rosalie Jones), recognized as the founder of the first Native contemporary dance company, illustrate how Indigenous dance can articulate Native familial and tribal narratives and worldviews. Writing about Native dances that take place on stage can be less controversial than discussing Indigenous ceremonies, which are often intended for Native people and insider participants. For *No Home but the Heart*, Daystar drew "from selective events in the lives of her great-grandmother, grandmother, and mother and tied them to historical events affecting the resettlement of [N]ative peoples in the late 19th century."[46] Similarly, Daystar's *Allegory of the Cranes* "is a personal ceremony of remembrance and revelation set in the cultural belief and storytelling of the Black-footed peoples of the North American high plains."[47]

Dreams can also be considered a form of narrative, and some stories attribute the origins of popular Native dances and regalia to dreams, which is apparent in narratives about the Grass Dance and Jingle Dress Dance.[48] Indigenous dreams inspiring Indigenous dances is a long practice. As Blue Bird described,

> A majority of our women in the old days came from Societies and strict upbringings. A lot of these women became dancers from personal dreams they had that told them what birds and creatures to follow, their colors and designs, where they should go, how they would dance, and to find other people with the same determination.[49]

Blue Bird also shared how this practice continues in the present day, even in the carceral context, "One of the Lakota prisoners [at the South Dakota State Penitentiary] is fulfilling a recent dream he had of making a blue traditional outfit and dancing. His outfit looks sharp."[50]

Decolonial Possibility #2: Resistance

Although neither Blue Bird nor Daystar's descriptions overtly address Indigenous resistance, this theme is implicit in contemporary Indigenous dance given the enduring presence of settler-capitalism. Native dance has defied and continues to defy assimilation and conversion policies; the

Ghost Dance, which is prominently interconnected to more-than-human well-being, provides perhaps the most well-known example. Nick Estes writes,

> In popular history books, the Ghost Dance appears briefly only to die at the Wounded Knee Massacre in 1890. The Ghost Dance in the revolutionary sense, was about life, not death; it was about imagining and enacting an anticolonial Indigenous future free from the death world brought on by settler invasion. It originated with Paiute prophet and healer Wovoka. In his vision, the Great Spirit's Red Son transforms the earth. This Red coming of the Messiah wipes away the colonial world, bringing back the animals, plants, and other-than-human ancestors destroyed by white men, and in turn, destroying the destroyers.... Its message of a coming Indigenous future spread like wildfire up the Western Canadian coast, down to the Southwestern United States and Northern México, and onto the Plains.[51]

Estes describes "life" as being the heart of the Ghost Dance, which counters "the death world brought on by settler invasion"—and the very structures of settler-capitalism.[52]

Yet, Native dance does not always express resistance in such overt ways; this practice can articulate Indigenous defiance while also being complicit in settler-capitalism. Like other movement modes, dance is a fluid practice, which can be analyzed in various and even contradictory ways. This mutable quality makes dance well-suited to subversively convey resistance. For instance, in the nineteenth and twentieth centuries, some Indigenous people elected to perform in Wild West shows, which were endeavors often run by non-Native people.[53] Although the Indigenous performers' participation in Wild West shows could be viewed as reifying settler-capitalism, it also allowed them to perpetuate and innovate their lifeways at a time when Native dance was legally prohibited. Moreover, the Indigenous actors in the Wild West shows at times reenacted battles, which depicted them in conflict with European American performers.[54] These battles could be viewed as a Native demonstration of Indigenous sovereignty and resistance to colonization and assimilation.[55]

Decolonial Possibility #3: "Dancing Sovereignty"

Indigenous resistance is further enacted in expressions of sovereignty in dance, which also demonstrate how different characteristics of Native dance intersect and overlap. Some scholars in Indigenous studies discuss Native "legal sovereignty" as separate from "visual sovereignty," while scholar Mique'l Dangeli has created the concept of "dancing sovereignty" to show how they interlock.[56] Dangeli defines "dancing sovereignty" as:

> [S]elf-determination carried out through the creation of performances (oratory, songs, and dances) that adhere to and expand protocol in ways that affirm hereditary privileges (ancestral histories and associated ownership of songs, dances, crests, masks, headdresses, etc.) and territorial rights to land and waterways among diverse audiences and collaborators.[57]

As Dangeli clarifies, Indigenous sovereignty, as enacted through dance, interconnects with more-than-humans.

In a recent essay, I draw on and expand Dangeli and other Indigenous studies scholars' discussions of sovereignty.[58] Through a close reading of the 1894 film, *Buffalo Dance*, one of the first to depict Native people, I define "sovereignty" as "Native expressions of agency and authority–rooted in Indigenous worldviews, languages, narratives, experiences, and practices–that relate to human and/or more-than-human collectives and promote Native well-being and futurities."[59] I argue that the Lakota dancers in the film express sovereignty through their choreography, which includes each actor signing in North American Hand Talk, the *lingua franca* prior to English in the partition of Turtle Island frequently referred to as the United States. As an articulation of Lakota/Indigenous sovereignty, *Buffalo Dance* also enacts relationships, community, and nation-building.

Considering assimilation policies have attempted to undermine Native self-determination, simply connecting to one's Native identity through Native dance can be a way of enacting sovereignty. Blue Bird observed,

> When men go out and dance for their first time, they like it and want to do it again, so I believe a lot of the men in prison want to have the connection. The men I know that want to find and understand their identity as Native Americans, but are afraid to take that step [...] grew up being about something else rather than growing up learning our way of life.[60]

Decolonial Possibility #4: Relationships, Community, and Nation-Building

Strengthening relationships, community, and Indigenous nations can combat settler colonial structures and practices, which have often

FIGURE 2.1 *Cante Tinza Itancan (Strong Heart Chief),* permanent ink and acrylic, drawing by George Blue Bird (Oglala Lakota), 2023. Permission of the artist.

sought to pit Native and other peoples against one another, portray humans as superior to more-than-humans, and relegate Indigenous peoples, practices, and sovereignties to the past. In contrast, Indigenous dance can nurture respectful and reciprocal human-to-human and human-to-more-than-human relationships in the past, present, and future, which could be considered "ecosomatic." For some Indigenous peoples, more-than-human relatives also constitute "nations."[61] There are a multitude of examples of how Native dance, on stage and off, can promote positive interdependencies. These interconnections are made visible through movement qualities which mimic and honor more-than-humans, dance regalia which can be fashioned from more-than-humans and procured according to Indigenous protocols that enact more-than-human sustainability, and the process of preparing for and enacting the dances and/or ceremonies. Blue Bird discussed how the movement qualities of a dance and its regalia can be related,

> Our Lakota women dancers follow the birds and many of our four-footed animals as leaders and motivators to inspire them and get them moving at our dances. In particular, they watch the eagles for their strength and great flying abilities. Our women want to be like them. They enjoy receiving their feathers, plumes, claws, and their spirit.[62]

Performances by Emily Johnson, a choreographer of the Yup'ik nation, illustrate how dance can forge new relationships and present possibilities for unifying humans and more-than-humans. Johnson's *Shore* (2014) expanded beyond the theater into the world in four key ways:

> COMMUNITY ACTION (volunteerism) in partnership with community organizations; STORY (a curated reading); PERFORMANCE that begins outdoors and moves into a theater; and a festive, culminating potluck FEAST to which the participants bring the stories and recipes of the dishes they bring to share.[63]

Native performances can challenge a hierarchy that makes on-stage performances more valuable than off-stage dance and can bring attention to the power of dancing to shape our futures. In Johnson's *Being Future Being* (2022), she focused on "the forces that brought this world into being," and presented a multilayered performance as "a site for transformation, ushering into focus new futures with the potential to reshape the way we relate to ourselves, our environment, and to the human and more-than-human cohabitants of our world."[64]

Scholar Cutcha Risling Baldy—who is Hupa, a dancer, and an insider to her community—similarly illuminates how ceremonies present possibilities for nurturing human-to-human and human-to-more-than-human relationships.[65] In the past, Hupa ancestors shared information with anthropologists about their dances, during a time when Hupa people refrained from practicing the coming-of-age ceremonies because of settlers targeting and violently attacking Native participants.[66] Risling Baldy argues that Hupa ancestors shared this information because they imagined a future time when Hupa people would be revitalizing their dances by drawing on prior Hupa practices to fortify the present and the future.[67]

Risling Baldy discusses the Hupa Flower Dance as a lived experience that, for women, "expanded their circles of support."[68] She also specifies that at the heart of

> The Xonsil-ch'idilye (White Deerskin Dance) and the Xay-ch'idilye (Jump Dance) [...] is the renewal of the world, and each time the dances are performed, the Hupa people are helping to renew, rebalance, and recenter the earth so that it will be safe and free of disease, death, and destruction.[69]

Notably, Risling Baldy's description resembles Estes' discussion of the Ghost Dance as "bringing back the animals, plants, and other-than-human ancestors."[70] Risling Baldy and Estes' descriptions clarify that these dances cannot be separated from the human-to-more-than-human connections that they enact, which can be intertwined with Native nation-building and healing.

Decolonial Possibility #5: Healing and Hope

Native dances can offer healing and hope, countering the detrimental impacts of settler colonialism on the holistic health of Native humans and more-than-humans. In Native movement practices, the refrain "movement as medicine" is well-recognized, and as Blue Bird highlights, "[t]o dance meant that many healings would come."[71] The Jingle Dress Dance emerged during the influenza pandemic of 1918–1919.[72] As the *Indian Country Today* staff describes:

> Known also as the Healing Dance, the Jingle Dance originated with the Ojibwe in the Great Lakes Region. The dress is said to have originated as a means of healing a medicine man's granddaughter. In the dream the elder was told to construct a jingle dress and have his granddaughter dance in it—and she was healed.[73]

Native practitioners trace the healing capacities of this dance to the color of the dancers' dresses and ribbons, as well as the sounds of the jingles.[74] The rhythmic patterns have been compared to the rustling of tree leaves and vibration of the Northern Lights, amplifying how more-than-human connections can heal human sickness.[75]

Native people continue to uphold the healing and hopeful qualities of the Jingle Dress Dance. LaRayne Woster, who is from the Rosebud Sioux Tribe and a Native American Studies teacher, shares:

> [The Jingle Dress Dance] is a celebration, but it's also the purpose of "I'm out here. I'm touching the earth. I'm dancing for the people. The People love this. It makes their hearts happy. If I can dance in prayer and help somebody feel better, that's the beauty of the dance."[76]

Human and more-than-human linkages are made visible by the Jingle Dress dancer who always keeps one foot on the earth, connected to the land at all times.[77] In addition, Susan Foster has shown how watching dancing can activate the same areas of the brain as dancing itself.[78] Therefore, even if someone cannot physically perform the dance, they can still benefit from watching the dance, which transmits healing benefits.

Acosia Red Elk (Umatilla), a ten-time Jingle Dress Dance champion and certified yoga instructor, created a fusion movement practice that she calls Powwow Yoga.[79] In late March and early April 2020, during the height of the COVID-19 pandemic, Red Elk shared two Powwow Yoga instructional videos on YouTube, which she made available open access.[80] Red Elk's Powwow Yoga II features four rounds of Jingle Dress Dancing with a yoga flow.[81] A water meditation that Red Elk created follows the final round of Jingle Dress Dancing and emphasizes human and more-than-human reciprocity, respect, and healing.

Decolonial Possibility #6: Well-being and Joy

Native dance—like dancing in general—can promote well-being and joy. Since Native bodies, identities, and dances interconnect with Native lands and more-than-humans, Indigenous human well-being and joy can be implicitly linked with that of more-than-humans.[82] Recent studies have shown that Native people accrue health benefits by engaging in cultural practices.[83] In the example of the Jingle Dress Dance, "healing" frequently implies sickness—or in mainstream discourses, trauma. "Well-being" infers the maintenance of good health, and Native well-being can challenge systemic structures that disproportionately and detrimentally affect the holistic health of Indigenous people. Likewise, Native joy defies the settler colonial stereotype of the sad and stoic "Indian." Movement connects to joy by producing endorphins, which "interact with the receptors in [the] brain that produce perceptions of pain" and "trigger a positive feeling in the body, similar to that of morphine [...] often described as 'euphoric.' "[84]

Research demonstrates that dance in particular can reduce the onset of dementia more so than reading, crossword puzzles, or other forms of exercise.[85] Notably, not all styles of dance produce the same cognitive benefits. According to neurologist Dr. Robert Katzman,

> Freestyle social dancing, such as foxtrot, waltz and swing, requires constant split-second, rapid-fire decision making, which is the key to maintaining intelligence because it forces your brain to regularly rewire its neural pathways, giving you greater cognitive reserve and increased complexity of neuronal synapses.[86]

Native dances are frequently improvised within a set structure, aligning with this definition of "freestyle" dancing.[87] Additionally, powwows are places where Indigenous people gather to "visit," meaning they are social events that bring together Native communities. Health columnist Jane E. Brody connects social events to well-being: "Social interaction is a critically important contributor to good health and longevity."[88] Brody cites *Harvard Women's Health Watch*: "Dozens of studies have shown that people who have satisfying relationships with family, friends, and their community are happier, have fewer health problems, and live longer."[89]

Native dance offers opportunities for Native people to express themselves, which is fundamental to surviving in societies where people are structurally marginalized. In a YouTube video titled, "Albuquerque's Native American Dances Unite Hip Hop and Pow Wow Culture," Randy L. Barton—"aka Randy Boogie, Dance, DJ, and artist" and "something of an icon in Southwest hip hop culture"—asserts that in hip hop: "All you're really seeing is the energy from oppression exploding. I'd rather explode than implode, cause if you implode that's suicide."[90] Notably, among Native young people, suicide is an epidemic.[91] Through "exploding" into hip hop dance, Barton sublimates the subversion he has experienced.[92]

Decolonial Possibility #7: Presence, Innovation, and Futurity

Indigenous dance—which, again, is interlocking with the land and more-than-humans—can communicate Indigenous presence, innovation, and futurity. In particular, live, in-person performances which feature Native people dancing can illuminate their enduring and embodied presences. Indigenous dance can also help make apparent Native spaces. This includes rendering visible settings which settler colonial discourses often do not associate with Indigenous peoples—such as showing urban landscapes as Native lands. In their performances, Native dancers may also utilize contemporary attire, props, and music, which can further contribute to audiences' understandings of Native peoples and practices as modern and dynamic.

Indigenous innovations in dance can occur in a multitude of ways. For instance, Native people can innovate their tribal and pan-tribal dances through varying their movement qualities or regalia. They can also fuse Native understandings or dance practices with other movement forms, such as yoga, which emerge from different sociopolitical contexts. Although yoga has Indigenous roots, Native people can also *indigenize* movement practices which are frequently considered to have non-Native origins, such as hip hop or skateboarding.

The PBS Voices video, "How These Native Dancers Blend Heritage and Hip Hop," weaves together many of the decolonial possibilities for Native dance.[93] The video is set in what is often referred to as Minneapolis, Minnesota, which the video identifies as "a huge starting ground for the American Indian Movement."[94] It features brothers Samsoche and Lumhe Sampson, who are who are Mvskoke Creek/Seneca Hoop dancers. Hoop dancing itself evidences Native innovations. Although the "practice is thousands of years old," it was not "until the 1930s that a young man named Tony White Cloud, Jemez Pueblo, played an instrumental role in its evolution and began using multiple hoops in a stylized version."[95]

Samsoche and Lumhe detail Native dancing as a means to facilitate human-to-more-than-human interconnections, such as with willow reeds–which Native people originally used to fashion the hoops–land, and water, in addition to human-to-human connections. Regarding human-to-human linkages, Samsoche shares, "To be able to keep these teachings [surrounding Native dance] alive is healing for us, but then it's also a very powerful tool for reaching people."[96] Samsoche's statement suggests that the impact of teaching and watching Native dance can be transformative for practitioners and audiences. Lumhe explains,

> What happened with colonialism and the genocide of Native Americans, we lost our connection to our culture. We were forbidden to sing our songs, our dances. And so, to have an opportunity to exercise it in this day and age to me is an ultimate act of sovereignty, resistance. And when I'm out there dancing, I'm representing myself, my family, my ancestors forevermore. I carry them all with me.[97]

Here, Lumhe references "ancestors," which according to some Native understandings, can include more-than-humans. Whereas this excerpt is not explicitly connected to Native futurity, in the video, Lumhe also

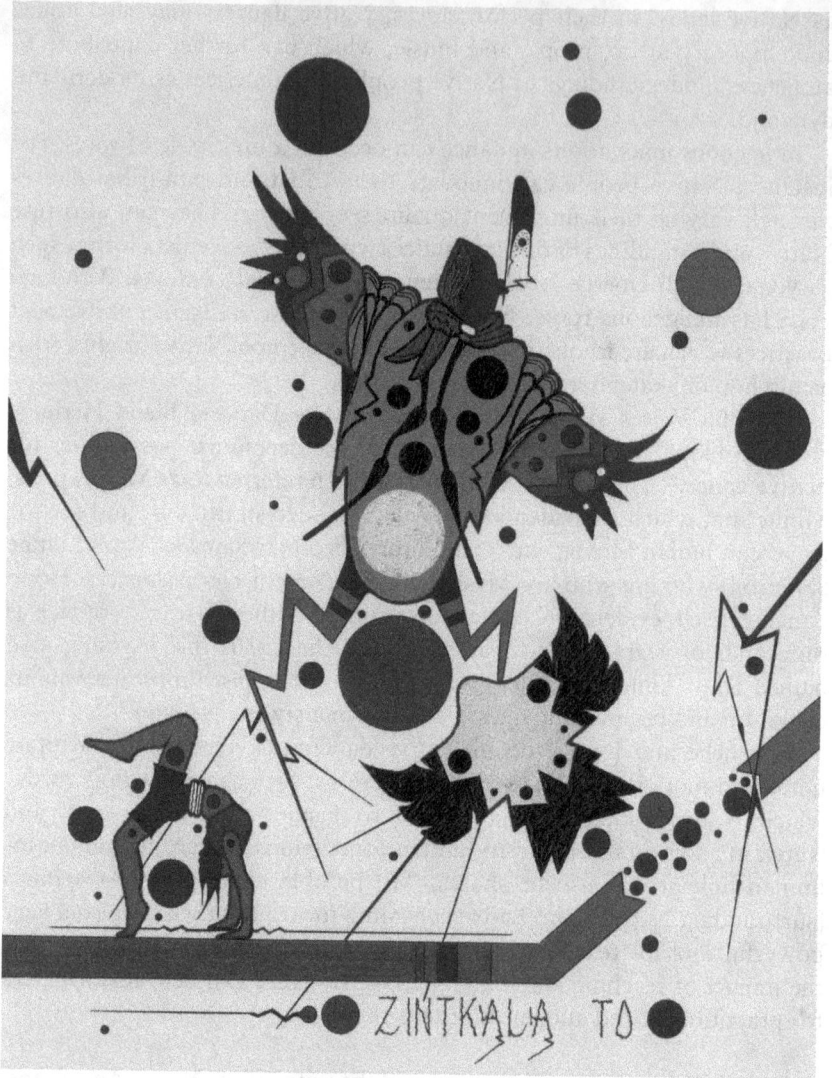

FIGURE 2.2   *Wanbli Kicopi Win (Calling Eagles Woman)*, permanent ink and acrylic, drawing by George Blue Bird (Oglala Lakota), 2023. Permission of the artist.

discusses, "My mother had us dancing from the moment we could walk."[98] Powwows commonly have a "Tiny Tots" category for Native toddlers and children. Indigenous babies and children are the embodied future of Native nations, and Indigenous dance—underpinned by Native logics—can be one way of socializing Indigenous children and nurturing a Native future into being. Blue Bird once wrote to me, "We need the spirit of our children. Their stories and their thoughts, along with the way each of them dances, are important."[99]

Conclusion: Land Back as a Powerful Ecosomatic Practice

For the reasons that I have outlined, Native dance can be considered an exemplary ecosomatic and decolonial practice. Yet, past and present, in de jure and de facto ways, settler-capitalist policies and practices have constructed obstacles that inhibit and outright prohibit Indigenous peoples from engaging in their movement practices, which are inseparable from bodies, lands, and more-than-humans. For instance, in the contemporary day, government agencies have destroyed sacred sites, disrupted tribal ceremonies, and prevented Native peoples from legally harvesting Native more-than-humans.[100] I have also spoken with several Native people in their early and mid-20s who are or were incarcerated at the South Dakota State Penitentiary and the South Dakota Women's Prison; they have shared with me that they participated in Native dance and other practices, such as beadwork, for the first time at the detention centers. At least in part, this indicates the structural hurdles—such as gas money and a working car— that some Native people are unable to overcome in the "free world" in order to access Native knowledge, which settler-capitalism has long sought to eliminate and marginalize.

The stakes of ecosomatics are that this emerging field can unconsciously reify settler-capitalism by obscuring Native peoples, practices, and/ or sovereignties; yet it can also work to challenge the violence of anthropocentrism and Cartesian dualism through discourse and practice. Ecosomatics, like dance studies, can make critical contributions to the academy by challenging Cartesian dualist assumptions. Although some scholarship in dance studies underscores human and more-than-human relationships, an ecosomatic approach to studying dance/movement highlights how the field of ecosomatics can simultaneously combat anthropocentrism and Cartesian dualism.

There is no single approach that all ecosomatic scholars and practitioners should use to challenge settler-capitalism; however, I advocate for centering Indigenous experts and their work and supporting the "land back" movement, which consists of much more than returning the land to Indigenous peoples. As Native, feminist scholar Lindsey Schneider (Turtle Mountain Chippewa) describes,

"Land back" [...] should be understood not as the return of title but rather as the full restoration of Indigenous land relationships. Title acquisition may indeed be part of the process, but cannot be its entirety. It is only through the restoration and flourishing of the complex web of Indigenous relationships with land, water, and our more-than-human-kin that we can hope to recover from the damage that settler colonial notions of land-as-property—with all their attendant conceptions of gender, heteropatriarchy, and domination—have done to the land and Indigenous peoples.[101]

Indeed, "land back" is a powerful ecosomatic practice that helps to ensure that the exemplary ecosomatic practices of Native peoples persist. Schneider's comments illustrate that ecosomatics can make an important intervention, even in Native and Indigenous studies discourses, by underscoring how bodies and bodily movements can vitally contribute to "the restoration of Indigenous land relationships."[102]

Schneider also highlights more-than-human "sites of Indigenous survivance" as sage teachers who "point us towards decolonization as an everyday imperative to nurture practices that support the flourishing of Indigenous life."[103] I would argue that prison too can be a "site of Indigenous survivance," and the Native humans who are and were caged there—such as George Blue Bird—are exemplary scholars, theorists, and practitioners of ecosomatics and decoloniality. In a recent message from Blue Bird sent via the online system that the South Dakota State Penitentiary contracts with, he provoked,

> I teach people to dance on a "get down right now" wherever we are at basis and this has changed prison's ugly demeanors into positive ones. Do you see dance as an effective method for change in Los Angeles and the world?[104]

I say, "Yes, George, I do. I absolutely do. Let's get down right now."

Notes

1   George Blue Bird, email to the author, October 11, 2020.
2   George Blue Bird, email to the author, July 1, 2020.
3   On "settler-capitalism," *see* Shannon Speed, *Incarcerated Stories: Indigenous Women Migrants and Violence in the Settler-Capitalist State* (Chapel Hill: University of North Carolina Press, 2019). Eve Tuck and K. Wayne Yang, "Decolonization Is Not a Metaphor," *Decolonization: Indigeneity, Education & Society* 1, no. 1 (2012): 12.
4   Tuck and Yang, "Decolonization," 5. To be sure, settler colonialism exists in other national and regional contexts that are outside the scope of this chapter, which focuses on Turtle Island.

5 Joseph Marshall III, "Lakota Itazipa: The Cultural Philosophy of Lakota Bows," *Lakota Wisdom Series* (YouTube, 2020), video. www.youtube.com/watch?v=wwVg1PTWBec&t=486s.

6 Jacqueline Shea Murphy, *The People Have Never Stopped Dancing: Native American Modern Dance Histories* (Minneapolis: University of Minnesota Press, 2007), 7.

7 Melanie Yazzie and Cutcha Risling Baldy, "Introduction: Indigenous Peoples and the Politics of Water," *Decolonization: Indigeneity, Education, and Society* 7, no. 1 (2018): 3. The term, "more-than-human," was coined by David Abram. *See* David Abram, *The Spell of the Sensuous: Perception and Language in a More-than-Human World*, 2017 ed. (New York: Vintage Books, 1996).

8 Elise Nuding, "Approaching Eco-Somatics: A Consideration of Potential Pitfalls and their Implications," *Journal of Dance & Somatic Practices* 13, no. 1–2 (2021): 30. https://doi.org/10.1386/jdsp_00034_1.

9 Ibid.; Kate Mattingly, "Choreographic Architecture and Vital Knowledge: Gaëtan Rusquet's *Meanwhile*," *Performance Research* 25, no. 2 (2020): 28.

10 Tuck and Yang, "Decolonization," 6.

11 Nuding, "Approaching Eco-Somatics," 33.

12 Ibid., 31.

13 This chapter is also indebted to feedback from Evangelina Macias, Sammy Roth, and Miya Shaffer. NB: This chapter uses the spelling "powwow" unless quoting someone who uses the spelling "pow wow."

14 George Blue Bird, email to the author, November 25, 2020.

15 Blue Bird, email to the author, July 1, 2020.

16 Linda Tuhiwai Smith, *Decolonizing Methodologies: Research and Indigenous Peoples* (London and New York: Zed Books, 1999), 79–80.

17 PBS.org. "Seven Generations—The Role of Chief." www.pbs.org/warrior/content/timeline/opendoor/roleOfChief.html.

18 Laura R. Graham and H. Glenn Penny, *Performing Indigeneity: Global Histories and Contemporary Experiences* (Lincoln: University of Nebraska Press, 2014), 3.

19 United Nations, "The United Nations Declaration on the Rights of Indigenous Peoples." (2007). Available at https://indianlaw.org/undrip/home.

20 UCLA Equity Diversity and Inclusion, "Resources on Native American and Indigenous Affairs: Native American and Indigenous Peoples FAQs," (2022). https://equity.ucla.edu/know/resources-on-native-american-and-indigenous-affairs/native-american-and-indigenous-peoples-faqs/#whoareind.

21 Teresa L. McCarty, "Education Policy, Citizenship and Linguistic Sovereignty in Native America," in *Language Policies and (Dis)Citizenship*, ed. Vaidehi Ramanathan (Bristol: Multilingual Matters, 2013), 116.

22 Courtney Cottrell, "NAGPRA's Politics of Recognition: Repatriation Struggles of a Terminated Tribe," *American Indian Quarterly* 44, no. 1 (2020): 67. https://doi.org/10.5250/amerindiquar.44.1.0059.

23 Eilis O'Neil, "Unrecognized Tribes Struggle without Federal Aid in the Pandemic," (April 17, 2021). www.npr.org/2021/04/17/988123599/unrecognized-tribes-struggle-without-federal-aide-during-pandemic.

24 Cottrell, "NAGPRA's Politics of Recognition," 67.

25 Mark G. Thiel, "Origins of the Jingle Dress Dance," *Whispering Wind* 36, no. 5 (2007): 17.

26 Benjamin R. Kracht, "Kiowa Powwows: Continuity in Ritual Practice," *American Indian Quarterly* 18, no. 3 (1994): 321. https://doi.org/10.2307/1184740.

27 Robin Prichard, "Masculinities and Performativities in Native American Dance," in *Masculinity, Intersectionality, and Identity*, ed. Doug Risner and Beccy Watson (London: Palgrave Macmillan, 2022), 146.

28 Leta Wise Spirit, interview by author, July 30, 2020.

29 Heidi Atter, " 'We Need to Be Visible': Two-Spirit Powwow Gives Dancers a Place to Be Themselves," *CBC News Saskatchewan* (June 16, 2019). www.cbc.ca/news/canada/saskatchewan/two-spirit-powwow-beardys-okemasis-june-pride-1.5177776.

30 Ibid.; Bay Area American Indian Two-Spirits, "About," (November 11, 2022). www.baaits.org/about; Gathering of Nations Powwow, "Two Spirit Grand Entry 2023." www.gatheringofnations.com/.

31 Evangelina Macias, "Dancing Defiance: From Native American Women's Fancy Shawl Dance to Burlesque and Pole Dancing" (Ph.D. dissertation University of California, Riverside, 2021).

32 Luana Ross, *Inventing the Savage: The Social Construction of Native American Criminality*, 1st ed. (Austin: University of Texas Press, 1998), 23, 42.

33 Shea Murphy, *The People Have Never Stopped Dancing*, 29–31.

34 *See* André Lepecki, "Introduction: Dance as a Practice of Contemporaneity," in *Dance*, ed. André Lepecki (Cambridge, MA: MIT Press, 2012), 14–23.

35 By Native peoples, I mean people of Native nations.

36 Shea Murphy, *The People Have Never Stopped Dancing*, 9.

37 George Blue Bird, email to the author, October 10, 2021.

38 Erin Tapahe, "This Morning at Standing Rock, Many Women Participated in a Sacred Ceremonial Women's Jingle Dance and This Was the First Time It Was Allowed to Be Recorded...," Facebook (October 29, 2016). www.facebook.com/erin.tapahe/videos/1220201151355408/.

39 Mishuana Goeman, "From Place to Territories and Back Again: Centering Storied Land in the Discussion of Indigenous Nation-Building," *International Journal of Critical Indigenous Studies* 1, no. 1 (2008): 24; Raechel Wastesicoot, "The Land Is Our Identity," *Land Life: The Nature Conservancy of Canada Blog*, June 21, 2019, www.natureconservancy.ca/en/blog/archive/the-land-is-our-identity.html.

40 George Blue Bird, email to the author, July 5, 2020.

41 K. Tsianina Lomawaima, *They Called It Prairie Light: The Story of the Chilocco Indian School* (Lincoln and London: University of Nebraska Press, 1994), 3.

42 Blue Bird, email to the author, July 1, 2020.

43 Alex Zahara, "Refusal as Research Method in Discard Studies," *Discard Studies* March 21 (2016). https://discardstudies.com/2016/03/21/refusal-as-research-method-in-discard-studies/.

44 Ibid.

45 Shea Murphy, *The People Have Never Stopped Dancing*, 199.

46 DAYSTAR Dance, "No Home but the Heart," (DAYSTAR Dance). https://daystardance.com/no_home_but_the_heart.html.

47 DAYSTAR Dance, "Allegory of the Cranes," (DAYSTAR Dance). https://daystardance.com/allegory-cranes.html.

48 Thiel, "Origins," 16. Adrian Primeaux, interview with the author, July 28, 2021.

49 Blue Bird, email to the author, August 5, 2020.

50 George Blue Bird, email to the author, August 26, 2020.

51 Nick Estes, *Our History Is the Future: Standing Rock versus the Dakota Access Pipeline, and the Long Tradition of Indigenous Resistance* (London and New York: Verso, 2019), 51.

52 Ibid.; Tuck and Yang, "Decolonization," 4.

53 Lester George Moses, *Wild West Shows and the Images of American Indians, 1883–1933*, 1st ed. (Albuquerque: University of New Mexico Press, 1996), 4–5.

54 Louis S. Warren, *Buffalo Bill's America: William Cody and the Wild West Show* (New York: Vintage Books, 2005), 93, 125.

55 Tria Blu Wakpa, "From *Buffalo Dance* to Tatanka Kcizapi Wakpala, 1894–2020: Indigenous Human and More-Than-Human Choreographies of Sovereignty and Survival," *American Quarterly* 74, no. 4 (December, 2022): 896.

56 Michelle H. Raheja, "Visual Sovereignty," in *Native Studies Keywords*, ed. Stephanie N. Teves, Andrea Smith, and Michelle H. Raheja (Tucson: The University of Arizona Press, 2015), 28–34.

57 Mique'l Dangeli, "Dancing Sovereignty: Protocol and Politics in Northwest Coast First Nations Dance" (Dissertation, University of British Columbia, 2015), 4.

58 Blu Wakpa, "From *Buffalo Dance* to Tatanka Kcizapi Wakpala," 900.

59 Ibid., 897.

60 George Blue Bird, email to the author, December 11, 2021.

61 Estes, *Our History Is the Future*, 7–8.

62 Blue Bird, email to the author, August 5, 2020.

63 Capitalizations in original. Emily Johnson: Catalyst, "Shore," (Emily Johnson: Catalyst). www.catalystdance.com/shore.

64 Emily Johnson: Catalyst, "Being Future Being," (Emily Johnson: Catalyst). www.catalystdance.com/being-future-being.

65 Cutcha Risling Baldy, *We Are Dancing for You: Native Feminisms and the Revitalization of Women's Coming-of-Age Ceremonies* (Seattle: University of Washington Press, 2018).

66 Ibid., 5–9.

67 Ibid., 8, 25–29.

68 Ibid., 144.

69 Ibid., 49.

70 Estes, *Our History Is the Future*, 51.

71 Kate Mattingly and Tria Blu Wakpa, "Movement as Medicine and Screendance as Survivance: Indigenous Reclamation and Innovation during Covid-19," *The International Journal of Screendance* 12 (2021): 150, https://doi.org/10.18061/ijsd.v12i0.7821. Blue Bird, email to the author, August 5, 2020.

72 Thiel, "Origins," 16.

73 ICT Staff, "Pow Wow Dance Style Origins: Jingle Dress Dance, Part 3," *Indian Country Today* (September 26, 2014; September 13, 2018). https://indiancountrytoday.com/archive/pow-wow-dance-style-origins-jingle-dress-dance-part-3.

74 St. Joseph's Indian School, "Jingle Dress Brings Hope and Healing," (St. Joseph's Indian School/YouTube, 2021), Video, 4′07″. www.youtube.com/watch?v=sMiNw_QVF0Y.

75 Tara Browner, *Heartbeat of the People: Music and Dance of the Northern Powwow* (Urbana: University of Illinois Press, 2002), 55. Acosia Red Elk, interview by the author, June 18, 2020.

76 St. Joseph's Indian School, "Jingle Dress Brings Hope and Healing."

77 Browner, *Heartbeat of the People*, 55.

78 Susan Leigh Foster, *Choreographing Empathy: Kinesthesia in Performance* (London and New York: Routledge, 2011), 7.

79 Tria Blu Wakpa, "Challenging Settler Colonial Choreographies during COVID-19: Acosia Red Elk's Powwow Yoga," *Critical Stages*, no. 23 (2021). www.critical-stages.org/23/challenging-settler-colonial-choreographies-during-covid-19-acosia-red-elks-powwow-yoga/.

80 Acosia Red Elk, "#1 Powwow Yoga with Acosia Red Elk," (YouTube, 2020), Video, 62′08″. www.youtube.com/watch?v=NSk0nDsiFxU; Acosia Red Elk, "#2 Powwow Yoga with Acosia Red Elk—focusing on Women's Jingle Dance," (YouTube, 2020), Video, 52′55″. www.youtube.com/watch?v=CMfAK-xxvqA.

81 Red Elk, "#2 Powwow Yoga with Acosia Red Elk—focusing on Women's Jingle Dance."

82 Goeman, "From Place," 25.

83 Kurt Schweigman et al., "The Relevance of Cultural Activities in Ethnic Identity Among California Native American Youth," *Journal of Psychoactive Drugs* 43, no. 4 (2011): 344. https://doi.org/10.1080/02791072.2011.629155.

84 Debra Fulghum Bruce, "Exercise and Depression," (April 1, 2022). www.webmd.com/depression/guide/exercise-depression.

85 Joe Verghese et al., "Leisure Activities and the Risk of Dementia in the Elderly," *The New England Journal of Medicine* 348, no. 25 (2003): 2512. https://doi.org/10.1056/NEJMoa022252.

86 Alanna Orpen, "Keep Dancing… It Turns Out It Is Good for the Brain," *BMC Series Blog*, April 4, 2016. http://blogs.biomedcentral.com/bmcseriesblog/2016/04/04/keep-dancing-turns-good-brain/.

87 Ibid.

88 Jane E. Brody, "Social Interaction Is Critical for Mental and Physical Health," *The New York Times* (June 12, 2017). www.nytimes.com/2017/06/12/well/live/having-friends-is-good-for-you.html.

89 Ibid.

90 KQED Arts, "Albuquerque's Native American Dancers Unite Hip Hop and Pow Wow Culture," (YouTube, 2020), Video, 6′3″. www.youtube.com/watch?v=4TzgHfELJ2s.

91 Mary Cwik et al., "Community Perspectives on Social Influences on Suicide within a Native American Reservation," *Qualitative Health Research* 32, no. 1 (2022): 16. https://doi.org/10.1177/10497323211045646.

92 KQED Arts, "Hip Hop and Pow Wow Culture."

93 PBS Voices, "How These Native Dancers Blend Heritage and Hip Hop," (YouTube, 2020), Video 5′27″. www.youtube.com/watch?v=ZYg1iW4-lmI.

94 Ibid.
95 Ibid. Dennis W. Zotigh, "History of the Modern Hoop Dance," *Indian Country Today* (May 30, 2007; September 12, 2018). https://indiancountryto day.com/archive/history-of-the-modern-hoop-dance.
96 PBS Voices, "Heritage and Hip Hop."
97 Ibid.
98 Ibid.
99 George Blue Bird, email to the author, January 16, 2022.
100 Meredith Privott, "An Ethos of Responsibility and Indigenous Women Water Protectors in the #NoDAPL Movement," *American Indian Quarterly* 43, Winter, no. 1 (2019): 75; Joe Szydlowski, "Winnemem Wintu Tribe Holds Coming-of-Age Ceremony at Lake Shasta," *Redding Record Searchlight* (Redding), July 4, 2012. https://archive.redding.com/news/winnemem-wintu-tribe-holds-coming-of-age-ceremony-at-lake-shasta-ep-375034843-354342 761.html.
101 Lindsey Schneider, "'Land Back' Beyond Reparations: Restoring Indigenous Land Relationships," in *The Routledge Companion to Gender and the American West*, ed. Susan Bernardin (London and New York: Routledge, 2022), 453.
102 Ibid.
103 Ibid., 461.
104 George Blue Bird, email to the author, January 21, 2023.

Bibliography

Abram, David. *The Spell of the Sensuous: Perception and Language in a More-than-Human World*. 2017 ed. New York: Vintage Books, 1996.
Atter, Heidi. "'We Need to Be Visible': Two-Spirit Powwow Gives Dancers a Place to be Themselves." *CBC News Saskatchewan*. (June 16, 2019). www.cbc.ca/news/canada/saskatchewan/two-spirit-powwow-beardys-okemasis-june-pride-1.5177776.
Bay Area American Indian Two-Spirits. "About." (November 11, 2022). www.baa its.org/about.
Blu Wakpa, Tria. "Challenging Settler Colonial Choreographies During COVID-19: Acosia Red Elk's Powwow Yoga." *Critical Stages*, no. 23 (2021). www.criti cal-stages.org/23/challenging-settler-colonial-choreographies-during-covid-19-acosia-red-elks-powwow-yoga/.
———. "From *Buffalo Dance* to Tatanka Kcizapi Wakpala, 1894–2020: Indigenous Human and More-Than-Human Choreographies of Sovereignty and Survival." *American Quarterly* 74, no. 4 (December, 2022): 895–920.
Brody, Jane E. "Social Interaction is Critical for Mental and Physical Health." *The New York Times* (June 12, 2017). www.nytimes.com/2017/06/12/well/live/hav ing-friends-is-good-for-you.html.
Browner, Tara. *Heartbeat of the People: Music and Dance of the Northern Powwow*. Urbana: University of Illinois Press, 2002.
Bruce, Debra Fulghum. "Exercise and Depression." (April 1, 2022). Accessed November 11, 2022. www.webmd.com/depression/guide/exercise-depression.
Cottrell, Courtney. "NAGPRA's Politics of Recognition: Repatriation Struggles of a Terminated Tribe." *American Indian Quarterly* 44, no. 1 (2020): 59–85. https://doi.org/10.5250/amerindiquar.44.1.0059.

Cwik, Mary, S. Benjamin Doty, Alexandra Hinton, Novalene Goklish, Jerreed Ivanich, Kyle Hill, Angelita Lee, Lauren Tingey, and Mariddie Craig. "Community Perspectives on Social Influences on Suicide within a Native American Reservation." *Qualitative Health Research* 32, no. 1 (2022): 16–30. https://doi.org/10.1177/10497323211045646.

DAYSTAR Dance. n.d. "Allegory of the Cranes." DAYSTAR Dance. https://daystardance.com/allegory-cranes.html.

———. n.d. "No Home but The Heart." DAYSTAR Dance. https://daystardance.com/no_home_but_the_heart.html.

Emily Johnson: Catalyst. n.d. "Being Future Being." Emily Johnson: Catalyst. www.catalystdance.com/being-future-being.

———. n.d. "Shore." Emily Johnson: Catalyst. www.catalystdance.com/shore.

Estes, Nick. *Our History is the Future: Standing Rock Versus the Dakota Access Pipeline, and the Long Tradition of Indigenous Resistance.* London; New York: Verso, 2019.

Foster, Susan Leigh. *Choreographing Empathy: Kinesthesia in Performance.* London; New York: Routledge, 2011.

Gathering of Nations Powwow. "Two Spirit Grand Entry 2023." (2022) Accessed November 11, 2022. www.gatheringofnations.com/.

Goeman, Mishuana. "From Place to Territories and Back Again: Centering Storied Land in the Discussion of Indigenous Nation-Building." *International Journal of Critical Indigenous Studies* 1, no. 1 (2008): 23–34.

Graham, Laura R., and H. Glenn Penny. *Performing Indigeneity: Global Histories and Contemporary Experiences.* Lincoln: University of Nebraska Press, 2014.

ICT Staff. "Pow Wow Dance Style Origins: Jingle Dress Dance, Part 3." *Indian Country Today* (September 26, 2014; updated September 13, 2018). Accessed November 11, 2022. https://indiancountrytoday.com/archive/pow-wow-dance-style-origins-jingle-dress-dance-part-3.

KQED Arts. "Albuquerque's Native American Dancers Unite Hip Hop and Pow Wow Culture." *YouTube*, 2020. Video, 6′3″. www.youtube.com/watch?v=4TzgHfELJ2s.

Kracht, Benjamin R. "Kiowa Powwows: Continuity in Ritual Practice." *American Indian Quarterly* 18, no. 3 (1994): 321–348. https://doi.org/10.2307/1184740.

Lepecki, André. "Introduction: Dance as a Practice of Contemporaneity." In *Dance*, edited by André Lepecki, 14–23. Cambridge, MA: MIT Press, 2012.

Lomawaima, K. Tsianina. *They Called It Prairie Light: The Story of the Chilocco Indian School.* Lincoln; London: University of Nebraska Press, 1994.

Macias, Evangelina. "Dancing Defiance: From Native American Women's Fancy Shawl Dance to Burlesque and Pole Dancing." Ph.D. dissertation, University of California, Riverside, 2021.

Marshall III, Joseph. "Lakota Itazipa: The Cultural Philosophy of Lakota Bows." *Lakota Wisdom Series.* YouTube, 2020. video. www.youtube.com/watch?v=wwVg1PTWBec&t=486s.

Mattingly, Kate. "Choreographic Architecture and Vital Knowledge: Gaëtan Rusquet's *Meanwhile.*" *Performance Research* 25, no. 2 (2020): 22–29.

Mattingly, Kate, and Tria Blu Wakpa. "Movement as Medicine and Screendance as Survivance: Indigenous Reclamation and Innovation during Covid-19." *The International Journal of Screendance* 12 (2021): 150. https://doi.org/10.18061/ijsd.v12i0.7821.

McCarty, Teresa L. "Education Policy, Citizenship and Linguistic Sovereignty in Native America." In *Language Policies and (Dis)Citizenship*, edited by Vaidehi Ramanathan, 116–142. Bristol: Multilingual Matters, 2013.

Moses, Lester George. *Wild West Shows and the Images of American Indians, 1883–1933*. 1st ed. Albuquerque: University of New Mexico Press, 1996.

Nuding, Elise. "Approaching Eco-Somatics: A Consideration of Potential Pitfalls and their Implications." *Journal of Dance & Somatic Practices* 13, no. 1–2 (2021): 29–39. https://doi.org/10.1386/jdsp_00034_1.

O'Neil, Eilis. "Unrecognized Tribes Struggle without Federal Aid in the Pandemic." (April 17, 2021). www.npr.org/2021/04/17/988123599/unrecognized-tribes-struggle-without-federal-aide-during-pandemic.

Orpen, Alanna, "Keep Dancing… It Turns Out it is Good for the Brain." *BMC Series Blog*, April 4, 2016. http://blogs.biomedcentral.com/bmcseriesblog/2016/04/04/keep-dancing-turns-good-brain/.

PBS. n.d. "Seven Generations—The Role of Chief." www.pbs.org/warrior/content/timeline/opendoor/roleOfChief.html.

PBS Voices. "How These Native Dancers Blend Heritage and Hip Hop." *YouTube*, 2020. Video 5′27″. www.youtube.com/watch?v=ZYg1iW4-lmI.

Prichard, Robin. "Masculinities and Performativities in Native American Dance." In *Masculinity, Intersectionality, and Identity*, edited by Doug Risner and Beccy Watson, 145–163. London: Palgrave Macmillan, 2022.

Privott, Meredith. "An Ethos of Responsibility and Indigenous Women Water Protectors in the #NoDAPL Movement." *American Indian Quarterly* 43, no. 1 (Winter, 2019): 74–100.

Raheja, Michelle H. "Visual Sovereignty." In *Native Studies Keywords*, edited by Stephanie N. Teves, Andrea Smith and Michelle H. Raheja, 25–34. Tucson: The University of Arizona Press, 2015.

Red Elk, Acosia. "#1 Powwow Yoga with Acosia Red Elk." *YouTube*, 2020. Video, 62′08″. www.youtube.com/watch?v=NSk0nDsiFxU.

———. "#2 Powwow Yoga with Acosia Red Elk—Focusing on Women's Jingle Dance." *YouTube*, 2020. Video, 52′55″. www.youtube.com/watch?v=CMfAK-xxvqA.

Risling Baldy, Cutcha. *We are Dancing for You: Native Feminisms and the Revitalization of Women's Coming-of-Age Ceremonies*. Seattle: University of Washington Press, 2018.

Ross, Luana. *Inventing the Savage: The Social Construction of Native American Criminality*. 1st ed. Austin: University of Texas Press, 1998.

Schneider, Lindsey. " 'Land Back' beyond Reparations: Restoring Indigenous Land Relationships." In *The Routledge Companion to Gender and the American West*, edited by Susan Bernardin, 452–464. London; New York: Routledge, 2022.

Schweigman, Kurt, Claradina Soto, Serena Wright, and Jennifer Unger. "The Relevance of Cultural Activities in Ethnic Identity among California Native

American Youth." *Journal of Psychoactive Drugs* 43, no. 4 (2011): 343–348. https://doi.org/10.1080/02791072.2011.629155.

Shea Murphy, Jacqueline. *The People Have Never Stopped Dancing: Native American Modern Dance Histories.* Minneapolis: University of Minnesota Press, 2007.

Smith, Linda Tuhiwai. *Decolonizing Methodologies: Research and Indigenous Peoples.* London; New York: Zed Books, 1999.

Speed, Shannon. *Incarcerated Stories: Indigenous Women Migrants and Violence in the Settler-Capitalist State.* Chapel Hill: University of North Carolina Press, 2019.

St. Joseph's Indian School. "Jingle Dress Brings Hope and Healing." *St. Joseph's Indian School/YouTube,* 2021. Video, 4′07″. www.youtube.com/watch?v= sMiNw_QVF0Y.

Szydlowski, Joe. "Winnemem Wintu Tribe Holds Coming-of-Age Ceremony at Lake Shasta." *Redding Record Searchlight* (Redding), July 4, 2012. https://arch ive.redding.com/news/winnemem-wintu-tribe-holds-coming-of-age-ceremony-at-lake-shasta-ep-375034843-354342761.html.

Tapahe, Erin. "This Morning at Standing Rock, Many Women Participated in a Sacred Ceremonial Women's Jingle Dance and This was the First Time it was Allowed to be Recorded...." *Facebook* (October 9, 2016). www.facebook.com/ erin.tapahe/videos/1220201151355408/.

Thiel, Mark G. "Origins of the Jingle Dress Dance." *Whispering Wind* 36, no. 5 (2007): 14–18.

Tuck, Eve, and K. Wayne Yang. "Decolonization is not a Metaphor." *Decolonization: Indigeneity, Education & Society* 1, no. 1 (2012): 1–40.

UCLA Equity Diversity and Inclusion. "Resources on Native American and Indigenous Affairs: Native American and Indigenous Peoples FAQs." (2022). https://equity.ucla.edu/know/resources-on-native-american-and-indigenous-affairs/native-american-and-indigenous-peoples-faqs/#whoareind.

United Nations. "The United Nations Declaration on the Rights of Indigenous Peoples." (2007). Accessed October 14, 2022. Available at https://indianlaw. org/undrip/home.

Verghese, Joe, Richard B. Lipton, Mindy J. Katz, Charles B. Hall, Carol A. Derby, Gail Kuslansky, Anne F. Ambrose, Martin Sliwinski, and Herman Buschke. "Leisure Activities and the Risk of Dementia in the Elderly." *The New England Journal of Medicine* 348, no. 25 (2003): 2508–2516. https://doi.org/10.1056/ NEJMoa022252.

Warren, Louis S. *Buffalo Bill's America: William Cody and The Wild West Show.* New York: Vintage Books, 2005.

Wastesicoot, Raechel, "The Land is Our Identity," *Land Life: The Nature Conservancy of Canada Blog,* June 21, 2019. www.natureconservancy.ca/en/ blog/archive/the-land-is-our-identity.html.

Yazzie, Melanie, and Cutcha Risling Baldy. "Introduction: Indigenous Peoples and the Politics of Water." *Decolonization: Indigeneity, Education, and Society* 7, no. 1 (2018): 1–18.

Zahara, Alex "Refusal as Research Method in Discard Studies." *Discard Studies* March 21. (2016). https://discardstudies.com/2016/03/21/refusal-as-research-method-in-discard-studies/.

Zotigh, Dennis W. "History of the Modern Hoop Dance." *Indian Country Today.* (May 30, 2007; updated September 12 2018). Accessed November 11, 2022. https://indiancountrytoday.com/archive/history-of-the-modern-hoop-dance.

3

ECOSOMATIC PERFORMANCE RESEARCH FOR THE PLURIVERSE

Daniel Ìgbín'bí Coleman

San Cristóbal de las Casas, México: 16.7370° N, 92.6376° W
Atlanta, Georgia, U.S.A.: 33.7532° N, 84.3853° W

Entrances

There is something majestic about being among natural formations that spiritually locate you in your inestimable smallness amid the vastness of nature. Every time I land in Tuxtla Gutiérrez, Chiapas, México, the final flight of three down into Chiapas, my body tingles with infinitesimal beingness. It is within these somatic sensations that my nervous system regains homeostasis and where I enter vibrational presence with everyone and everything around me. I have spent years listening to and learning to trust the sensations that arise in my body, enlivening every element of my soma, noticing what it is to feel this and the places that trigger a deep sense of embodied attunement with my spiritual self and the people and worlds around me. Chiapas, México, and the Southern United States are both places that do that for me.

When you land at the airport in Tuxtla Gutiérrez, set apart from the rest of the city, you feel the oppressive humidity engulf you. From there, it is just over an hour drive, by a bus or van, to San Cristóbal de las Casas; the humidity and the temperature drop as you wind your way up through the mountains into cooler and higher altitudes. Gazing through the windows of the van and down into the shadowy ravines marked by the clouds that hover over the mountains below, I revel in feeling small—moving away from the noisiness of anthropocentric hubris. I let my memories of this place float and recollect all the ways that it has shaped me as I settle into silence to be with the fullness of this land. The lull of the ride and the power of the green mountains ground me. I begin to feel the hum of this land once

DOI: 10.4324/9781003390985-5

again. I know I am approaching the city by the presence of fog, if it has not greeted the van sooner, followed by the many colors of the buildings and their lights up ahead that beg the attention of my line of sight.

This is the place—San Cristóbal de las Casas—that deepens my understanding of the spiritual access possible through engaging in ecosomatic practice. Ecosomatic practice, for me, includes the exercises, actions, and performance choices and processes that guide embodied listening and attunement to ecological interrelatedness of place. My use of ecology here is in the most capacious sense that considers social, cultural, and political ecologies. My penchant for this practice came from the embodied messages I found by witnessing some of what a land holds and *why* I could feel certain lands deeply. Through the enlivening vibrational sensations that have always coursed through my body when I have encountered and lived in this place, I realize that my somatic responses teach me about the imbrications of spiritual and physical porosity. It is from here that I cultivate ecosomatic methodology and practice. More on this after first situating a more comprehensive picture of its emergence.

My spiritual and physical work and movement in the world(s) is marked by various "we's." I disidentify with the individualism of liberal humanism that has fomented late capitalist and anthropocentric extractivism. Instead, I center becoming and un/becoming through collective entanglements that make ecosomatic sensing with place possible. Overlapping and entangling collectivities are how we co-constitute selves, one another, and place itself. The communities that I exist inside of are those that mainstream society cannot affirm or oftentimes, even comprehend. We are queer, trans, and gender non-conforming, differently abled, neurodivergent, Black, brown, multiracial and multinational, Native/Indigenous, undocumented, kinky, militant, poor, squatters, mentally ill, revolutionary, organizing/provoking, fat, hairy, current and former sex workers, tattooed and pierced, transfeminist, survivors of myriad traumas, disciplinarily disobedient intellectuals, artists, healers and healing-incoherent creatures fed up with the ills of modernity and wanting something more (which is to say, not *this*). Together we love, organize, create, and re-create, fail messily and tragically, create both harm and repair, generate entirely new and healing relational paradigms, triumph with unspeakable brilliance, break-up, fall out, build undamaged worlds, make several ways out of no way at all, and continue to choose life when slated for death and/or non-being.[1] We were doing mutual aid before it was called that.[2] We feel a response-ability to political movement (whether that looks like making sure our kin has home and food or whether that means street activism) by virtue of being alive.[3] We get in where we fit in. No matter what state in what country I have taken up residence or sojourn in, I find these people—my people. My fully human people.

We are drawn together not by identity, *sensu stricto*, but more by affinity—by a knowledge that what we have been presented with in a hegemonic and socially scripted way does not fulfill the desires of our souls, our innards, or our dearest and most precious longings. We understand that the nations and societies that formed us are founded by injustice and sustained by an endless list of -obias and -isms. We have a lasting understanding that reform will never bring the sustained transformations that we desire, but we help with strategic political goals as necessary or when required. Within these systems, rather than repressing that which arises from within (or without), only to risk having it come out in the most harmful of ways, we prefer to simply live it. To live it all. And even then, we find recesses of our selves where the detritus of the script lives. When this happens, we learn the painful lessons that live there, excavate what remains, and try again. When we are hated and ostracized for choosing to live as we are, we continue the healing work of not internalizing the vitriol, instead knowing that these responses are demonstrative of the tremendous power we hold within—in the most untouchable of places. We choose to embody, relate, and create in ways that defy what we are told we are supposed to be, know, and want; not defiance for the sake of defiance but for the sake of wanting to *actually live*, in this life, in these bodies, right now. These impetuses are what have driven my way of moving and being in the world(s).

My approach to ecosomatics, in its merciless porosity, names sociopolitical situatedness formed through "we's" as part of how we come to apprehend world(s). The place from which I experience world(s) is as a mixed-Black transmasculine nonbinary and queer person born in the United States. My Blackness taught me about the intersections of gender and racialization in a childhood spent in predominately white institutions. My experience as a girl and woman of color has shaped how I have come into masculinity and *trans*gressing its confines, choosing always to operate from a transfeminist life practice. The shape of my body, whether due to my femininity, my Blackness, and/or my transness, has had many ways of "getting me in trouble" whether this was through fetishization, anti-Black racism, tokenization, homophobia, transphobia, and an inability to experience "proper" racial or gender passing in most places through which I move. I express this situatedness as part of what has shaped my experience of living in my body, even if I am only able to name a partial element of my existence due to the limitations of social categorization and the Western over-emphasis on the field of visuality.

Ecosomatic Practice, San Cristóbal

It is a crisp 55°F (12°C) outside, as is typical for summers in San Cristóbal. It is mid-afternoon and the grass is wet from the intermittent rains. I move listening to and feeling the seduction of the energy of rebellion that seems to roll off the majestic mountaintops and emanate up from the ground itself. My movement studies the place, as my body imagines the views of various animals throughout this landscape. There is a solemnity to the cold that begins to envelop my body. I move my body imagining what a bird's-eye view sequence might see and hear as various struggles co-exist in this place. Pictured in the image below (*see* Figure 3.1), my body leans into the brisk air and I imagine what it is to soar—flight as grounded presence.

Like a kaleidoscope running through my head, I enter into a trance-like space, watching gathering points of resistance against various forms of gendered violence; organizing for the lack of basic infrastructure within and beyond the city's limits; struggles against the extreme austerity measures placed upon public schools and the blockades raised in protest; the displacement of nearby Native/Indigenous communities by national and multinational corporations holding encampment in the Plaza de la Resistencia; and the increasingly visible queer population making itself known.

FIGURE 3.1   Daniel Ìgbín'bí Coleman in a video still from *La Envergadura: Ecologías Trans*, ©2019, permission of the author.

I have positioned my body to play with memory in this open space because from here I am witnessed by a massive mountain on one side and can hear and see the city from the distance on the other side. The red dirt reminds me of a time before the penetration of capital and of other Southern places of red dirt. Another memory emerges.

When the rain pours down hard during the hurricane season of the summer, everything momentarily stops while people hide under the makeshift covering of their stalls, stand under the overhangs of shops, and hold newspapers over their heads while moving quickly toward shelter. From this vantage point near the mountain, I can feel the push and pull of the calmness and the surge, characteristic of the rhythm of the city's life where social stratification and gentrification is revealed. The pause and contemplation that the rain brings is significant because it is often truly torrential. It is both quickly burdensome and quickly dries up, returning life to its normal hustle and bustle.

This rhythm strikes me as a heartbeat: one that races and slows regularly, sometimes daily. There are so many Native/Indigenous people who make the trek to town every day to sell their goods and merchandise—the quantity of goods sold representing their livelihood. They are competing with the internal markets of the city—the appropriation of Native/Indigenous dress for *coleto* (European-mestizo people) consumption, for example—and live their lives in this day-by-day gamble.[4] The neoliberal reality has sedimented itself into the fiber of the land, even after 1994. There are still *coletos* who expect Native/Indigenous people to move aside if they walk by them.

All these images glide across my mind's eye as I surrender my body to place. As distant smells of petrol filter into my nostrils, I feel the wide-reaching effects of neoliberalism and its attendant economies of extraction. Some of the people who had sole dominion over these highlands for millennia are now relegated to the outskirts of the city struggling to make a subsistence living. Much is to be paid for the presence of worlds who believe in a singular world here.

I lay back into the fog and smoke that are accompanying me through these sensorial recollections—copal and condensation furl around my arms and upper body. My body tells me that the ancestors of this land never left it and never will. The smoke speaks of ceremony and knowing. I tap into the movement of my arms, spreading them like wings and again imagine what it is to be a creature in flight. My limbs imagine the air is water and I let the movement take me. I wait for the movement to tell me when it is ready to still and come to a close.

In this vignette of ecosomatic movement/performance practice, what happens in the mind as the body explores a space is part of what the work brings out of the archive. Rather than focusing on literal representations of a space, ecosomatic work aims to lean into the fullness of the body's capacity to take a place in, to recall, and story layers of co-creation. By moving in a site, I create site again and again, in collaboration with everything around me and in all the memories contained within me. Though I am moving as a solo artist and sentient being, I am ever-collectively intwined, entangled, and co-constitutive of all that has formed me and the layers of life around and within me. It is here that we might come to understand a place anew, time and again.

Ecosomatics as Method

I have developed an ecosomatic methodology for research practice in the pluriverse that comes from the desire to trust the memories of embodied knowledge that become archived in the body through engaging in different worlds. I will focus on my work in San Cristóbal de las Casas here. I participated in and alongside struggles for the livelihood of women, queer, and trans people who are Black and/or brown (brown understood primarily as having Native/Indigenous ancestry from the Americas and being racialized thusly), and against state forms of terror and violence like feminicide and disappearance. I have moved as an artivist and organizer because I am spiritually driven to participate in fighting against the extermination of the ones I love and those in interconnected communities of affiliation and kinship.[5] In this ecosomatic approach, the goal is to create containers for the metabolic process of tracing through time and participation in lifeworlds—a pluriversal negotiation away from singular narratives that attempt to create easy histories of place.

Ecosomatics is a performance as research (PAR) methodology and practice that asks us to live sensitively in our bodies, acknowledging their porosity, precarity, resilience, and shifting strength and sensibility.[6] Ecosomatics is a performatively archeological endeavor where we use our physical, spiritual, social, and political bodies as a reading practice of the praxes we have been inside of and alongside of, in our endeavors to build other worlds, live in other worlds, and build toward liberation. Within the formal elements of this practice, modalities like installation, visual art and objects, sound and music, duration, and forms of movement and/ or stillness can serve as artistic mediums for engaging in processes of re-memory, archival tracings, and ecopoetic witnessing of processes born in communities of struggle. Ecosomatics, for me, is all about the process, though it can certainly have a product, as my work has.[7] This means that though formal artistic practices and elements might be used, the ecosomatic method is in the process of creation itself, before, during, and after any archive of a consolidated "piece." The ecosomatic research is all of the "primary material" that is sifted through the body as memory receptable for community and kinship labor toward something else.

Theoretical Approaches for Ecosomatic Work

A couple of underlying theoretical prerogatives shape my approach to ecosomatic work. In my praxis of living and moving, these truths have brought me great clarity about the epistemological and ontological priorities of my intellectual work: (1) the singular idea of a universe understood as a totalizing epistemic field we (our species) like to call *The* world is not singular at all. Further, I would venture to say the present culture wars and their crises of imagination that make the resistance to human plurality violently clear are a sign and symptom of the unnatural homogenizing and colonial projects that seem to be running ablaze in the blood of those with sociopolitical power, like incurable epigenic pandemics.[8] Given this, I am driven by pluriversal thought and consciousness of the existence of many worlds. (2) Within the manifold crises manufactured by homogenizing anxieties, as a species many continue to be terrified and terrorized by the body, our bodies, other people's bodies, and their infinite wisdom, capacity, desires, and possibilities for connecting to everything and everyone around us. Despite our hyper technologically-mediated lives and human-made compounds that now reside within all of us, we remain in a material and fleshly form having what some call a "human existence."[9] It is holding these two truths and intellectual priorities that grounds my ecosomatic work.

Approaching ecosomatics from here inspires my pluriversal commitment to acknowledging the cosmological roots of knowledge production. I am invested in how cosmological roots of thought are essential to both Black and Afro-descendant and Native/Indigenous worlds in the hemispheric Americas. I begin with spiritual cosmology that culls Indigenous West African understandings of cosmogony and cosmology, the body, life, and all our relations to forces of nature through its recovery in the diaspora. The living protocols of Afro-diasporic practices demonstrate their transformations through the trans-Atlantic slave trade and natural adaptations to a world not built for plurality, so that descendants in the diaspora might continue to have spiritual technologies to survive in the wake.[10] Returning to Chiapas from and through multiple cosmological perspectives, namely, Afro-diasporic practices, brings me to more profound engagements with the Native/Indigenous cosmologies and spiritual understandings of life. Returning to Chiapas to engage in ecosomatic performance increases my understanding of these cosmologies as already ecologically interdependent. These spiritual cosmologies have directed some of the most consequential political movements in Chiapas—an entire population of people driven by the ancient *and* modern wisdom their ancestors passed down for millennia and adapted to meet the time.[11] People who have always understood that to destroy nature is to destroy ourselves.

Pluriversal politics round out my ecosomatic approach, allowing ecosomatic methodologies to lead us toward something else. Following the direction of Arturo Escobar, Marisol de la Cadena, and Mario Blaser, I understand pluriversal politics as modes of relating to the social that recognizes as sovereign manifold stories of the origin of the cosmos, our planet and all its creatures (including us), and how nature operates.[12] Pluriversal modes of thinking are part of larger decolonial thought formations and practices that, rather than engage in cultural relativism, challenges the very concept of "cultural" as it is deployed by the dominant groups of any given society. Pluriversal ways of knowing understand something much more radical: we literally live in different worlds. The great tension in and gaslighting of the political spheres we are located inside and the parties that claim to represent us is that they are incapable of recognizing the many cosmologies that guide our lives. The singular idea of a world must be created and re-created, with tremendous amount of waste and bloodshed, to convince itself of the necessity of its existence. All this is giving us is a dying planet. Global Western hegemony, supremacy, and militaristic imperialism have been powerful forces in this attempt at a homogenized conception of what our lives are and should be. A pluriversal decolonial understanding of others and their worlds would mean that we might just have to recognize other humans as part of the same species and their relationships to other forms of living and non-living beings and things.[13] Ecosomatic practice allows my pores to grasp why one of the primary Zapatista emblems is *"un mundo donde quepan muchos mundos"* or "a world in which many worlds fit."[14]

Practices for Ecosomatic Attunement

The following practices are meant to help the body/your bodies tune in with place. They can be modified or translated as needed to relate to your specific location.

Practice 1—Walking

Spend the greater part of a day walking in different parts of the location you are engaging in ecosomatically. Be intentional about silence, even if you greet others along the path. Take in the place with your full senses. Record in body and mind, alone. Walk/move until your body is tired and your limbs and feet feel sore. This will mean different things for different people depending on your body's capacity. Turn inward as you walk, while taking in every sensation that you can. Return to your resting place and prepare a nice meal for yourself and get your body into hot water. Drink mugwort tea before lying down to help open your dreams. When you awaken, write down any symbols and messages that your dreams presented. This practice can be repeated as many times as needed and can be broken into several days to spend time in different parts of place.

Practice 2—Water

Immerse yourself in a natural body of water in the site of choice. If you do not go alone, be sure to spend time with the water, alone. Move your body in water until you bring it to a point of saturation/exhaustion. Exit the water and rest on land near it. Drink and nourish your body. Sit and let the tiredness wash over you in an open- or closed-eyed meditation. Listen to the stand-out images, sensations, sounds, smells that come to you. Record them in some way.

Practice 3—Mountain

Sit in a place where you feel small, ideally in a place where you can see a mountainous region surrounding you (or at least ahead, flanking, or behind you). You do not need to sit in any specific position, but just sit. Bring a notebook and something to write by hand with you. Bring hydration. Keep your eyes open. Choose several hours of silence. Bring your awareness to the sounds around you and inside of you. Notice where your thoughts go, without judgment, and sit with how they are shaping your experience (or not). Write when you are moved to record information that comes to you. Close your eyes intermittently (but never for more than 30 seconds) and tune in to the "noise" of "silence." Find vibration (be creative about what this means) and tune into its source. Hum to the sound of the vibration and let your body tell you when your sound/song should stop. Find silence again. Remain here for at least 2 hours. Slowly transition out. It is advisable to remain relatively silent for what remains of your day, while integrating what messages have arisen.

Notes

1 It is in transfeminist community organizing that I have found kinship with people across these identity vectors and social positions, and then some. What has most excited me by transfeminist organizing and thought is the willingness to be queer (in the queer theory sense of a joyfulness in being different and

non-hegemonically situated), and/or in the context of Abya Yala (the Kuna word that replaces "Latin America" in decolonial thought) where communities invite one another into digressions from gender and sexual normativities to free up otherwise orientations to the world that do not rely on or turn to respectability in our life practices. For a very consequential text in transfeminist thought, *see* Elena Urko and Miriam Solá, eds., *Transfeminismos: Epistemes, Fricciones, y Flujos* (Tafalla: Txalaparta, S.L.L., 2013).

2 "Mutual aid" is a term for a form of community organizing that became quite literally indispensable during the COVID-19 pandemic. Mutual aid is also a critical position of barter and support that does not engage in saviorship or relationships of charity. *See* the work of Dean Spade, here: www.deanspade.net/mutual-aid-building-solidarity-during-this-crisis-and-the-next/.

3 Response-ability, or the ability to respond, comes from the work of Kelly Oliver. *See* Kelly Oliver, *Witnessing: Beyond Recognition* (Minneapolis: University of Minnesota Press, 2001), 135.

4 The appropriation of Native/Indigenous dress for coleto consumption can be seen in the series of designer stores that have women's blouses with embroidery patterns particular to those communities selling for top dollar. And, with no connection to what the embroidered symbols represent.

5 "Artivist" is a word that bridges artist and activist meaning those who work at this intersection, making intentionally and explicitly political art. "Political" not as in electoral politics, but "political" as in responding to the charged nature of our lived experiences as they enter the field of power relations.

6 During my Master's program at San José State University, under Riley's mentorship and tutelage, I was introduced to how to make performance as a way of thinking about the world. In this work, I follow in the vein of Riley and Hunter. *See* Shannon Rose Riley and Lynette Hunter, "Introduction," in *Mapping Landscapes for Performance as Research: Scholarly Acts and Creative Cartographies*, eds. Shannon Rose Riley and Lynette Hunter (Houndmills and Basingstoke: Palgrave Macmillan, 2009), xv–xxiv.

7 The "product" of the ecosomatic work in Chiapas is available for viewing here: www.danielbcoleman.com/7432010-2019-ecosystem-service-trans-ecologies#1.

8 The culture wars in both the United States and México are endless. I understand "culture wars" to be struggles for what is ethically and morally acceptable in any given society, often creating a binaristic electoral politics approach to much of human behavior and struggling to legislate accordingly.

9 *See* Yusof's commentary on the result of nuclear fallouts on our bodies: Kathryn Yusoff, *A Billion Black Anthropocenes or None* (Minneapolis: University of Minnesota Press, 2018).

10 For more on the "wake," *see* Christina Elizabeth Sharpe, *In the Wake: On Blackness and Being* (Durham: Duke University Press, 2016).

11 Aymara/Bolivian feminist sociologist Silvia Rivera Cusicanqui explains that Native/Indigenous people would have always had their own "modernity," and the colonial encounter that led to what we presently know as modernity does not foreclose on the reality that there would have been an entirely different conceptualization of the modern without this violence. *See* Silvia Rivera

Cusicanqui, "*Ch'ixinakax utxiwa*: A Reflection on the Practices and Discourses of Decolonization," *South Atlantic Quarterly* 111, no. 1 (2012). https://doi.org/10.1215/00382876-1472612.

12 *See* Marisol de la Cadena and Mario Blaser, *A World of Many Worlds* (Durham: Duke University Press, 2018). *See also* Arturo Escobar, *Pluriversal Politics: The Real and the Possible* (Durham: Duke University Press, 2020).

13 I am a Wynterian scholar of race (following Sylvia Wynter) and proceed in her footsteps with this comment—one I see as part of her primary argument in Sylvia Wynter, "1492: A New World View," in *Race, Discourse, and the Origin of the Americas*, eds. Vera Lawrence Hyatt and Rex Nettleford (Washington and London: Smithsonian Institution Press, 1995).

14 To read how the Zapatistas continue to construct "a world in which many worlds fit," *see* https://enlacezapatista.ezln.org.mx/2018/01/01/palabras-del-comite-clandestino-revolucionario-indigena-comandancia-general-del-ejercito-zapatista-de-liberacion-nacional-el-1-de-enero-del-2018-24-aniversario-del-inicio-de-la-guerra-contra-el-olvi/.

Bibliography

de la Cadena, Marisol, and Mario Blaser. *A World of Many Worlds*. Durham: Duke University Press, 2018.

Escobar, Arturo. *Pluriversal Politics: The Real and the Possible*. Durham: Duke University Press, 2020.

Oliver, Kelly. *Witnessing: Beyond Recognition*. Minneapolis: University of Minnesota Press, 2001.

Riley, Shannon Rose, and Lynette Hunter. "Introduction." In *Mapping Landscapes for Performance as Research: Scholarly Acts and Creative Cartographies*, edited by Shannon Rose Riley and Lynette Hunter. Houndmills; Basingstoke: Palgrave Macmillan, 2009.

Rivera Cusicanqui, Silvia. "*Ch'ixinakax utxiwa*: A Reflection on the Practices and Discourses of Decolonization." *South Atlantic Quarterly* 111, no. 1 (2012): 95–109. https://doi.org/10.1215/00382876-1472612.

Sharpe, Christina Elizabeth. *In the Wake: On Blackness and Being*. Durham: Duke University Press, 2016.

Urko, Elena, and Miriam Solá, eds. *Transfeminismos: Epistemes, Fricciones, y Flujos*. Tafalla: Txalaparta, S.L.L., 2013.

Wynter, Sylvia. "1492: A New World View." In *Race, Discourse, and the Origin of the Americas*, edited by Vera Lawrence Hyatt and Rex Nettleford. Washington; London: Smithsonian Institution Press, 1995.

Yusoff, Kathryn. *A Billion Black Anthropocenes or None*. Minneapolis: University of Minnesota Press, 2018.

4

MATERIAL/MATERIAL

Thousandfold Somas and Poetry of Emergence

Sondra Fraleigh
St. George, Utah, U.S.A.: 37.0941° N, 113.5749° W

In Vertical Time
We were water
questing land
then
Just beneath the skin—
we vanished and assembled morphology.

Our somas sleep
now
and return to shore.

Hyletic Interplay

Matter matters. Material nature matters, and humans are part of this mattering. There is no one irrefutable truth about human nature or mattering. The body of the Earth and our human bodies are materially linked. The nature of being human is constantly emerging intra-actively and collectively with other kinds of nature. Our lived bodies are not stable biological essences; rather, they are materially in process. They are not representations; they are ongoing performances and enactments of possibilities. Somatic experience emerges through the hyletic interplay of the cultural body with ethical/ecological potentials. Sense perception is not passive but shapes and is shaped by our sentient openness to phenomena—including tangible objects, kinesthetic occurrences, events, images, memories, and lived experiences. Such openness (responsiveness)

DOI: 10.4324/9781003390985-6

is built throughout life in somatic materializations of human consciousness and the world.[1]

Watsuji Tetsurō's phenomenology of nature demonstrates that "perceptions–even the simplest ones–are always of whole forms in particular contexts."[2] Self and nature structure subjectivity relationally and have a history completed through culture.[3] Similarly, through material feminism and the physics of Niels Bohr, Karen Barad explains the performative engagement of material nature with the human body and non-human world as intra-action, wherein phenomena are not represented things but ongoing relations that make embodied concepts meaningful.[4] Heidegger puts the ethos of emergence this way: "Greater than actuality stands possibility."[5] This chapter follows such philosophies where the things-in-themselves of phenomenology become expressive and performative rather than positional.

The Text

The text describes entanglements between material nature and cultural happenings, moving toward lived and material nature pioneered in feminist theory and spearheaded by thinkers such as Karen Barad and Vicki Kirby.[6] The ecological phenomenology of Edward S. Casey, Glen A. Mazis, and Watsuji Tetsurō also supports the whole. New materialism has emerged from philosophy, science studies, and cultural theory, cutting across human and natural sciences. The nature of being human is constantly emerging intra-actively with other kinds of nature. Our lived bodies are not stable biological essences; rather, they are materially and ecologically in process. *This work develops a somatic, performative phenomenology of corporeal materialization and transformation from these inspirations.*

Phenomenology is intra-active at root, taking place as a performance that underscores the kinetic, emergent nature of consciousness. As a phenomenological performance unfolds, it flows, creating and re-creating itself in analysis of experience and insights. Phenomenology materializes in text, media, or live events: setting aside common assumptions, taking risks, and following interconnected investigations. As a method, phenomenology folds inward to question taken-for-granted knowledge and opens outward toward discovery and form. The method of this chapter pursues questions of transformation through articulations of somatic processes and poetry. Likewise, it resources Watsuji Tetsurō's Japanese phenomenology of nature where being-in-the-world is relational as being-with-others, and the *space of self* is a network of nature and culture.[7] The founder of phenomenology, Edmund Husserl, describes the *emergent kinetics* of phenomenological methodology. The phenomenological-constitutive emergence takes unity in the flowing live present. Expanding toward an ecosomatic gaze, Husserl says: "Consciousness of the world [...] is in constant motion."[8]

The chapter surfaces with a brief prologue on butoh and its political materiality in the aftermath of World War II in Japan. I return to World War II and my dance studies in Germany not long after the war, eventually reflecting on the "everydayness" of body politics and how the body of the Earth and our human bodies are materially and spiritually linked. I develop *matching* as a somatic strategy from these reflections, as the text turns away from bodily perfection and mastery.

Surfacing with Butoh

Butoh is an avowedly material form of dance and apt example of affective/ideological facets of bodily materializations. Rooted in mid-twentieth-century Japan, butoh emerged as the "dance of darkness" and "ancient dance" after World War II.[9] Tatsumi Hijikata originated butoh as a performative critique of the Western colonialization of Japan, conspicuous materialism, and the conservative immobility of Japanese society. Today, residues of Hijikata's radical dance continue to migrate across cultures through dance translations individually crafted and comprehended.

Butoh is not a progressive art, sinking, as it does, toward mud and elemental ecology. Its movements are based on images and atmospheric changes, morphing from elegance to anguish through corporeal aesthetics. Most significantly, for Japan after the war, butoh brought marks of suffering into dance, sublimating the body while extending its indeterminate states. This dance genre seldom lands anywhere: it keeps changing. The experience of *ma*, the space between in Japanese, or liminality in psychology, generates butoh aesthetics. Butoh dancers explore psychological shadows, often relative to concerns for the suffering of nature.[10] Aesthetics of emergence and morphology allow dancers to disappear or intensify, just as nature changes with surprises. The human materials of butoh intersect with environmental materials: insects, trees, lakes, stones, and much more, while the space between these materials creates an improvised precarity. What humans make of their experience and how they perform connectivity is an individual matter of emergent consciousness. This is true for both the dancer and the audience if an audience is present.

"Inner Material/Material" (1960) is the title of Hijikata's first short essay in which he decries materialist production entering Japan after World War II.[11] His crisis concerning capitalist production and the "bleeding" of nature is even direr in his following essay, "To Prison" (1961).[12] He would rather go to prison and "get caught smack in the middle of a mistake" than take "a bad check called democracy."[13] While he rails, he depicts his surrealist dance tactics in view of "materials," inner materials: ideals, ideas, happenings, and emotions. I take my cue in stating the somatic materials of performance from Hijikata's inward glance.

Part 1. Movement Material/Performed Material/Everyday Material

As material, dance is not just any movement; it is performed on purpose—sometimes (but not always) for an audience. Autotelic performance is done for intrinsic purposes, explored in somatic transformative processes for performer benefit. This kind of performing is not for audience engagement; instead, it explores dance and movement for participants. Performances always exist in context—in acting, dancing, cooking, conversation, and more. As an attentive way of doing, performance might be a simple enactment of a task, or it might link to achievement. To perform in the arts is to orient actions in expert ways that are practiced every day—to sing and dance as one might like. Affective influences, feelings, and emotions circulate through such performed materials and extend beyond the human in ecosomatic orientations and events. Affectivity is everywhere embodied and conveyed in the lifeworld, as this book holds. Concerning expressions of affectivity and the body, Merleau-Ponty says: "Emotion is not a psychic internal fact but rather a variation of our relations with others and the world, which is expressed in our bodily attitude."[14] Similarly, Casey's recent work, Turning Emotion Inside Out, holds that emotions surpass the human subject and are not simply generated or held as self within.[15]

The emerging field of ecosomatics investigates relationships between the direct experience of the body with the larger field of living beings in which human life is embedded and too often imperious. Ecosomatic performances focus on the ecological self relative to specific events and places in the environment. I call my dances in the environment Place Dances in light of Casey's phenomenology of nature and place-worlds, his view of "finding soul in place."[16] These dances are ongoing performances of perceiving and remembering the soul of particular places. "Canyon Consciousness" delineates Place Dances that engage the rugged beauty of Utah's canyonlands.[17]

Indeed, we orient everyday according to place and nature. The body acts on, intends, anticipates, and relates to its environment. Husserl described the manifold nature of intentionality in orientations of consciousness toward the environing lifeworld of nature and the intuited and affective worlds of sense and culture. He envisaged the human as materially and subjectively a part of this multilayered lifeworld. Lifeworlds are experienced with complexity, united with the body as "a single psycho-physical thing."[18]

In a word, humans are never separate from nature, except they separate themselves psychologically and somatically.

Later, Watsuji's Japanese phenomenology (in the wake of Husserl and Heidegger) confirmed intentional continuity as "interactional exchanges" with otherness in belonging to nature and the world.[19] This study asks how transformative experience can occur in our relationships to nature and each other, with what shifts of intention and interactional exchanges? Human attitudes to nature matter in how we position ourselves as *within* or *above* nature. What kind of performances, including those right in front of us and still arriving every day, settle human postures toward nature? I don't have immediate answers to these questions, but I trust I can write my way into them.

Intentionality makes all the difference in ecosomatic orientations. We might pursue performance as an everyday occurrence, one that holds transformative possibilities through a shift in consciousness and intention. The *everyday* could mean many things; thus, we pause on this term. In the phenomenology of Heidegger, everydayness is shown through the repetitiveness of daily life. He says "how" one "lives into the day" can take many turns and even be forced into "burdensome" tasks. Everydayness might bog down into "dullness" and even "suffering." Heidegger writes that as a concept, the everyday is only possible in frameworks of being as ontologically disclosed. In his study of being and time, he describes being as temporal, material, and relational. "Being-with-others" is known in the everyday, "facing-forward" in time while "leaving behind" a history of all that has been. How one "lives into the day" is everydayness. His emphasis on how we live suggests method, process, and performance. Existence has possible variations and is not always dull or doomed in the everyday, as Heidegger also outlines in *Being and Time*: "In the moment of vision," and indeed often just "for that moment," existence can even "master" the everyday, "but it can never extinguish it."[20] I also seize this moment, but I question the onus of mastery as it tackles the everyday.

Your bodied soul sounds
When in an instant, death comes with hands
in his pockets, appears with a pale complexion
assured in violence. Why didn't he stop
weighing down your neck?
In gasps then silence,
darkness abjured—and breath taken,
everyone
saw your need to breathe,
while his need
was to go on hurting.

The killing of George Floyd in Minneapolis on May 25, 2020 captured the world's attention and re-exposed racial crises in our time. Floyd's death spurred nationwide protests against police brutality and a reckoning over everything from public relics suggesting the shames of slavery to sports team names that disrespect ethnic identities.

Matter matters; it returns historically and manifests intentions. Heidegger, who supported Nazi causes as Hitler rose to power in Germany, suggests mastery, struggle, overcoming, and control as matters of comportment in the everyday.[21] Instead of mastering the everyday in a posture of overcoming and control, we might better practice *somatic matching*, or *conscious attunement*. Somatic matching rests on a belief that agency and action can listen, reciprocate, and not dominate. I have written of this previously through the concept of *matching not mastery* and a revaluation of darkness.[22] Butoh arose partly as a revaluation of darkness and was called "the dance of darkness," not because it was considered evil, but because it advanced the power of morphology and exposed the problems of light/dark binaries.[23] "Start with your handicap" was Hijikata's advice about dancing. Privilege in normalization of perfect bodies and perfect movement is oppressive; let's consider the "perfect state" with "perfected bodies" that the Nazis envisioned.

I studied dance in Germany in 1965 in the wake of World War II and prepared for this by reading about Adolph Hitler's rise to power in several sources, particularly William Shirer's *The Rise and Fall of the Third Reich*.[24] It was harrowing to learn about Hitler's political hold on so many millions of people and to read Shirer's graphic depictions of the Nazi objectification and mutilation of human bodies in Josef Mengele's medical experiments on Jews. I studied with Mary Wigman in her studio in West Berlin barely acknowledging the elephant in the room—all that had happened to countless people in the war, the killing of millions of Jews, the war dead on all sides, but I was there to study dance (wasn't I?).

Expressionist art and dance had already come to fruition long before my time in Berlin. Wigman's first *Witch Dance* was in 1914, and Hitler invaded Poland in 1939, the year I was born. Eventually, I was able to see similarities between butoh and expressionism, mainly through their surrealist tactics and admission of ugliness. Neither art form expected perfect bodies nor purified movement. That came through Nazi beliefs in a superior Aryan race of Northern European descent. Hitler's monstrous goal was to perfect that race.

A year and a half of studying expressionist forms in Germany helped shape my consciousness, which is still emerging. Berlin was rebuilding when I studied there, but much was still in ruins in 1965; no one had really recovered, most certainly not nature nor the soul of Germany. I returned in 2003 after the hated wall between East and West Berlin went down, but I was still holding the memory of ghost somas rising from the streets. Today, I appreciate regenerative leadership in German politics and former Chancellor Angela Merkel's courage, which led to descriptions of her as the de facto leader of the European Union and the most powerful woman globally. It is notable that in her rise to power, she became Minister for the Environment, Nature Conservation, and Nuclear Safety in 1994. As I write, I sense a pearl of cultural wisdom in the zeitgeist and body politics of Germany, one that braves immigration and otherness, however tenuous. Of course, the Neo-Nazis are also on the move, but they have much to push against this time.

When I studied in Berlin, Frau Mathilde Thiele was the primary teacher at the Wigman school aside from Wigman herself. Many years later, Frau Thiele lived 20 minutes from me in upstate New York, and we visited often. I was in mid-life, and she was toward the end of her life. She told me her personal stories of living through two world wars, and incredibly how she survived the devastating bombing of Dresden. I have recorded her stories and photographs of her key dance on artist Käthe Kollwitz and World War I on my website.[25] Thus, I understand something of personal struggle during wartime. I also see the importance of self-reflection in body politics, that autocrats can't take power that isn't prepared for them from the collective body politic. Hitler succeeded in large part because antisemitism was prepared for him, already shaping cultural minds in the field of his times.

Sew
into my heart,
uncover what's missing
in the materials I seam every day.
What fabrics
folly
hatreds
and envy.
Etch the stitching.
Let me see it, and hear more clearly.

Listening is a sense demonstration of matching. Attention itself is a kind of matching. We match that which we attend to carefully. In her work on ethics, philosopher Iris Murdoch views attention as moral training and virtue, "looking carefully at something and holding it before the mind."[26] She says that "moral change comes from an attention to the world whose natural result is a decrease in egoism through an increased sense of the reality of [...] other people, but also other things."[27] Everyday performances might shift somatic attention from exclusive self-awareness to inclusive-awareness of others and rescue of nature.

In her recent phenomenology, Audrey Lane Ellis studies how dance improvisation can magnify corporeal attention through listening rather than accomplishment, tying corporeal listening to *ethics*.[28] Experience does not happen to people; instead, it involves direction of attention, making decisions, self-moving, moving with and alongside others, and shaping (or stumbling into) everyday and special performances. Heidegger is correct; the way we live into the everyday is important, but he misses the somatic point of attunement, a quality of attention and intention that one can cultivate. Experiences are not simply "had"; we act or not, and experience accrues through individual and collective intentions. Such agency involves how we see ourselves and others, imagine and hope, and especially how we move into everyday life. *How* indicates a way of orienting performance. *How* is there in how we roll over to sit up, stand, and walk; how we breathe into material activations of everyday being—tasks, emotions, plans, and disappointments—

All eventualities
My aging husband is in a wheelchair now,
But he still holds the residue of all he has done and been.
His material-spiritual being sustains him still as he adapts,
matching his precious few steps to the affordances around him.

Every evening, we remember the places we have visited
and the people we have known. All of it presences again.
Our acts together and apart are present in fading materials,
floating non-solid liquids wrapped up in encounters
with people and animals—not to forget our pets—
and not to mention the weather.

How the weather moves us,
and so do sounds of crickets and frogs.
Our little dog loves us—
as naturally as whales sing to their young
in the moving waters.

We soak in songs and several worlds of matter,
matching or resisting things: stuff, materials,
canyon weeds, ragweed cactus, manzanita, melting snow,
red dirt, real and imagined people.

Materials of somatics appear in several ways, often through unspectacular, everyday movement in simple dance explorations or performances. Somatics is a field of practices not well-understood; thus, this chapter includes definitional aims. Somatic movement approaches in their educational and therapeutic settings are based on experiential, exploratory methods, not on accomplishment; their precarity teaches us to respect material entanglements. Respect for material life appears as ethical contemplation in somatics. Outcomes of somatic explorations do not yield scientific truths. More to the point, they engender present-centered attention and listening. Like the one-pointed attitude of Zen, somatic explorations teach the substance of awareness in the moment of action, which is precisely the project of performance, whether everyday-commonplace or expert.

Part 2. Somatic Empathy/Transformative Practices

The field of somatics is concerned with lived contexts of perception. As Gaston Bachelard and Maurice Merleau-Ponty critique it, perception conceived as sense data is analyzed in abstract outcomes removed from lived contexts. Sense experiences are highly mental data lost to the continuity of poetic emotions.[29] Soma-psyche is the lived space and depth of feeling in what perception uncovers. Soma is another word for the body organism, inseparable from the psyche as the sentient basis for feeling and moving. In soma-psyche, the physical and psychological link, and they flourish or shrink relative to nature. Casey's essays in philosophical psychology suggest connective spaces of nature, the ground underfoot and the sky above, archetypal presences in image-forms. Specific environments bring us into their poetics of soul and history.[30]

The neuroscientific psychology of Antonio Damasio conceives soma as the silent background of extended consciousness. In his pathbreaking work, *The Feeling of What Happens: Body and Emotion in the Making of Consciousness*, he shows how the somatic biologic stratum of life, what he terms the "proto-self," is not available to consciousness or capable of perception or control.[31] It is, however, the organic foundation for feeling and consciousness.[32] Consciousness is global and organismic, not a local brain property. Damasio argues that feelings are nature's evolutionary solution for preserving life. Conscious awareness of feeling is connective. Change in body state and cognition and subject-object relations activates somatosensory felt life.[33] "You simply cannot escape the affectation of your organism, motor and emotional most of all."[34]

Somatically oriented movement and dance explorations call attention to qualities of feelings that emerge in consciousness, especially as markers of change. What we care about and are afraid of appear, and through the feeling that movement awareness engenders, we can reflect on habits of heart and mind. In asking into somatic phenomena in the first place, this work is anchored in considerations of somatic empathy and transformation. Identifying humans with nature and each other, Husserl's philosophy of lived body and lifeworld explains human empathy as "intersubjective mutuality."[35] This study is committed to the reality and mattering of nature, and the manifold orientations of somatic practices harken to this commitment. Nature enfolds humans, and our fate is tied up in our care for each other and the Earth. By now, however, you may be wondering if we have a problem with this.

> *Our unrequited misery*
> Longs for what is missing
> In acoustic imagination
> parts of liquid memory
> and severances.
> Nature has a hold on us—
> So, what will we do
> when the ocean's gone missing?

Wrapped in longing, empathy with nature is vital to our collective bodily health and sustainable emergence. The survival of the planet depends on our feeling and care for nature and our felt belonging to environments beyond four walls. Movement approaches of somatic practices demonstrate material entwinements of the human body in a broader aesthetic (affective) collective and unsettle the discursive/material, intellectual divide that material feminism also questions.[36]

> *We are more than words*
> Even light gleams belong to materials
> in bright ontologies of being-more-than-self.
> We change daily and become
> more-or-less
> forward and in retrograde.
>
> The day wanes and sheds
> while we experiment and dance,
> feet up to head down then horizontal
> with many voiced mysteries.
>
> > *Yikes!*
> > *Is it too late to seek a better world?*
> > Much as waters flow and trickle

humans flow and trip,
steeped in particulate matters.

My work
pray enters the living mystery
when I lie down in the grass
and look up to merge with all that is birthing
darkness and light—
the calm cosmological ethics of *all*.

> *Am I doing this the wrong way!*
> *Probing potentials of human-nature?*
> In the way somatic moves
> explore transformations
> of walking, reaching, falling,
> and responding?
> In how we evidence
> collective materiality?

We who might otherwise be peaceful in nature
Engage in second-guessing and aggression,
mingling
with polyester in crowded figures,
listening
through rustles of silk and fragrance
flora and fauna
as animals ourselves.

Words don't replace the material world, nor the materiality of human life. The hope of critical theory belies the discursive side, failing to unsettle the material science it is up against.[37] Critical theory has eschewed nature as a category of concern but now has pause to reassess.[38]

> *Hold on a second! What is nature?*
> Is nature defined in *letting be* (*seinlassen*)
> as Heidegger scripted?
> Or in leaving white to the light,
> while setting all else aside as dark animus?

Is darkness not nuanced and beautiful in variety?
Uncovering the sinews and tints of nature,
will justice not be justice for all?

I can be present with my attention in tune with my own nature,
matching sky-blue with my gaze as given, but not to me alone.

I share in a somatic politics of well-being for all-beings.
And I opt for *letting be*: it seems right to me
we let everything sing
as our innate potentials flex and grow.

Critiques of dance history have discounted nature and now need to resettle
this. When I broach nature as a topic in performance studies, the usual
retort is that nature is not real; it is a romantic ruse and trope. This attitude
arose primarily through Ann Daly's critique of Isadora Duncan's dances
regarding nature.[39] Romance, however, is not one thing, nor is nature.

Like long muscles
Romance striates in furrows and streaks.
The impracticality of romance courts mind
much as humans and other animals
fall in love and like.

Romance bubbles and
barely does it seem to rain
but the fog eludes.

Today I saw the White Rain Lily
and felt the exotic protection
of the giant California Fan Palm—
here in the desert where it snows in winter
and thunder shakes the towering canyons.

I have danced some of the Duncan repertory related to waves and the
rushing of the sea. My body remembers the material surge of Isadora's dance
amid the brilliance of harps. It is inspiring to embody the feeling of breath
motivating her movement. Her dance convinces me that nature is a reality.

Humans and other animals
are bone,
flesh, and nerve,
and human tastes are entangled
in the material swelling sea:
its majesty now afloat
with plastic bottles and debris.

When I wonder how movement can be material
How human nature
embodies material minds,
I stumble, fall, and find out—in quick-sparked
feelings of falling, gliding like a plane,
then landing in a heap of mind.

In the public garden
My puppy and I meet people of every ilk.
Today—a man beaming—out-of-his-mind
about getting married in the park's picnic rotunda.
We talk about his bride as he unloads chairs
From the bed of his truck.

Unfortunately
I lose my furry one in the wood chips and need to call her back.
My human is barking—"can't have that," says puppy,
happy for not being attached.

Life is already here, emerging, becoming, going, and arriving. No matter how old we become; we are still arriving as our material ethics and aesthetics evolve. Matter, movement, and emotion are inseparable. As somatic markers, sensing and feeling surface through intentions and actions. It matters that we stop to listen and feel. Body and emotion emerge in the making of consciousness. Damasio teaches that feelings are "guides to the future," and harbingers of ethics and culture.[40]

My heart is a portal and a performance
All the material goods I touch, ingest, and apply to my body matter to my inner material/material becoming. The materials and people I contact change my body and self-understanding. My environment, daily habits, and the objects around me shape my consciousness in complex ways that connect to my responsivity.

In what direction do I want the contents of my consciousness to nudge me when I am unaware? What quality of heart do I want to cultivate? How will I perform everyday encounters? With care, if I care, and with fairness, if I am fair?

Part 3. Material Glances

In this part, I build on two questions accruing from the first parts of this chapter: (1) how somatic movement explorations provide experiential avenues toward perceptual transformations, and (2) how somatic explorations evidence collective materiality of nature and humans. Below, I answer these questions in brief glances, and from the position of a performer-participant. I keep in mind that these are evolving questions of process and developmental emergence.

First Glance: Somatic Explorations Orient Movement and Dance toward Perceptual Transformations

When we move intentionally, we materialize movement and embody the feelings inherent in the movement. When I dance, I am not expressing myself. Instead, *I seek to surpass myself*. I wish to attune to the dance I'm doing and to its movement materials. Like a text that moves me past my present circumstance and limits, dancing can renew and carry me beyond limitations.[41] This is my hope. But does anyone move beyond habitual limitations? Sometimes they do, and taking this chance makes the somatic venture worthwhile. People want to change the status quo, transform their pain, sorrow, and loss, and unfold their untapped possibilities. Movement and memory have a bond through the body. They are linked in the material substance of mind, heart, and character, not set in stone, but carried forward in potential.

Second Glance: Movement Materializes Culturally and Somatically

Human movement materializes as a journey with somatic and cultural entwinements. The cultural means of somatic study begins with the body and evokes the feeling mind. The embodied cultural journey is not easy because each body has a complex history. Everyone has a history. Further, the body is not a stable material acted upon by outside forces, political or social. In materializing, bodies carry somatic histories, specific cultures, and generate unpredictable corporeal forces and cultural minds.

The field of somatics is originally defined through *soma as self-perception* by philosopher Thomas Hanna.[42] I look behind this definition to conceive *soma* more fully as a process of organic gestalt depth perception, silently pervading all of our actions and cultural expressions. Soma as self-perception is incomplete; *soma-psyche* is the lived context and depth of feeling that renders perception possible in the beginning. Movement is activated from gestalt somatic sources and can connect back to them. This silent resource (or innate potential) can eventually give rise to creative cultural actions or proffer depressive states and cultural nihilism. Soma-psyche is not predictable, but it can flower and respond well to care.

To consider just one consequence of the somatic cultural frame: somatic empathy in warmness toward others, or intra-subjective mutuality, is paramount, particularly for people who are culturally oppressed. Simone de Beauvoir expressed this well in *The Ethics of Ambiguity*.[43] She characterizes oppression that aims to reduce oppressed minorities to the status of an object (a statistic), one that excludes them from the authority to make meanings and establish values. A just society is one that values the freedom of all. Marginalized and oppressed groups use their power to transform negative values ascribed to them into those they can seize as beautiful and experience the freedom to invent themselves in their own best image.[44]

Letting be
As nature lets everything be
with room to grow.
Vines connect as vines do.
Trees leaf and some flower.

I lie in the grass
on my favorite
slope in the park,
watching the children
learn how to play tennis,
hearing them laugh and crash,
and I look in wonder
at the patchwork of people and colors,
old, bent, young, giddy, tan,
black, and eggplant, all sharing tints—
these are the colors I am in love with.

I lie here as the sun warms everything.
I know my back will feel better
from the softness of grass: letting be
in the daily performance of the grass, trees,
and birds, the sound of firmly-grasped rackets
hitting across the net in the onrushing
sweet spot of play—where
nature and culture meet.

Breathe
To ease knots, allow meaning an opening
like highly contagious wet weeds;
even they must learn how to hide discreetly.

Inhale, hold
then sigh to find a bodied way, following today, tomorrow.
Imagine how a form emerges, maybe a bounce or stretch,
finding with ease the shapes in front and behind.

Energy stores
wait and tune in, revitalizing urges
to expand on pause, not straining effort.

Breaking
running to my side through the cool dust,
you in your arrival and silence.
Poetic flesh, implicit choice,
your morphing shape and color—
not threatening nor with the face of a judge.
You make the difference that bodies me intact.

The significant issue of soma is posited through phenomenology in its reductions of self, bracketing the notion that the self can be known, envisioning the self as precarious. When I consider how otherness is formative in social and philosophical ontologies, I understand that being is more extensive than self. Materializations of self enfold others and other materials. There is more to the self than "self," if you will. There is the whole of being, the ontology of "being-in-the-world" that Husserl and Heidegger explained at the heart of their works.[45] *The cultural environing world* matters, increasingly so, since we are a crucial part of the material world of feeling, object clarity, and muddiness we come to know through all of our senses. We humans touch the materials of the world, and in turn they touch and entwine us.[46]

Cultural minds
They scatter like rhizomes
in process—intervening, interruptive, and connective.
No mind escapes everyday familiarity, the seeping frame
of relatedness, whether friendly or threatening.
Soma is not a singularity, but a radix.

Somatic values of bodily life inform culture at every moment, seen and unseen. Soma is a thousandfold, as active and vast as global politics. The personal body is part of a larger body. Somatic material is part of the material nature that permeates human life and the cultural ethos that sweeps through and interrupts it. Long before we have any conceptions of freedom and agency, soma foregrounds experiential and cultural values.

Third Glance: Somatic Movement Explorations Influence Material Body Identity

Soma folds
into life and spreads out—
as the root of flesh, soil, and relativity—
body of space, time, and place.

For you, for now
In the way nature designed you to have joy in motion, in options to sit on a towel or big blanket, to sit tall or slouch. Perhaps not to sit at all. Wherever you are, relax your shoulders, close your eyes, allow your hands to rest on your knees or lap. With softness, take your attention to the bottom of your spine. Attune to your imagination ruminating at the base of the spine, in the pelvic bowl, and the hips. Notice what comes up for you, any fidgeting or distractions.

No need to be perfect. Most important is that you take time to bring your attention inward. No worries. Invite being present; be present with what is—then if you please, allow images of energy to swirl in the belly bowl. Move any awareness that comes from the heart to the base of the spine and up the spinal column—now draw this energy into your forehead. Suppose your posture changes. With your fingers, tap the middle of your forehead. Here is the pituitary gland that informs all the other glands. Let your body change in any way it does. To finish, perhaps the breath wants to reach into the chest. Rest a minute or two; then scan your body-sense for possible insights.

Move an image of a tall tree along your spine
Lie down on your back, or stand simply. From where you are, move an image of a tall tree along your spine, not gripping your toes, fingers, or jaw. Tap lightly a few times with one finger the space between your brows. When you are ready, bring your palms together at the chest, or pause them at the belly.

Sit, and let your head hang forward; feel the curve and length of the back of the neck. *Choose... Choose...* What? Inhale through the nose, then exhale through the mouth as you return the head to balance on the neck and spine. Twist easily and slowly without stress; twist to the right and the left—and then toward whatever inner material emerges. It might be something or nothing. Let go. Sit or lie down to rest a while if you choose to.

Matter matters when the body renews. You might feel more in tune with a new and improved home base (homeostasis). This is the hope, but of course, it is a precarious hope.

Transformation is fickle and flawed
it does not like perfection because change is lazy;
it likes to wait and worry a little,
but when transformation takes up the mantle of expression,
bloodstreams make way—minds become spacious—
and materials catch up.

Forth Glance: Age Does Not Deter Somatic Ttransformation

Students in my somatic yoga class are over 60 for the most part. One might think all their transformations have already taken place. But no. Many are ready for significant changes, and by now, they know how to apply themselves in learning. They listen, try out new things, and (as I remind them), they don't stress toward results. They trust in their material bodies and transformative natures—at least, this is the trust between us. The human body is constantly changing and emerging into old age and letting go.

In somatic yoga classes, I describe the material meanings of the energy centers of yoga or chakras in Sanskrit. The feet stand on the ground and relate to grounding and support. The legs and pelvis come next, including the enteric nervous system or belly chakra. This embodied complex gives credence to creativity and is generative and prone to guilt. Giving attention to the belly bowl, we can let it relax consciously. In the belly, we are also reminded that the material body is self-generating. Humans, like other animals, reproduce.

Continuing to move upward through the body's energy systems, we encounter the primary breathing diaphragm and the heart closely allied. The powers of agency, decision, caring, and grieving metabolize throughout the body, but they are symbolized and imagined through breath and soul. Above the heart lies the throat with matters of eating and speaking as part of its meaning. The throat is where we say, "I sing; I speak my truth; I have a voice." Words have material sources in the throat. Many animals sound, sing, and speak.

The third eye in the middle of the forehead is substantial in India and in yoga. This eye prompts insight and is far-seeing. It is both material and ethereal. Lastly, the crown chakra is at the crown of the head. The top of the head is envisioned as a lotus flower in yoga. In humans, it symbolizes white-silver hair and the wisdom of age.

Soma snows material
I cut my hair today,
as drifting, floating down, its thousand snowfalls play
such magic tricks while flying proud away.

Fifth Glance: Movement Is Lived through Material Body History

Movement always exists through material, rocks sliding down a hill, or dancers on a stage. As lived, movement is experienced; as material, it is complex and thousandfold. We move toward things and away, with and around, under and over, stepping through the world in agony and joy. Movement humanizes its material comportment: engaged in caring contact or performed with malice. Affectivity is part of movement, not detachable, but a property of movement itself, both motional and emotional, as dancers learn.

In making dances, choreographers commonly refer to movement as material. Movement materials of dance are spatially defined by where the dancer travels in space. When she dances in concert, she carries such spatial matters of body through design and in the collective emergence of the whole. The material of movement is also temporal; it has time and takes time. It carries the specificity of rhythm and personal history. Laban Movement Analysis (L.M.A) studies and records these material matters of dance and movement.

Here we are concerned with connective somatic properties, how movement and dance connect humans to others and the world. Merleau-Ponty's phenomenology of reciprocities explains such connectivity, particularly through *chiasm*. Whatever we touch touches us.[47] As he puts it poetically and systemically: "Our own body is in the world as the heart is in the organism... and with it forms a system."[48] Perception is not passive; it has intentional connective properties. It begins with the world of the perceiver and is constituted in engagement with the world; there is co-determination and a somatic organic source. Otherness is involved from the beginning.

The infant gains knowledge of the world through contact with human others and all the affordances they provide, which is a partnership. The healthy infant learns to attune with others and participates in the unfolding of her potentials. A distinct phenomenology of soma and perception deepens the intersubjective and ethical dimensions of such development.

Somatic ethics arrives early in childhood as movement makes ethics visible and felt. Will the growing child learn how to engage others in play? Will she learn how to share or look only to her own needs and wants? Movement carries ethical dimensions that ground early in the lived body and become habitual, but they are also renewed consciously in adult life in cultivations of self and community. These are somatic materials of movement and care, as reflecting an aesthetic history of care.

Sixth Glance: Sensing and Feeling Aid Material Transformation

Spending time in the natural world
Feeling at ease there, integrated with edges and enclosures, enacting the body at home with itself, untangling psychic skeletons, developing rootedness and sensing place as an adventure; enjoying food, communing with people in green places near lakes and rivers, lily ponds, deep canyons, dense forests, and red desert sands; noticing bird songs, insect buzzing, and the sheerness of dragonflies; feeling resourceful and confident, inviting security and bravery; facing, as ever, an unpredictable future.

Whither did you go body?
Looked away for a minute, and you were gone.
See where you were on the worn couch
and the half-eaten pear on the table.

Where did you go, agile dancer?
Wherefrom your arched foot?
Ahh, there you are,
sitting in your writing chair near the piano,
listening to the air conditioner with television
moaning in the background.
And there you are again, gazing out the picture window
into the summer-bleached landscape.[49]

If you are discontent body
I forgive you
because you are still here
with your warmth and loneliness.

As movers gain confidence in movement, they understand themselves as movers, not as bystanders. In somatic terms, the power of perception is strong, and healing begins with this. People are often drawn to somatic practices through pain and a desire to heal. It sounds odd that one might listen to pain or make friends with sorrows and heartaches. Do pain and suffering have anything to say? Of course, they speak, but not always in words. They speak intra-actively through sense and material, and they know their pesky entanglements can dissolve because the material body is kinetic and malleable.

Closure along with change and emergence is the hope and confidence of movers in somatic practices. Acknowledgment of grief (for instance) anticipates renewal and transformation. Feelings matter; emotions matter. Feelings are guides to the future, and affectivity is fundamental to cultural and ethical progress.[50]

Seventh Glance: Body Politics Are Somatically Linked to Ecology and Social Issues through Ethics

In the beginning of the chapter, we said that somatic experience emerges through *hyletic interplay* of the cultural body with ethical/ecological potentials. Current research shows that issues of ecology and social justice are somatically linked through ethics. Respect for biodiversity conservation is reflected in social issues where protection of nature and social justice mirror each other. This is the view of contributing authors of *Contested Nature*.[51] Body politics directly permeate individual and collective bodies because the body of the Earth and human bodies are materially and spiritually linked. The way the powers of society regard our bodies materializes somatically, immanently in feelings of security or neglect to begin with, then more outwardly in politics of care or dismissal. Regarding ecological ethics, we observe that tyrannical regimes care more about power than protecting human communities and nature, and tyranny takes root easily when we ignore its signs.

How then do individuals intersect with body politics somatically—and for the good of nature and the social collective? We can turn in several directions for inspiration, especially toward the arts. Theater and dance practices broach civil discourse and disobedience through example and experiment; theater, dance, and movement practices curry nonverbal dialogue and implicit politics. Audrey Ellis puts this well: "Dancing does not bracket the body from social, cultural, and ideological forces, but rather registers these forces as corporeal and experiments with the tensions and pressures they create."[52] Attention to the inner material of movement, especially its affective powers, affords connective tissue to further investigate the body's political pressures in social transformation. We have only begun to learn how to harness such innate corporeal sources.

Speaking concretely, our responsibility to the Earth's body redounds to our human bodies and traumas. Most recently, shifting demographics in the American body politic sway and buffet the collective. The national census of 2021 shows that America is increasingly diverse. How will our emergent collective navigate a multiracial power-sharing democracy, if indeed America remains a democracy? Nothing is guaranteed in an emergent collective. Political tensions are at work in testing the American democratic fabric. As I write in July of 2021, much of the Republican Party leadership coalesces around the false belief that Donald Trump won the 2020 election and is the rightful president, not Joseph Biden. The spreading of this untruth deeply disturbs the material body of democracy in America and has already spread death and destruction in our nation's capital through the January 6th insurrection.

Body politics are matters of life and breath. They are somatic issues that tear at the fabric of our material bodies, bodies that are already charged with meaning. Merleau-Ponty puts it well. This is what it means to be in a field: there is a "belonging" that defines the field for us.[53]

> *My body is already defined*
> in the field of politics and pandemics,
> charged with social meanings
> and dismissals,
> my aging sensibilities
> deemed insignificant.
>
> But not to me. Are my feet
> not still meant for walking
> mouth for talking
> eyes crying
> and nose for breath?
>
> Yes, I say to life,
> don't take it for granted.

And dance?
I can always dance my truth
or cry for more space
and breath.

Glancing Back: Somatic Materials Show Up in Dreams and Are
Affective

I often evidence the somatic imaginal material of soma and psyche in dreams. I dance my dreams with other dancers in dream sharing and dancing with Jungian teacher Jeanne Schul. This weekend upcoming, I will share and dance the dream I had last night. As I look at my quickly scribbled script of the dream, I see several materials and anticipate performing bound forceful movements. At some point, I will ask about the meaning of my dream as I interpret it myself. Meanings will emerge through the circumstances of telling and dancing, and the dance could change them.

Housetop Dream

Chapter One: I am a child wearing a simple dress in the dream. My father or someone like him puts me on a round metal house decoration, a disk, and places me on top and right in front of our modest house, standing off by itself in the country. The metal disk has other metal pieces inside, making a curved inner design. I am happy to be placed on the disk; it seems a privilege, even though I know I'm not a weathervane and will be stuck on top of the house for an unknown purpose, maybe forever.

Chapter Two: A great wind comes along and blows me off the house, buffeting me about on the disk, and I don't know if I will make it back to the safety of my perch. I remember the feeling of being pasted to the disk and that the wind takes me along with it in a flurry. By now, I am part of the metal material and the pervasive wind.

Chapter Three: What does it mean? My first thought is that it is about leadership and resilience. But I will let my dancing of the dream with others inform the meaning. For now, I have questions. Why does the wind visit and dislodge me? I wake up while still in the flurry of wind. Is the spell and burst of the wind like life? Where do I feel the dream in my body? At first blush, I feel my body as placed, perched, and expanded all at once. Then I think of the break and squall. Some of my body floats. Some of it touches and sticks to the metal, very emplaced and immobile. And I ask when I can get off. I also wonder if I really want to unfasten my body, or would I even be able to. I resolve to let the process of dancing my dream, and witnessing someone else mirror it, transform my associations. The somatic and bodily enactment will teach and change me. It always does.

Dancing the Dreambody

When I dance my dream in our somatic gathering, I make a vivid association, remembering that my childhood bedroom in my family home was upstairs and at the front of the house. I felt safe there, but my parents quarreled violently downstairs, and I worried about them. Later on, and especially in the second half of my life, I understood their emotional turmoil. As I dance this housetop dream, my feelings for my background turn gently toward gratitude, and I move softly and unstick from the disk. I can move from high to low, whirl brightly, and be okay with it all. As my dance continues, it places me in a field of flowers and tall grasses, and I become curious about how they might taste. I remember becoming an adventurous eater, tasing unknown foods in my travels. As part of the material-me, the taste of otherness becomes part of my inner material.

Ecosomatic Materialization

I have attempted to show how somatic dance and movement involve direction of intentional consciousness, difference, orientation, imagination, ethics, and meaning—first with Hijikata's politically motivated butoh, shifting to reflections on darkness, race, and war, and lastly through examples of embodied politics and dreams. The inner connections of this chapter aim to show how materials of dance and movement draw upon central goals of human life beyond that of art for audience reception—and that somatic modes of activity provide rich methods toward human flourishing. Body and nature are of the same emergent substance, and they share a common ecology and somatic ethos. The material of human movement is our vulnerable and affective body entwined with environmental nature, culture, and ecology. Body and nature are of the same substance; they share ecologies. Matter and movement entwine each other in somatic performances that carry corporeal intention and implicit ethics, just as our body is continually emerging with the bodying momentum of the world.

In Lateral Time
Water blends time's inner materials—ecosomatic treasure—
ancient music I long for—in figures without measure.

Notes

1 Glen A. Mazis, *Merleau-Ponty and the Face of the World: Silence, Ethics, Imagination, and Poetic Ontology* (Albany: SUNY Press, 2016), 255–260.
2 David W. Johnson, *Watsuji on Nature: Japanese Philosophy in the Wake of Heidegger*, Northwestern University Studies in Phenomenology and Existential Philosophy (Evanston: Northwestern University Press, 2019), 166–167.
3 Ibid., 5.
4 Karen Michelle Barad, "Posthumanist Performativity: Toward an Understanding of How Matter Comes to Matter," in *Material Feminisms*, eds. Stacy Alaimo and Susan Hekman (Bloomington: Indiana University Press, 2008), 133.
5 Martin Heidegger, *Being and Time (1927)*, trans. John Macquarrie and Edward Robinson (New York: Harper & Row, 1962), 62–63.
6 Karen Michelle Barad, *Meeting the Universe Halfway: Quantum Physics and the Entanglement of Matter and Meaning* (Durham: Duke University Press, 2007), 3.
7 Johnson, *Watsuji on Nature*, 14.
8 Edmund Husserl, *The Crisis of European Sciences and Transcendental Phenomenology: An Introduction to Phenomenological Philosophy*, trans. David Carr (Evanston: Northwestern University Press, 1970), 109.
9 Sondra Fraleigh, *Butoh: Metamorphic Dance and Global Alchemy* (Urbana: University of Illinois Press, 2010).
10 Sondra Fraleigh, "Butoh Translations and the Suffering of Nature," *Performance Research: A Journal of the Performing Arts* 21, no. 4 (2016). https://doi.org/10.1080/13528165.2016.1192869.
11 Tatsumi Hijikata, "Inner Material/Material," *TDR* 44, no. 1 (2000a). Originally published in 1960 as *Naka no sozai/sozai*, a pamphlet for Hijikata DANCE EXPERIENCE *no kai* (recital).
12 Tatsumi Hijikata, "To Prison," *TDR* 44, no. 1 (2000b). Originally published in 1961 as *Keimusho e* in *Mita Bungaku* (The Mita Literature): 45–49.
13 Ibid.
14 Maurice Merleau-Ponty, "The Film and the New Psychology (1945)," in *Sense and Nonsense* (Evanston: Northwestern University Press, 1964).
15 Edward S. Casey, *Turning Emotion Inside Out: Affective Life beyond the Subject* (Evanston: Northwestern University Press, 2022).
16 Edward S. Casey, *Spirit and Soul: Essays in Philosophical Psychology*, 2nd ed. (Putnam: Spring Publications, 2004), 319–341.
17 Sondra Fraleigh, "Canyon Consciousness," in *Dance and the Quality of Life*, ed. Karen Bond (Cham: Springer, 2019). *See also* Fraleigh's practice pages, "Enworlding Place Dances and Potatoes" in Chapter 20.
18 Edmund Husserl, *Ideas Pertaining to a Pure Phenomenology and to a Phenomenological Philosophy, Second Book: Studies in the Phenomenology of Constitution*, trans. Richard Rojcewicz and André Schuwer (The Hague: Kluwer Academic, 1989), 383–390; Edmund Husserl and Eugen Fink, *Sixth Cartesian Meditation: The Idea of a Transcendental Theory of Method*, trans. Ronald Bruzina (Bloomington: Indiana University Press, 1995), 164–165.
19 Johnson, *Watsuji on Nature*, 119.
20 Heidegger, *Being and Time*, 421–423.

21 Fred R. Dallmayr, *The Other Heidegger,* Contestations: Cornell Studies in Political Theory (Ithaca: Cornell University Press, 1993).

22 Sondra Fraleigh, *Dancing Identity: Metaphysics in Motion* (Pittsburgh: University of Pittsburgh Press, 2004).

23 Fraleigh, *Butoh.*

24 William L. Shirer, *The Rise and Fall of the Third Reich: A History of Nazi Germany* (New York: Simon and Schuster, 1960).

25 Sondra Fraleigh, "*Das Warten*: Mathilde Thiele, A Life in Dance and War," Eastwest Somatics, 2021. www.eastwestsomatics.com/sondra.

26 Iris Murdoch, *Metaphysics as a Guide to Morals,* 1st American ed. (New York: Allen Lane, Penguin Press, 1993), 3.

27 Ibid., 52.

28 Audrey L. Ellis, "From Animation to Activation: Improvisational Dance as Invitation and as Interruption" (Ph.D. Doctoral, Stony Brook University, 2021).

29 Mazis, *Merleau-Ponty and the Face of the World,* 259.

30 Casey, *Spirit and Soul,* 299–307.

31 Antonio Damasio, *The Feeling of What Happens: Body and Emotion in the Making of Consciousness* (San Diego, New York, and London: Harcourt, Inc., 1999).

32 Ibid., 172.

33 Ibid., 80, 169.

34 Ibid., 148.

35 Husserl, *Pure Phenomenology,* 179.

36 Vicki Kirby, "Natural Conversations: Or What If Culture Was Really Nature All Along," in *Material Feminisms,* eds. Stacy Alaimo and Susan Hekman (Bloomington: Indiana University Press, 2008), 224–227.

37 Barad, "Posthumanist Performativity," 130.

38 Kirby, "Natural Conversations."

39 Ann Daly, *Done into Dance: Isadora Duncan in America* (Middletown: Wesleyan University Press, 1995).

40 Antonio R. Damasio, *The Strange Order of Things: Life, Feeling, and the Making of the Cultures* (New York: Pantheon Books, 2018).

41 Sondra Horton Fraleigh, *Dance and the Lived Body: A Descriptive Aesthetics* (Pittsburgh: University of Pittsburgh Press, 1987/1995), 22–35.

42 Thomas Hanna, *Somatics: Reawakening the Mind's Control of Movement, Flexibility, and Health* (Reading: Addison-Wesley, 1988), 20–21.

43 Simone de Beauvoir, *The Ethics of Ambiguity (1948),* trans. Bernard Frechtman (New York: Philosophical Library, 1992).

44 There is much scholarship on how agency and resistance may be possible within systems of oppression and marginalization. bell hooks has spoken of the power of the margins in her groundbreaking 1989 essay, bell hooks, "Choosing the Margin as a Space of Radical Openness," *Framework: The Journal of Cinema and Media* 36 (1989). Daphne A. Brooks describes "Afro-alienation acts," as "tactic[s] that the marginalized seize on" in order to convert the "condition of alterity" into "cultural expressiveness and a specific strategy of cultural performance." Daphne A. Brooks, *Bodies in Dissent: Spectacular Performances of Race and Freedom, 1850–1910* (Durham: Duke University Press, 2006), 4. *See*

also Patricia Hill Collins, *Black Feminist Thought: Knowledge, Consciousness, and the Politics of Empowerment* (London and New York: Routledge, 2008); Shaun R. C. Wallace, "Fugitive Voices," in *American Cultures as Transnational Performance: Commons, Skills, Traces*, eds. Katrin Horn et al. (Abingdon; New York: Routledge, 2021).
45 Husserl and Fink, *Sixth Cartesian Meditation*; Heidegger, *Being and Time*.
46 Maurice Merleau-Ponty, *The Visible and the Invisible*, trans. Alphonso Lingis (Evanston: Northwestern University Press, 1968).
47 Ibid.
48 Maurice Merleau-Ponty, *Phenomenology of Perception*, trans. Colin Smith (London and New York: Routledge, 1962), 235.
49 Poem for Alison East, using Husserl's words for *enworlding*: "a being-tendency" of *whither* and *wherefrom*. A *being-tendency* leading out of the world, *whither*, and returning, *wherefrom*, is connective and rooted. Husserl and Fink, *Sixth Cartesian Meditation*, 21, 99.
50 Damasio, *The Strange Order of Things*.
51 Steven R. Brechin, *Contested Nature: Promoting International Biodiversity with Social Justice in the Twenty-first Century* (Albany: State University of New York Press, 2003).
52 Ellis, "From Animation to Activation," 112.
53 Merleau-Ponty, *Phenomenology of Perception*, 4.

Bibliography

Barad, Karen Michelle. *Meeting the Universe Halfway: Quantum Physics and the Entanglement of Matter and Meaning*. Durham: Duke University Press, 2007.
———. "Posthumanist Performativity: Toward an Understanding of How Matter Comes to Matter." In *Material Feminisms*, edited by Stacy Alaimo and Susan Hekman, 120–154. Bloomington: Indiana University Press, 2008.
Brechin, Steven R. *Contested Nature: Promoting International Biodiversity with Social Justice in the Twenty-First Century*. Albany: State University of New York Press, 2003.
Brooks, Daphne A. *Bodies in Dissent: Spectacular Performances of Race and Freedom, 1850–1910*. Durham: Duke University Press, 2006.
Casey, Edward S. *Spirit and Soul: Essays in Philosophical Psychology*. 2nd ed. Putnam: Spring Publications, 2004.
———. *Turning Emotion Inside Out: Affective Life beyond the Subject*. Evanston: Northwestern University Press, 2022.
Collins, Patricia Hill. *Black Feminist Thought: Knowledge, Consciousness, and the Politics of Empowerment*. London; New York: Routledge, 2008.
Dallmayr, Fred R. *The Other Heidegger*. Contestations: Cornell Studies in Political Theory. Ithaca: Cornell University Press, 1993.
Daly, Ann. *Done into Dance: Isadora Duncan in America*. Middletown: Wesleyan University Press, 1995.
Damasio, Antonio. *The Feeling of What Happens: Body and Emotion in the Making of Consciousness*. San Diego; New York & London: Harcourt, Inc., 1999.

Damasio, Antonio R. *The Strange Order of Things: Life, Feeling, and the Making of the Cultures*. New York: Pantheon Books, 2018.

de Beauvoir, Simone. *The Ethics of Ambiguity (1948)*. Translated by Bernard Frechtman. New York: Philosophical Library, 1992.

Ellis, Audrey L. *From Animation to Activation: Improvisational Dance as Invitation and as Interruption*. Ph.D. Doctoral, Stony Brook University, 2021.

Fraleigh, Sondra. *Dancing Identity: Metaphysics in Motion*. Pittsburgh: University of Pittsburgh Press, 2004.

———. *Butoh: Metamorphic Dance and Global Alchemy*. Urbana: University of Illinois Press, 2010.

———. "Butoh Translations and the Suffering of Nature." *Performance Research: A Journal of the Performing Arts* 21, no. 4 (2016): 61–71. https://doi.org/10.1080/13528165.2016.1192869.

———. "Canyon Consciousness." In *Dance and the Quality of Life*, edited by Karen Bond, 23–44. Cham: Springer, 2019.

———, "*Das Warten*: Mathilde Thiele, a Life in Dance and War," Eastwest Somatics, 2021. www.eastwestsomatics.com/sondra.

Fraleigh, Sondra Horton. *Dance and the Lived Body: A Descriptive Aesthetics*. Pittsburgh: University of Pittsburgh Press, 1987/1995.

Hanna, Thomas. *Somatics: Reawakening the Mind's Control of Movement, Flexibility, and Health*. Reading: Addison-Wesley, 1988.

Heidegger, Martin. *Being and Time (1927)*. Translated by John Macquarrie and Edward Robinson. New York: Harper & Row, 1962.

Hijikata, Tatsumi. "Inner Material/Material." *TDR* 44, no. 1 (2000a): 36–42.

———. "To Prison." *TDR* 44, no. 1 (2000b): 43–48.

hooks, bell. "Choosing the Margin as a Space of Radical Openness." *Framework: The Journal of Cinema and Media* 36 (1989): 15–23.

Husserl, Edmund. *The Crisis of European Sciences and Transcendental Phenomenology: An Introduction to Phenomenological Philosophy*. Translated by David Carr. Evanston: Northwestern University Press, 1970.

———. *Ideas Pertaining to a Pure Phenomenology and to a Phenomenological Philosophy, Second Book: Studies in the Phenomenology of Constitution*. Translated by Richard Rojcewicz and André Schuwer. The Hague: Kluwer Academic, 1989.

Husserl, Edmund, and Eugen Fink. *Sixth Cartesian Meditation: The Idea of a Transcendental Theory of Method*. Translated by Ronald Bruzina. Bloomington: Indiana University Press, 1995.

Johnson, David W. *Watsuji on Nature: Japanese Philosophy in the Wake of Heidegger*. Northwestern University Studies in Phenomenology and Existential Philosophy. Evanston: Northwestern University Press, 2019.

Kirby, Vicki. "Natural Conversations: Or What if Culture was Really Nature All Along." In *Material Feminisms*, edited by Stacy Alaimo and Susan Hekman, 214–236. Bloomington: Indiana University Press, 2008.

Mazis, Glen A. *Merleau-Ponty and the Face of the World: Silence, Ethics, Imagination, and Poetic Ontology*. Albany: SUNY Press, 2016.

Merleau-Ponty, Maurice. *Phenomenology of Perception*. Translated by Colin Smith. London; New York: Routledge, 1962.

————. "The Film and the New Psychology (1945)." Translated by H. L. Dreyfus and P. A. Dreyfus. In *Sense and Nonsense*, 52–53. Evanston: Northwestern University Press, 1964.

————. *The Visible and the Invisible*. Translated by Alphonso Lingis. Evanston: Northwestern University Press, 1968.

Murdoch, Iris. *Metaphysics as a Guide to Morals*. 1st American ed. New York: Allen Lane, Penguin Press, 1993.

Shirer, William L. *The Rise and Fall of the Third Reich: A History of Nazi Germany*. New York: Simon and Schuster, 1960.

Wallace, Shaun R. C. "Fugitive Voices." In *American Cultures as Transnational Performance: Commons, Skills, Traces*, edited by Katrin Horn, Leopold Lippert, Ilka Saal and Pia Wiegmink. Abingdon; New York: Routledge, 2021.

5

DECENTERING THE HUMAN THROUGH BUTOH

Lani Weissbach

Indianapolis, Indiana, U.S.A.: 39.7684° N, 86.1581° W

From its inception in the late 1950s, the Japanese avant-garde dance form known as "butoh" was a radical departure from traditional dance forms. Butoh's founders Tatsumi Hijikata and Kazuo Ohno endeavored to strip away any regulation of the body, whether aesthetic, social, or cultural to allow the body to "speak for itself."[1] They developed specific techniques to foster a state of consciousness that opened the dancer to be moved by inner and outer stimuli without a predetermined plan.

Rather than using mimicry or mirroring, butoh dancers allow their bodies to morph into, or become, images or qualities of the world around them. Butoh scholar Sondra Fraleigh explains that this act of becoming "is a metamorphic matter, typically, of becoming creature, becoming other animals and objects, becoming elements of earth or atmospheres, and becoming nature as inhering in nature."[2] Critical to this process is the intention to empty oneself of human attachments akin to the emptiness of Zen. The concept of emptiness in Buddhism points to the insubstantiality and impermanence of phenomena, and butoh explicitly applies this to our human form. As a method of inducing this state of mind in her workshops, butoh teacher SU-EN would repeat as we worked on embodying such elements as slime, a twisted branch, or a black hole, "These are not my arms and my legs. This is not my beautiful face."[3]

The act of becoming that is unique to butoh is also an act of decentering the human, which is a radical paradigm shift in how we can experience ourselves in an interconnected web of life. To me, decentering the human means that I consciously *let go of my attachment to myself* and open the

DOI: 10.4324/9781003390985-7

lens of my conscious and somatic presence to include and be moved by the wider world around me. In this receptive, fluid, and permeable state, I am not passive or in any way denying that there is a self that is present.[4] Rather, I temporarily expand this sense of self to include the world around me. In essence, I *become* more-than-human.

This process of becoming is supported by another butoh method for decentering—the practice of literally taking oneself off-balance, often in barely perceptible shifts in one's position. Using micromovements to take the body into asymmetry allows the dancer to attune to more subtle internal and external sensations and stimuli, which contributes to the sense of detaching from one's human form.

In the early 1990s, I was a principal dancer in the butoh company Iona Pear Dance Theatre based in Hawaii, and training and performing in a variety of outdoor sites was an integral part of our work. I remember feeling a profound shift within myself as I emerged from these dances, and for some time afterward, a deep sense of aliveness, peace, and openheartedness permeated my life. I felt an unequivocal sense of connection and belonging to the world. But the world includes ugliness, weakness, and entropy, and butoh, like Buddhism, teaches us to be with life just as it is, not how we want it to be. By decentering my humanness through butoh, I matched the world and discovered the great power of being in a state of non-resistance to the messiness of mud, the stench of rotting fruit, and cigarette butts lining dirty city sidewalks. It was here, in the thick of it all, that I discovered empathy, or "feeling with" the other/world—perhaps a crucial first step toward overcoming polarization and preserving every endangered species on our planet. Indeed, as Fraleigh states, "When we dance in gratitude for the earth, we are not likely to exploit it," and I believe this gratitude arises not simply by giving thanks, but by experiencing ourselves as part of the totality of life.[5]

Decentering Practices

Facilitate these practices with a partner who can read them and leave space for your exploration. Alternatively, you can record the narratives yourself and leave space and time for movement in the recording. If you are a teacher, use these practices freely with your students. Read them aloud and wait for everyone in the class to come near a resolution or close. Trust your instinct concerning the progress of the group as a whole. Let everyone know it is almost time to finish by ringing chimes or using verbal cues and give them a short while to end as they wish.[6]

Rolling Meditation

Lie on your back with knees bent and the soles of your feet in standing. Notice the sensations of your breathing. On the next exhale, make one small movement—a micromovement—and notice the subtle change in the position of your body, which is now slightly asymmetrical, imbalanced. Stay here and feel this new form in your body as you breathe in. On the next exhale, make another micromovement, and pause again to feel this new shape. As you continue this process, it will eventually take you onto your side, belly, opposite side, and finally back onto your back. Your roll will not move you across the floor; rather, you will try to stay in about the same place in space, as you revolve around yourself one micromovement at a time. In awkward moments, soften your body and enliven your *hara*—the center of your body. Feel the specific part of body that is initiating the micromovement, and experiment with moving from parts of the body that don't normally initiate movement.

Wide Vision Meditation in Stages

Sit, stand, or lie down. Begin by feeling your body, here and now. Sense your weight and the parts of your body touching the ground. Soften your skin and feel your natural breathing. Begin to expand your awareness following each stage, below. Take at least 30 seconds to stay with each stage of the meditation before progressing to the next stage. As you progress to each wider view, always go back to the beginning before going to the next step. It would sound like this: (1) see yourself in this room; (2) see yourself in this room, and in this town; (3) see yourself in this room, in this town, in this city, etc.

Note: A leader can verbally guide each stage, or this can be done in the imagination.

Stages: The first few stages can be adjusted to your physical/geographic location.

See yourself in this...
Room
Town
City
State
Country
Continent
Planet
Solar System
Galaxy
Universe

Walk of Emptiness: Butoh Walking Meditation

Stand with your feet aligned under your hip sockets, in a parallel position. Bring attention to the soles of your feet touching the surface you are standing on. Open your eyes and see your horizon, allow your gaze to be soft, and practice seeing yourself in your surroundings with wide vision. Balance the energy exchange through your eyes—allow yourself to see the world, and sense the world also seeing you. Keep the knees soft and slightly bent, but not enough to change your height. Begin your walk by slowly shifting your weight to your left foot, sliding the right foot only about one centimeter off the ground, and placing it down so your right heel is only about an inch or less in front of your left toes. Continue to walk in this fashion, with your arms relaxed at your sides. With each step, empty yourself of your attachments and your identity as a person. Clear and open your energy field with this walking meditation for at least 8–10 minutes.

Ascending/Descending

Start in standing. Begin to lower yourself down to the floor without a predetermined plan. This practice does not have to be performed in a smooth fashion, simply go down to the ground and find a way to come back to standing. Continue this process; the task is to find new pathways down and up each time. Lower your body down as far as is possible for you. Start at a medium speed, or a comfortable speed. Gradually, begin to slow your movement down in increments. Eventually, the process of ascending and descending will comprise many micromovements. Pay close attention to the moments that feel awkward or are difficult to move through—such as when you are half-way up or down (*see* Figure 5.1). At these points, notice how your body may want to shift into a comfortable, balanced position. Notice what is coming up in your mind. Soften your being and see if you can stay with the dilemma a bit longer. Continue to use micromovements to find a way through.

Becoming More-than-Human

Go out to a place that draws your attention. It can be a wild, natural space, a human-made environment, or have elements of both. Begin with walking slowly—choose either the butoh walking mediation or your natural walk—opening all your senses to the environment. Be aware of your feet on the earth, step by step. Go barefoot if possible.

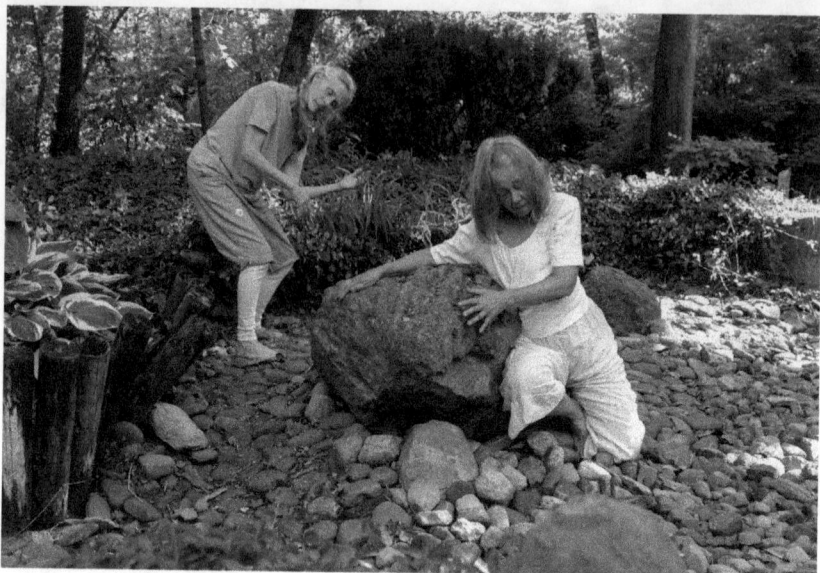

FIGURE 5.1 Mary Jo DeMyer (left) and Nina Ryan ascending and descending in micromovements. Photo courtesy of Charles Borowicz.

Allow your body to be called to a particular location in this environment, and when you arrive, find a specific form in your body that intuitively reflects something about the shapes, energies, and textures of your starting place. If possible, place your belly-center (*hara*) close to the ground. Stay quiet and still for a while before moving (*see* Figure 5.2).

Become present to any upwellings of movement that may begin to emerge in your body. The movement might be very faint, even a twitch or tremor. Allow the movement to grow, evolve, or dissipate on its own, and stay aware of how your body's form is changing in the process. As your dance evolves, allow a full spectrum of dynamics to emerge organically and in direct response to your environment. Movements can range from imperceptibly small and subtle to rapid and/or seismic shifts. Allow your body to be moved to new forms and places. Surrender any good ideas or "fixing" your body's position and stay with what is happening in each moment, as you become more (*see* Figure 5.3).

FIGURE 5.2 Mary Jo DeMyer reflecting the environment in stillness. Photo courtesy of Charles Borowicz.

FIGURE 5.3 Nina Ryan becoming Crooked Creek. Photo courtesy of Charles Borowicz.

Notes

1 Jean Viala and Nourit Masson-Sekine, *Butoh: Shades of Darkness* (Tokyo: Shufunotomo Co., Ltd., 1988), 17.
2 Sondra Fraleigh, "We Are Not Solid Beings: Presence in Butoh, Buddhism, and Phenomenology," *Asian Theatre Journal* 37, no. 2 (2020): 4. https://doi.org/10.1353/atj.2020.0037.
3 SU-EN butoh workshop, San Francisco (Summer, 2005).
4 Gretel Taylor, "Empty? A Critique of the Notion of 'Emptiness' in Butoh and Body Weather Training," *Theatre, Dance and Performance Training* 1, no. 1 (2010): 85. https://doi.org/10.1080/19443920903478505.
5 Fraleigh, "We Are Not Solid Beings," 5.
6 I am deeply grateful for the students and creative collaborators I have been so fortunate to work with over the years and for the teachers who have influenced and inspired these practices. Special thanks to Cheryl Flaharty, SU-EN, Denise Fujiwara, Gerry Trentham, and Sondra Fraleigh.

Bibliography

Fraleigh, Sondra. "We Are Not Solid Beings: Presence in Butoh, Buddhism, and Phenomenology." *Asian Theatre Journal* 37, no. 2 (2020): 464–489. https://doi.org/10.1353/atj.2020.0037.
Taylor, Gretel. "Empty? A Critique of the Notion of 'Emptiness' in Butoh and Body Weather Training." *Theatre, Dance and Performance Training* 1, no. 1 (2010): 72–87. https://doi.org/10.1080/19443920903478505.
Viala, Jean, and Nourit Masson-Sekine. *Butoh: Shades of Darkness.* Tokyo: Shufunotomo Co., Ltd., 1988.

6

SHAKY ISLANDS AND RISING SEAS

Dancing Entanglements in the Global South

Karen Barbour

Kirikiriroa (Hamilton), Aotearoa (New Zealand): 37.7869° S, 175.3185° E

Shifting Narratives

In this essay, I investigate entanglements with the more-than-human world through moving and dancing in the Global South, most particularly in Aotearoa New Zealand. To provide a theoretical background for my discussion of local geographies, embodied relationships to place, and ecosomatic dance, a section of this chapter focuses on discussion of "entanglement," a concept drawn from the transdisciplinary theorizing of feminist scholar Karen Barad.[1] Throughout, I offer *narrative vignettes* that juxtapose my academic writing through theory with the nuanced lived experiences of moving with shaky islands and oceans; *each of the more personal sections will be in italics*. These vignettes function to create a conversational assemblage of the multiple narratives of my embodied entanglements in the more-than-human world.[2] Further, the different styles of writing reflect a methodological entanglement, supporting an alternative re-reading and re-writing of texts toward transdisciplinarity.

Responding to the theme of "geographies of us," I move from a perspective of the Global South, particularly *Te Moana-Nui-a-Kiwa* (the Pacific Ocean) and the islands of *Aotearoa* (New Zealand), while negotiating flows between the local and global. With an embodied basis in ecosomatic dance, my lived experiences embrace entanglements with agentic shaky islands, vast expanses of ocean, and rising tides. At this time, significant embodied shifts are underway, complicating and giving rise to new expressions of agency. And so it is that I dance and write into entanglement.

DOI: 10.4324/9781003390985-8

Faults

Squished in the back seat of the car alongside my sisters, I'm soaking in the view of the hills, trees, cliffs, rivers, farmland, bush, and occasional houses passing by my window as we journey the narrow country roads. I yearn for exploring these places, breathing in the scents of earth and weather, searching for fossilized shells in the limestone cliffs, and brushing leaves overhanging the rivers, instead of the constraining car. Learning my place in the world enthusiastically, rivers, bush, and cliffs are my favorite places.

"Can you see out there Karen, where the river is?" My father interrupts my thoughts. "See where the hills and cliffs rise up higher on the other side of the farmland? Rivers often flow in the valleys showing us that fault lines may be under the surface." Dragging myself out of my reverie, I follow my father's gaze.

"What is a fault line Dad?"

"Well, underneath the surface of the earth..." my father begins, flexing his teacherly muscles,

there are large rock masses, called tectonic plates that cover the planet and where these plates meet, the rock from one plate grinds against the other. Sometimes one plate slides under the other and the higher plate creates hills, cliffs, and mountains. Our Southern Alps sit over the two huge plates called the Australian and Pacific tectonic plates and all of New Zealand is affected by the movement of these plates deep under our islands. That's why we have mountains and valleys. Water runs from the mountains into streams, lakes, and rivers in the valleys and on to the sea. Plate movement continues today and that's also why we have earthquakes and volcanoes.

"But why is it called a fault line?" I persist, my understanding of the word "fault" being limited to my use in the sentence, "It's not my fault," uttered in defense when I failed to look after my younger sisters properly. My dad continues,

In social science, people sometimes apply words differently. Geographers and geologists talk about rock formations and when rock formations are broken or interrupted or displaced along lines of weakness, they call this a "fault," a "fault line," or a "fault zone." If you want to be a social scientist Karen, you will need to learn to use lots of words differently according to context and you will learn to observe the land and people to understand the world around you.

No one else in the car is entering the conversation, my sisters' sleepy heads lolling to the side and my mum quiet in front of me. My dad was finishing his university degree as well as teaching at the high school in our small rural community. I knew he was interested in geography and he was writing something called a thesis. I wasn't really sure what it meant—writing a thesis or being a social scientist—but I was fascinated by earthquakes and volcanoes.

Thinking about fault lines and what remains of the past lives of people and sea creatures, the world seemed so much more powerful to me than people, capable of creating islands and destroying even the biggest stone buildings. It seemed to me that we were all mixed up with everything on the Earth, and what got left behind after big movements of the Earth were just traces, fossils, and partial remains. Imagining the land of my home changing around fault lines and lives from fragments in ruins, I was learning histories of the world.

Entanglements

The concept of entanglement and the theory of agential realism arise from the connections between quantum physics and social theories, as articulated through quantum physicist, Karen Barad's, feminist new materialist lens.[3] Barad draws specifically on the work of Niels Bohr, who concluded that entities do not exist in themselves but instead come into focus as phenomena when we use certain apparatuses to observe them; there is an inseparability between ways of observing the world and the matter of the world. In this sense, "we do not uncover pre-existing facts about independently existing things as they exist frozen in time like little statues positioned in the world. Rather, we learn about phenomena–about specific material configurations of the world's becoming."[4] The world thus interpreted is not made of things but of phenomena—of ontological entanglements. As a consequence, Barad argues that meaning and matter are inseparable.[5] Instead of staying within the quantum world in their thinking, Barad engages with social theories and feminist arguments in understanding the matter of the more-than-human world, and in understanding what matters.

I note here that a focus on the fleshy and organic matter of the world, including embodiment, is not new in feminist theory and praxis, or in dance research.[6] Nor is the idea of "becoming" or the deep embodied interrelationships in the "lifeworld" new in phenomenology.[7] In this chapter, I draw from eco-phenomenology as I have in previous writings and also explore the potential of feminist new materialism.[8] As Astrida Neimanis argues, "*[m]aterial feminism is thinking with matter. Matter here is lively; it destabilizes anthropocentric and humanist ontological* privilege."[9] The language of feminist new materialism, namely, Barad's work, develops and extends concepts of "intra-action," "agential agency," "entanglement," and "performativity," discussed in turn below.

Barad's work is valuable to a range of social science researchers, applied in understanding our embodied material experiences as humans in relationship with non-human materiality, and particularly by feminist writers.[10] Barad's theory of agential realism offers a transdisciplinary understanding of the world as dynamic, continually intra-acting materiality. This inherent inseparability and agency of both human and non-human matter gives rise to what we experience as phenomena in the world. These intra-actions are entanglements involving "material-discursive forces–including ones that get labeled 'social,' 'cultural,' 'psychic,' 'economic,' 'natural,' 'physical,' 'biological,' 'geopolitical,' and 'geological'–that may be important to particular (entangled) processes of materialization."[11] Barad explains,

> To be entangled is not simply to be intertwined with another, as in the joining of separate entities, but to lack an independent, self-contained existence. Existence is not an individual affair. Individuals do not preexist their interactions; rather, individuals emerge through and as part of their entangled intra-relating.[12]

The concept of entanglement expresses agency in material relationships, agency shared by all creatures and assemblages in the more-than-human world.[13] Each person's material agency is entangled with many agencies in the more-than-human material world and "feminist materialisms invites [sic] us to consider how non-human bodies or matters might contribute to their own actualization."[14] Agency is understood not as an attribute that one possesses but instead emerges as part of the intra-action and entanglement through which phenomena are revealed.[15] All phenomena are held in the memory of the world and come into being through entanglements of materiality. As Barad describes,

> Phenomena are not located in space and time; rather, *phenomena are material entanglements enfolded and threaded through the spacetimemattering* [sic] *of the universe... Memory—the pattern of sedimented enfoldings of iterative intra-activity—is written into the fabric of the world. The world "holds" the memory of all traces; or rather, the world is its memory (enfolded materialization).*[16]

To clarify further, "First, there is no separation of agentic entities prior to their intra-action. Second, and similarly, all dynamic agentive beings are able to act, which disputes the notion that human beings are the only ones acting."[17] New materialist understandings of intra-action thus have methodological implications for researchers, including requiring researchers to acknowledge their own responsibility in making the particular agential cuts they make, and that there are also non-human others who act in research contexts.

Feminist new materialists understand methodological choices as "agential cuts," one of many expressions of agency to be found in the matter of the more-than-human world. When I research, the methods I use create "cuts" that momentarily stabilize what I experience of the world so I can see phenomena.[18] Thus, researchers must consider ethics as they express agency in the decisions about what to research and how to create understandings of the world. Barad's feminist new materialist theorizing is described as an "ethico-onto-epistemology and agential realism in which she [sic] redefines connections to the shared world by attuning to the entangled matter that is created within intra-actions."[19]

Performance

Developing the theory of agential realism further, agency expressed through our researcher actions becomes performativity.[20] As understandings arise in my entanglements in the matter of the more-than-human world, these experiences may be understood as performances. As an embodied practitioner, I am attuned to the possibilities of understandings arising in embodied entanglements: with other dancers, with the surfaces and structures upon which I move, with other creatures of the more-than-human world, and especially through site dance and ecosomatic dance practices.[21] According to site dance researcher Victoria Hunter,

> Dance and movement knowledge is [...] a vital and important contributor to new materialist discourse as the dancing, moving body not only unearths and explores human-nonhuman entanglements but also illustrates, expresses and articulates the outcomes of such intra-actions through non-verbal means.[22]

As a dancer and embodied practitioner, I might describe the entangling of research with the materiality of the world as ethically responding, dialoguing, interpreting, expressing agency and performativity. In Figure 6.1, I am immersed, entangled, as salt water, flesh, and bone—alternately resting as the sea washes into rock pools, and then clinging desperately alongside plant and shellfish kin, exposed as the water rushes away. I am in these moments, merely materiality of the world.

FIGURE 6.1   *Entanglements 1*, West Coast, Whaingaroa Raglan. Photograph by Rodrigo Hill 2022.

Shaky Islands

Aotearoa New Zealand, my home, may be described as a volatile and geologically new group of three main islands (and 600 much smaller islands), isolated in the southwest of the largest ocean, *Te Moana-Nui-a-Kiwa* (the Pacific Ocean). Eons ago, the islands I call home were part of the supercontinent of Gondwana, which then separated into the continents of Australia, India, Antarctica, and *Te Riu-a-Māui* (Zealandia).[23] Where the Pacific and Australian tectonic plates converge, parts of Zealandia (and ancient Gondwana) were uplifted and re-exposed, forming and re-forming the islands in an ongoing process.[24] However, the movements of these "tectonic plates are not fixed but, like gears, rotate in relation to each other although the centers of rotation slowly migrate over time," thus also creating tilting and folding.[25] Today, most of the continent of Zealandia is submerged, my island home being mere lonely fragments that currently rise above the sea. Graham described the islands as "adrift in the South Pacific Ocean, separated from the rest of the world by vast distances and blessed with some of the most varied and spectacular landscapes on Earth."[26] Within the islands' rocky and eroded structure are the remnants of both marine and terrestrial pasts, "a physical expression of the deformation of Zealandia caused by plate tectonics" and a consequence of uplifting, tilting, slipping, faulting, and folding movements.[27]

The islands of Aotearoa are continually becoming, rising over the fault lines created by the relentless movement of tectonic plates pressing, grinding, and sliding under each other and rupturing repeatedly. At any time, the islands express this geological performativity as earthquakes shake and grumble, and active volcanoes breathe and erupt. Landslides, continual erosion, hydrothermal eruptions, and tsunamis also occur, and as a consequence, Aotearoa has been known as the "shaky islands."[28] Geologists measuring the islands' performativity regularly record up to 15,000 tremors annually.[29]

Sometimes my home shakes hard: the South Island city of Christchurch rocked and groaned through major earthquakes in 2010 and 2011 (185 people died), and coastal areas of Kaikoura were traumatized in a 2016 earthquake, leaving thousands of *Paua* and other shellfish dead. The islands are always rising and falling—breathing, grumbling, and resting. Remnants of earlier performances are expressed in the many fault zones as well as dormant volcanic cones that memorialize past movement of these islands.[30] Active volcanos, including the large volcanos nearest to my home, may blow and spit, erupting with little or no warning. Mount Ruapehu in the central North Island erupted most recently in 2007. Whakaari/ White Island near the eastern coast of the North Island erupted in 2019, tragically killing 22 people involved in volcanic site-seeing.[31] Reflecting on experiences of the shaky islands led cultural studies researcher Nigel Clark to comment that:

Where we live, and the forces that shape the land around us—the restlessness of the earth, the vagrancies of weather—are a part of the "context" of our everyday lives, part of the way we view ourselves and our world.[32]

Thus, "shaky islands" is an apt description for Aotearoa and makes explicit the entangled significance of earthquake and volcanic experiences, along with oceanic isolation and island living.

My everyday life entangles with the living and moving of the islands, variously uneasy and restless, and at other times quiet, creating wonder in the spectacular landscape of past performances. Alison East expresses that through attuning to the land and embodiment, dancers may perform as nature, and "rhythms derive from an internal pulse and from the visual and felt rhythms of the landscape."[33] Expressing a sense of a shared life of entanglement, Sondra Fraleigh writes,

Elemental nature is inspirational, even in thunder and hail. We are not separate from nature's elements; rather, they dwell within, fluid as water and murky. Nature is not simply in surroundings. *Nature is in us* [...] we can become aware of what we share with others, with deserts and canyons, with lakes and oceans. Shared life (we-life) can be brought forward consciously in dance and other arts.[34]

My everyday living and my dancing are inspired by and expressed through performing the more-than-human world. Island performances are the agency of the more-than-human world, past performances containing the memory of the world.

Dancing with Ruaimoko

Resting, softening, and allowing, my organs releasing from their habitual skeletal structure with the pull of gravity, descending toward the wooden floor and deeper toward the earth, I give up the effort of containment. Releasing my weight, I accept the grounding and support of the earth. In these moments, everything slowing and easing, I am finding and re-finding and re-finding again and again, becoming with each breath and heartbeat.

"Rū nui, rū roa, rū nui, riu roa, rua nui, rua roa...," the lyrics of the waiata *(song) with which I am moving, sing of the god of earthquakes, Ruaimoko, moving in the belly of the earth.*[35] *Sensing through aching internal shifting, rolling, reorienting, I am feeling with Ruaimoko, creator of earthquakes, caverns, and valleys, tattooer of our islands.*

"Te rū i riu, te rū i te rua, te rū i mokotia..." the song continues. I too feel the surface cuts of tattooing, my own agency carved into my skin, and the deeper shifts and changes as Ruaimoko carves my home islands.

Sensing my shifting bones, organs releasing anew in each quietening, resting moment, I feel as Ruaimoko moves in the belly of the islands. Memories stirring in my embodiment of my own child moving in my belly: gentle rhythms and stretching, and the unexpected performances of testing limits.

As a child I learned that all the Earth is mother, Papatūānuku, forever separated from her lover in the sky, Ranginui.[36] *In their separation, in the light created between them, their children grew, expressing as trees and plants, birds and people, oceans and fish and fresh water, winds, storms, and conflict. The tears of Ranginui fall on his lover as rain, and her desire rises from the earth as steam and heat in a never-ending shared breathing. While their children live and grow and argue in all their performing, in the belly of Papatūānuku lives their last child, Ruaimoko. Moving in the belly of Papatūānuku, Ruaimoko continues carving the earth, performing unexpected agential cuts, shaking, groaning, and rolling in earthquakes.*

In this moment and requiring exhalation, full movements allow my pelvis to rise and then ride the descent with gravity. Moving is a performance of surfacing memories and a scattering of the debris of words and fears. Seeping sweat and tears from my crevices, I am releasing, allowing permeability and porousness in my transforming. In my quietest moments I'm ever rolling, seeking to rest my gaze in the crook of my elbow, wrapping arms to calm my heart, exchanging breath with the world, cradling hope in my breast, riding the descent of my pelvis.

Dancing with Ruaimoko *is a reverent, sensual, intimate moment, expiring then and awakening anew now. Together rolling as earthly bodies, rising, folding, sliding, dancing, grounding, awakening again and again, entangled and alive.*

Stories and Matter

In order to represent my embodied experiences with shaky islands and rising seas, I create a conversational assemblage of the multiple stories of entanglements in the more-than-human world in which I dwell. These stories are agential cuts (in the language of feminist new materialisms); methods of momentarily stabilizing the becoming of the more-than-human world so that phenomena may be revealed. In this undertaking I aim to foreground the complexities of entanglement, resisting the production of a singular argument and a merely human story.[37] Thus, my methodology draws on narrative inquiry, sharing key moments of affective and embodied understandings through stories of people and islands. Considering that matter is agential, this is a lively and performative writing; in this sense, it is an effort to reveal "storied matter," which

> describes the idea that from its deepest lithic and aquatic recesses to the atmospheric expanses, and from subatomic to cosmic realms, matter is capable of bringing forth a display of eloquence, which can be explained as the "ontological performance of the world in its ongoing articulation."[38]

The stories of non-human matter emerge in our human ways of storying our lives, and in this thorough entanglement our human embodiment also emerges. Dwelling in the more-than-human world, we are "within a community of expressive presences," requiring us to be attentive to more-than-human stories and meanings.[39] Oppermann writes so clearly of this task,

> If we make storied matter part of our storytelling culture, it can play an important role in bringing the state of the world more to public awareness, and we can impart new ideas and insights about our experiences and

perceptions of the planet. Because storied matter induces "out-of-the-box" thinking, which is exactly what is needed today to develop solutions to our current problems and to build post-anthropocentric discourses. Hence, giving matter access to articulation by way of stories that co-emerge with the human is not only a way to emancipate matter from silence and passivity, but also to liberate ourselves from the images, discourses and practices of our own Cartesian dreamworld.[40]

Accepting this challenge to make storied matter a part of this story, I write through multiple agential cuts, creating a kind of assemblage throughout the chapter, engaging the insights of dance and somatic scholars, feminists, scientists, and social scientists.

The Fish of Māui

As a child at school, I listened to the ultimate fishing story, a description of the origins of my home islands.[41] *The description began with* Māui-Pōtiki, *the youngest son, stowing away on his older brothers'* waka *(canoe).*

Eager to join his brothers' fishing trip and to prove his worth, Māui *revealed himself to his surprised brothers only after the* waka *had traveled far south in* Te Moana-Nui-a-Kiwa. *Always clever,* Māui *had prepared for success in his adventure, crafting an enchanted a hook made from his grandmother's jaw bone.*

With the help of this hook, Māui *caught a giant* ika *(fish) on his line and hauled it to the surface. After* Māui *pulled the* ika *from the ocean, however, his brothers began cutting into it. This* ika *became the main northern island of my home, named* Te Ika-a-Māui *(the fish of* Māui*). The mountains, valleys, and caverns are the traces of the brothers' cuts.*[42] *And our southern island is sometimes called* Te Waka-a-Māui *(the canoe of* Māui*).*

For me, this fishing story suggests why our southern island rocks and groans in the ocean, and the northern island continues to dive, roll, and rise in the water, moving constantly like a fish struggling on the line. The head of the fish hosts our capital city Wellington, nearest the waka. *In the belly of the fish is the largest lake, Taupo, with its ever-watching volcanoes, and the tail of the fish curves northward away from the* waka, *as the fish struggles defiantly against the hook.*

Rising Seas

In the current geological epoch, we humans are collectively recognizing the impact of our actions in the world, particularly now with increasingly rapid climate changes occurring around us. Climate change impacts are occurring in what is described as the Anthropocene: the concept that humans have recently become a "geological force" in our activity on the planet.[43] More than people just leaving marks on the Earth's surface, the Anthropocene "proposes that we have become geologic or geophysical agents: that we have impacted upon the working of the Earth as a whole."[44] We understand that our industrialization and land-use produces carbon dioxide and other gasses—described as greenhouse gas emissions—leading to climate change across the planet. Our greenhouse gas emissions are influencing Earth's climate system, causing warming that is irrevocably changing global climate and all of the more-than-human world.[45] Understanding the Anthropocene, it has become obvious that global climate change is directly entangled with local environmental challenges, and correspondingly, local concerns may be described within global planetary change. This recognition is certainly occurring in Aotearoa and *Te Moana-Nui-a-Kiwa*.

A major impact of anthropogenic climate change is rising sea levels, already a significant concern as glaciers and polar ice sheets melt and thermal expansion increases sea levels.[46] For many of us in the South Pacific Ocean, sea level rise is already obvious. While the average sea level rise globally might be calculated as 3.5 mm per year, local sea level rise is experienced at varying rates. The performativity of the islands of Aotearoa means that we also observe regular vertical land movements from large seismic events like earthquakes, as well as more subtle continual changes resulting from plate tectonics and erosion. In areas in which our coastlines are subsiding, the annual rate of sea level rise is likely to double and the timeline for direct impacts to be observed is greatly reduced.[47] Comprehensive mapping of our island coastlines leads to alarming projections,

We know that global sea-level rise of 25–30 cm by 2060 is baked in and unavoidable regardless of our future emissions pathway. But what may be a real surprise to people is that for many of our most populated regions, such as Auckland and Wellington, this unavoidable rise is happening much faster than we thought. Vertical land movements mean that these changes in sea-level may happen 20 to 30 years sooner than previously expected.[48]

Our capital city in Wellington, the largest city of Auckland, and the west coast beach areas in Whaingaroa nearest my home are likely to experience this doubling effect of rising sea levels and subsiding land. According to James Schulmeister, "New Zealand is both one of the most sensitive locations for examining future environmental change and conversely one of the most geologically active."[49] Again, this complexity creates unique entangled performances of the more-than-human world.

The impact of sea level rise on our islands is calculated "optimistically" on the basis of meeting targets for CO_2 emission reductions.[50] Our overall ecological well-being in Aotearoa and in *Te Moana-Nui-a-Kiwa* is under threat. Aotearoa may be described as a first-world country, but we nevertheless share an inevitably and increasingly changeable environment, along with many islands and Global South nations. Rising sea levels literally erode our coastlines and are already submerging some islands, creating challenges to the survival of the more-than-human world. There is no disentangled way to live separately from our environment, and yet the capitalist practice of using up the resources of the natural world continues. Right now, in Aotearoa, the discussions are shifting from plans for "managed retreat" from areas threatened by rising sea levels, to working on cooperative planned relocations.[51] These are necessary policy-based planning processes and are predominantly focused on human concerns. There are many wider matters of ecosystem survival and relocations, adaptations and regenerations of the more-than-human world that need to be addressed.

Consequently, we need new ways of describing our world through which matter gains access to articulation by way of co-emerging stories with us people.[52] We need new concepts that force us to humbly recognize that we arise as phenomena only through entanglements with other material agents. For me, I exist only through entanglements with extraordinary shaky islands and rising seas. And so I need research methods and practices for living creatively that support such recognitions and revelations in the more-than-human world.

Ecosomatics

In living ecosomatics, I appreciate my dancing with the more-than-human world, participating in entangled performances in the everyday and particularly in site dance.[53] I am reminded of how eco-phenomenologist David Abram expresses his everyday experience as a dance: "Whenever I quiet the persistent chatter of words within my own head, I find this silent or wordless dance already going on—this improvised duet between my animal body and the fluid, breathing landscape that it inhabits."[54]

This silent dance always already going on is the dance of matter and the dance that matters: the traces of my dancing already done and my dancing yet to be are held in the more-than-human world. In ecosomatics, I experience and share practices of developing awareness and moving consciously.[55] Through learning from and respecting the world, I appreciate how all is entwined in ecological wellness, as well as in crisis.[56] Fraleigh writes,

Cultivating attention can make a difference to the world and our place in it. Paying attention to earth and the more-than-human-world, I experience myself moving toward health and wholeness, not just for myself, but for the planet [...] We are not in this life alone: ultimately we are not separate from each other, and neither are we separate from non-human nature.[57]

The practices of ecosomatics encompass deepening awareness and the underlying concept is that "body sensibility and conscious action can facilitate planetary awareness."[58] From this planetary awareness may grow care, empathy, and action. Robert Bingham expresses that his "physical practice became the ground for aesthetic reflection on the relationship between human bodies and the earth in the context of crisis," providing an "effective tool for waking up viscerally."[59] However, the challenge for me is to move beyond the awareness of the complex entanglements and crises of the Anthropocene, and the sense of overwhelming helplessness.[60] As many scientists, activists, and both Fraleigh and Bingham express, naming the climate crisis explicitly is "a call to action: a circumstance requiring response."[61] More than awareness is required but this is the beginning. "Individually and collectively, we need to pay attention in all the ways that we can. Our way here is through arts activism, making it more visible in performance, word and image. Every attempt to make a difference counts."[62]

Thus, I may describe my living in ecosomatics as a practice of dancing, writing, and sharing: practices of storying matter on the page and in performance, through which new understandings of grief and loss, survival, relocation, adaptation, and regeneration arise in the more-than-human world. Ecosomatic dance, entangled as and in salty water (Figure 6.2), is the silent dance in which I experience new understandings and through which insights are revealed.

Toward Hope

Stepping uncertainly into the dance studio, my students seem at a loss without chairs or projected lecture notes, and unnerved by their lecturer lying on the wooden floor. I rise to greet them, padding quietly about and offering gentle reminders to remove shoes. "Kia ora koutou, welcome. Please come and sit on the floor in a circle with me." With more encouragement, the first-year students of our university's new Bachelor of Climate Change degree gather together.[63] Most are majoring in Biology, Earth Sciences, Geology and Ecology, or in Environmental Planning and Geography. Suffice to say, none of them likely imagined sitting on the wooden floor of a dance studio with their lecturer.

FIGURE 6.2 *Entanglements 2*, West Coast, Whaingaroa Raglan. Photograph by Rodrigo Hill 2022.

Already I am observing the weight of the climate change science they are carrying from their learning in other classes. Climate anxiety is a common experience for young people living in the era of the Anthropocene and these students are absorbing more and more scientific evidence of crises every day.[64] *In classes with me they have been learning about the role of arts activism in our world of anthropogenic climate change; necessary learning and a strategy designed to alleviate climate anxiety that the Dean of Science and I have discussed with care.*

Inviting the students into the dance studio today, I am offering from my heart, intentionally nourishing hope in my own cells too. Talking quietly together, I facilitate the students sharing their stories of personal practices that support relaxation and resilience—walking in the trees, drawing, listening to music, playing with siblings, and moving meditatively. I encourage them gently as we move together through the studio space, pausing to notice how we feel in the moment. As the speed of moving slows and I feel the energy shifting in the room, I invite awareness of rising and falling breath. In the sensations of everyday moving and breathing, there is both familiarity and potential for new revelations. Taking time to descend toward resting on the floor, I offer the suggestion that there is safety in moving inward as we fold and bend, just as there is confidence expressed in the risks of reaching outward in the world. I describe what we are doing to the students using simple everyday movement cues, but for me, I observe the dance.

Speaking more quietly, I guide the students as we move, tracing backward through the movement patterns we learned as babies. From moving on land, we are coming closer to the earth, passing through crawling as we did as babies, into sitting and lying on our bellies, and to resting curled on our sides as though floating in the waters of our mother's womb. This is the gentle accessible practice of Land to Water Yoga, offered as an ecosomatic practice.[65] My voice and tone shifts to match the students' quietening. Breathing deeply and more attentively becomes their only activity for some time.

When sensing the students' readiness to move again, I invite rolling and changing position with each exhale. As they find their way into simple movements of their own choosing, we share in the nourishment of movement, fostering attentiveness to well-being in the moment. Breathing and moving becomes a shared practice, an option that may sustain and support personal resilience as the students engage in the extraordinary task of not only surviving our climate crises, but also of becoming leaders in positive change and regeneration for the future. A personal and pedagogical practice of hope....

Shaky Moments

Living many shaky moments and writing through my own eco-changes, I am nonetheless finding moments of renewal, of hope in carving with *Ruaimoko*, in sensing into my islands of home, in sharing ecosomatics with students, in acknowledging the struggle, like a fish hooked on the line of human striving. Now is merely a moment, and in each moment things are ceasing to be—unique species, political movements, fresh drinking water, whole islands. Now is a moment in which I record in dancing what is ceasing to be. I can, as Fraleigh writes, "move more consciously to live more consciously, including our relationship to a world in ecological turmoil."[66]

Moving to live more consciously in relationship to our environments is an offering I make to my students too. For those learning about climate change, there is value in ecosomatics as a personal practice for well-being and managing increasing climate anxiety. For my dance students, ecosomatic experiences offer a creative means to re-connect within local environments and places of significance, and even to participate in hopeful performances as ecosomatic activism.

For myself, I am continually seeking ways to understand "the mutual constitution of entangled agencies," in order "to come to terms with the staggering transformations we are witnessing" in the Anthropocene.[67] I move to witness and also to begin tracing into what is becoming. I, too, am "coming to terms with the present, in order to intervene in it and transform it."[68] Moving and writing, I also remember times in which being active, expressing agency, was too full of anxiety, pain, hopelessness. (These memories help me empathize with my students' experiences of climate anxiety.) Now I am saying yes in and with the more-than-human world. As a dancer, I am always in a sway, this rising, falling, and rolling dance, mapping in my cells what I am ceasing to be and what I am in the process of becoming. So dancing and writing becomes an assemblage, a story of entanglements in which some things happen, acknowledging grief and moving with it, dancing with… feeling into… and toward… some kind of hope in the more-than-human world.

Notes

1   Karen Michelle Barad, *Meeting the Universe Halfway: Quantum Physics and the Entanglement of Matter and Meaning* (Durham: Duke University Press, 2007), 3; *see also* Karen Michelle Barad, "Posthumanist Performativity: Toward an Understanding of how Matter comes to Matter," *Signs: Journal of Women in Culture and Society* 28, no. 3 (2003).
2   Assemblage theory or "assemblage thinking" is a posthuman philosophical approach to the ontological diversity of agency, which developed from the work of Bruno Latour, Gilles Deleuze, Félix Guattari, and others. *See* "The Agency of Assemblages," in Jane Bennett, *Vibrant Matter: A Political Ecology of Things* (Durham: Duke University Press, 2010), 20–38. The term, "more-than-human" was coined by David Abram; *see* David Abram, *The Spell of the Sensuous: Perception and Language in a More-than-Human World*, 2017 ed. (New York: Vintage Books, 1996). In particular, I refer to Kathleen Stewart's practice of assemblage in writing. *See* Kathleen Stewart, *Ordinary Affects* (Durham: Duke University Press, 2007), 5.
3   Barad's PhD is in Physics, with a focus on theoretical particle physics and quantum field theory. Barad, *Meeting the Universe Halfway*, 3.
4   Ibid., 90–91.
5   Ibid., 90; Brandon Jones, "Mattering," in *Posthuman Glossary*, eds. Rosi Braidotti and Maria Hlavajova (London and New York: Bloomsbury, 2018), 244.
6   For a discussion of embodiment and feminist research in the context of dance, *see* Karen N. Barbour, "Dancing Epistemology, Situating Feminist Analysis," in *Back to the Dance Itself: Phenomenologies of the Body in Performance*, ed. Sondra Fraleigh (Urbana: University of Illinois Press, 2018); Karen Barbour, *Dancing Across the Page: Narrative and Embodied Ways of Knowing* (Bristol: Intellect Books, 2011).

7 For a discussion of lifeworld and phenomenology in the context of dance, *see* Sondra Fraleigh, ed., *Back to the Dance Itself: Phenomenologies of the Body in Performance* (Urbana: University of Illinois Press, 2018).

8 Karen N. Barbour, Vicky Hunter, and Melanie Kloetzel, *(Re)positioning Site Dance: Local Acts, Global Perspectives* (Bristol: Intellect Books, 2019), 17; Barbour, *Dancing Across the Page: Narrative and Embodied Ways of Knowing*, 49.

9 Italics in original. Astrida Neimanis, "Material Feminisms," in *Posthuman Glossary*, eds. Rosi Braidotti and Maria Hlavajova (London and New York: Bloomsbury, 2018), 242.

10 For related feminist research, *see* Marianne Clark, "Reimagining the Dancing Body through Barad," in *Sport, Physical Culture, and the Moving Body: Materialisms, Technologies, Ecologies*, eds. Joshua I. Newman, Holly Thorpe, and David L. Andrews (New Brunswick: Rutgers University Press, 2020); Victoria Hunter, *Site, Dance and Body: Movement, Materials and Corporeal Engagement* (Cham: Palgrave Macmillan, 2021); Allison Jeffrey, Karen Barbour, and Holly Thorpe, "Entangled Yoga Bodies," *Somatechnics* 11, no. 3 (2021). https://doi.org/10.3366/soma.2021.0364.

11 Barad, "Posthumanist Performativity: Toward an Understanding of how Matter comes to Matter," 810.

12 Barad, *Meeting the Universe Halfway*, iv.

13 Mary P. Sheridan et al., "Intra-Active Entanglements: What Posthuman and New Materialist Frameworks Can Offer the Learning Sciences," *British Journal of Educational Technology* 51, no. 4 (2020): 1278, https://doi.org/10.1111/bjet.12928. Bennett, *Vibrant Matter*, 23.

14 Neimanis, "Material Feminisms," 243.

15 Barad, *Meeting the Universe Halfway*, 141.

16 Italics in original. Karen Michelle Barad, "Quantum Entanglements and Hauntological Relations of Inheritance: Dis/continuities, SpaceTime Enfoldings, and Justice-to-Come," *Derrida Today* 3, no. 2 (2010): 261. https://doi.org/10.3366/E1754850010000813.

17 Sheridan et al., "Intra-Active Entanglements: What Posthuman and New Materialist Frameworks Can Offer the Learning Sciences," 1278.

18 The concept of a "cut" is different from the phenomenological practice of "bracketing" a lived experience, although both offer opportunities to engage with phenomena. *See* Barbour, "Dancing Epistemology, Situating Feminist Analysis," 233–246.

19 Sheridan et al., "Intra-Active Entanglements: What Posthuman and New Materialist Frameworks Can Offer the Learning Sciences," 1277.

20 Barad, *Meeting the Universe Halfway*, 136.

21 Jeffrey, Barbour, and Thorpe, "Entangled Yoga Bodies," 354.

22 Hunter, *Site, Dance and Body: Movement, Materials and Corporeal Engagement*, 16.

23 GNS Science Te Pū Ao, "Te Riu-A-Māui" Our continent, www.gns.cri.nz/our-science/land-and-marine-geoscience/te-riu-a-maui-our-continent/. Paul W. Williams, ed., *New Zealand Landscape: Behind the Scene* (Amsterdam: Elsevier, 2017), 5.

24 James Schulmeister, ed., *Landscape and Quaternary Environmental Change in New Zealand* (Amsterdam: Atlantis Press, 2017), vii.

25 From "The Creation of Zealandia" in Williams, *New Zealand Landscape: Behind the Scene*, 1.

26 Ian J. Graham, ed., *A Continent on the Move: New Zealand Geoscience into the 21st Century* (Wellington: Geological Society of New Zealand, 2008), back cover.

27 From "The Creation of Zealandia" in Williams, *New Zealand Landscape: Behind the Scene*, 15–16.

28 For example, Nigel Clark, "Cultural Studies for Shaky Islands," in *Cultural Studies in Aotearoa New Zealand: Identity, Space, and Place*, eds. Claudia Bell and Steve Matthewman (South Melbourne: Oxford University Press, 2004).

29 GeoNet, 2022. www.geonet.org.nz/.

30 Nick Mortimer and James M. Scott, "Volcanoes of Zealandia and the Southwest Pacific," *New Zealand Journal of Geology and Geophysics* 63, no. 4 (2020): 371. https://doi.org/10.1080/00288306.2020.1713824.

31 GeoNet, 2022 www.geonet.org.nz/about/volcano/whiteisland.

32 Clark, "Cultural Studies for Shaky Islands," 6.

33 Alison East, "Dancing Aotearoa: Connections with Land, Identity and Ecology," *Dance Research Aotearoa* 2, no. 1 (2014): 101, https://doi.org/10.15663/dra.v2i1.24; Alison East, "Performing Body as Nature," in *Moving Consciously: Somatic Transformations through Dance, Yoga, and Touch*, ed. Sondra Fraleigh (Urbana: University of Illinois Press, 2015), 165.

34 Italics in original. Sondra Fraleigh, "Canyon Consciousness," in *Dance and the Quality of Life*, ed. Karen Bond (Cham: Springer, 2019), 30.

35 Maisey Rika (vocalist), *Ruaimoko* with TeKahautu Maxwell (lyrics), Anika Moa (vocalist), and Mahuia Bridgman-Cooper (taonga puoro), released 28 September 2012. www.youtube.com/watch?v=yvFD3aILwZU.

36 In sharing my fragmentary understandings of Māori Indigenous knowledges and traditions, I recognize that much depth of meaning is unavailable to me as a Pākehā (non-Indigenous) woman. In contrast, I witness my son's growing depth of understanding and expression as a young Indigenous man of Te Ao Māori (the Māori world). While such depth of knowledge is not mine to access, I nevertheless grew up with common understandings as were embedded in mainstream schools, and readily available through literature, art, and popular culture in Aotearoa. It is to these commonly known understandings that I refer. *See* Te Ahukaramū Charles Royal, "Story: Māori Creation Traditions," *Te Ara: The Encyclopedia of New Zealand* (February 8, 2005). https://teara.govt.nz/en/maori-creation-traditions; Eileen McSaveney, "Story: Historic Earthquakes," ibid. (June 12, 2006). https://teara.govt.nz/en/historic-earthquakes.

37 Stewart, *Ordinary Affects*, 5.

38 Barad qtd. in Serpil Oppermann, "Storied Matter," in *Posthuman Glossary*, eds. Rosi Braidotti and Maria Hlavajova (London and New York: Bloomsbury, 2018), 411. For original, *see* Barad, *Meeting the Universe Halfway*, 149.

39 David Abram, *Becoming Animal: An Earthly Cosmology* (New York: Pantheon Books, 2010), 173.

40 Oppermann, "Storied Matter," 414.

41 As noted above, I again refer to my own fledgling understandings of *whenua* (land) and *Te Ao Māori* developed through encounters with *whānau* (family), school, literature, art and popular culture. *See* Te Ahukaramū Charles Royal, "Story: Whenua—How the Land was Shaped," *Te Ara: The Encyclopedia of New Zealand* (June 12, 2006). https://teara.govt.nz/en/whenua-how-the-land-was-shaped.

42 There are many versions of Maui's adventures told by peoples of *Te Moana-Nui-a-Kiwa* (the Pacific Ocean). My narrative does not reflect the full depth of Indigenous knowledges that are accessible to peoples of Te Moana-Nui-a-Kiwa. Ibid.

43 Paul J. Crutzen, "Geology of Mankind," *Nature* 415, no. 6867 (2002): 23. https://doi.org/10.1038/415023a.

44 Nigel Clark and Bronislaw Szerszynski, *Planetary Social Thought: The Anthropocene Challenge to the Social Sciences* (Cambridge: Polity Press, 2020), 5.

45 The Intergovernmental Panel on Climate Change (IPCC), *Climate Change 2021: The Physical Science Basis. Contribution of Working Group I to the Sixth Assessment Report of the Intergovernmental Panel on Climate Change*, The Intergovernmental Panel on Climate Change (IPCC), (2021) www.ipcc.ch/report/ar6/wg1/.

46 NZ SeaRise Te Tai Pari o Aotearoa, "New Zealand Sea-Level Rise Projections," www.searise.nz/new-zealand-sea-level-rise, 2022.

47 NZ SeaRise Te Tai Pari o Aotearoa, 2022.

48 Levy, qtd. in "Sea Level Is Rising Faster than We Thought," *NZ SeaRise*, 2022, www.searise.nz/blog/2022/5/3/sea-level-is-rising-faster-than-we-thought.

49 Schulmeister, *Landscape and Quaternary Environmental Change in New Zealand*, ix.

50 The Intergovernmental Panel on Climate Change (IPCC), *Climate Change 2021*; Te Kāwanatanga o Aotearoa New Zealand Government and Te Hau Mārohi ki Anamata, "Towards Productive, Sustainable and Inclusive Economy: Aotearoa New Zealand's First Emissions Reduction Plan," Te Kāwanatanga o Aotearoa New Zealand's First Emissions Reduction Plan, (2022). https://environment.govt.nz/publications/aotearoa-new-zealands-first-emissions-reduction-plan/.

51 Christina Hanna, Iain White, and Bruce C. Glavovic, "Managed Retreats by Whom and How? Identifying and Delineating Governance Modalities," *Climate Risk Management* 31 (2021). https://doi.org/10.1016/j.crm.2021.100278.

52 Oppermann, "Storied Matter," 414.

53 On-site dance, *see* Barbour, Hunter, and Kloetzel, *(Re)positioning Site Dance: Local Acts, Global Perspectives*; Hunter, *Site, Dance and Body: Movement, Materials and Corporeal Engagement*.

54 Abram, *Spell of the Sensuous*, 53.

55 Sondra Fraleigh, *Moving Consciously: Somatic Transformations through Dance, Yoga, and Touch* (Urbana: University of Illinois Press, 2015), 55; Karen Barbour, "Embodied Values and Ethical Principles in Somatic Dance Classes: Considering Implicit Motor Learning," *Journal of Dance & Somatic*

Practices 8, no. 2 (2016): 190, https://doi.org/10.1386/jdsp.8.2.189_1; Karen Barbour, Marianne Clark, and Allison Jeffrey, "Expanding Understandings of Wellbeing through Researching Women's Experiences of Intergenerational Somatic Dance Classes," *Leisure Studies* 39, no. 4 (2020): 3. https://doi.org/10.1080/02614367.2019.1653354.

56 Sondra Fraleigh, "Body and Nature: Quest for Somatic Values, East and West," *Journal of Dance & Somatic Practices* 13, no. 1–2 (2021): 118. https://doi.org/10.1386/jdsp_00040_1.

57 Fraleigh, "Canyon," 24.

58 Sondra Fraleigh and Robert Bingham, eds., "Performing Ecologies in a World in Crisis," Special Issue, *Choreographic Practices* 9, no. 1 (2018): 3.

59 Robert Bingham, "In the Shadow of Crisis: Dance and Meaning in the Anthropocene," in *Dance and the Quality of Life*, ed. Karen Bond (Cham: Springer, 2019), 68.

60 On the impact of climate change on our emotions, *see* Panu Pihkala, "Toward a Taxonomy of Climate Emotions," *Frontiers in Climate* 3 (2022): 1–22, https://doi.org/10.3389/fclim.2021.738154; Sarah Jaquette Ray, *A Field Guide to Climate Anxiety: How to Keep your Cool on a Warming Planet* (Oakland: University of California Press, 2020).

61 Fraleigh and Bingham, "Performing Ecologies," 6.

62 Ibid., 5.

63 The University of Waikato, 'Bachelor of Climate Change', www.waikato.ac.nz/study/qualifications/bachelor-of-climate-change.

64 Glenn Albrecht, *Earth Emotions: New Words for a New World* (Ithaca: Cornell University Press, 2019), 11; Pihkala, "Toward a Taxonomy of Climate Emotions," 7; Ray, *A Field Guide to Climate Anxiety: How to Keep Your Cool on a Warming Planet*, 5.

65 I am describing a version of the Shin Somatics Land to Water® yoga practice developed by Sondra Fraleigh at the Eastwest Institute. *See* Sondra Fraleigh, *Land to Water Yoga: Shin Somatics Moving Way* (New York and Bloomington: iUniverse, Inc., 2009).

66 Fraleigh, "Body and Nature," 120.

67 Barad qtd. in Rosi Braidotti, *The Posthuman* (Cambridge: Polity Press, 2013), 96. Originally in Barad, *Meeting the Universe Halfway*, 33.

68 Rosi Braidotti, "Affirmative Ethics and Generative Life," *Deleuze and Guattari Studies* 13, no. 4 (2019): 464. https://doi.org/10.3366/dlgs.2019.0373.

Bibliography

Abram, David. *The Spell of the Sensuous: Perception and Language in a More-than-Human World*. 2017 ed. New York: Vintage Books, 1996.

———. *Becoming Animal: An Earthly Cosmology*. New York: Pantheon Books, 2010.

Albrecht, Glenn. *Earth Emotions: New Words for a New World*. Ithaca: Cornell University Press, 2019.

Barad, Karen Michelle. "Posthumanist Performativity: Toward an Understanding of How Matter Comes to Matter." *Signs: Journal of Women in Culture and Society* 28, no. 3 (2003): 801–831.

———. *Meeting the Universe Halfway: Quantum Physics and the Entanglement of Matter and Meaning*. Durham: Duke University Press, 2007.

———. "Quantum Entanglements and Hauntological Relations of Inheritance: Dis/continuities, SpaceTime Enfoldings, and Justice-to-Come." *Derrida Today* 3, no. 2 (2010): 240–268. https://doi.org/10.3366/E1754850010000813.

Barbour, Karen. *Dancing Across the Page: Narrative and Embodied Ways of Knowing*. Bristol: Intellect Books, 2011.

———. "Embodied Values and Ethical Principles in Somatic Dance Classes: Considering Implicit Motor Learning." *Journal of Dance & Somatic Practices* 8, no. 2 (2016): 189–204. https://doi.org/10.1386/jdsp.8.2.189_1.

Barbour, Karen, Marianne Clark, and Allison Jeffrey. "Expanding Understandings of Wellbeing through Researching Women's Experiences of Intergenerational Somatic Dance Classes." *Leisure Studies* 39, no. 4 (2020): 505–518. https://doi.org/10.1080/02614367.2019.1653354.

Barbour, Karen N. "Dancing Epistemology, Situating Feminist Analysis." In *Back to the Dance Itself: Phenomenologies of the Body in Performance*, edited by Sondra Fraleigh, 233–246: Urbana: University of Illinois Press, 2018.

Barbour, Karen N., Vicky Hunter, and Melanie Kloetzel. *(Re)positioning Site Dance: Local Acts, Global Perspectives*. Bristol: Intellect Books, 2019.

Bennett, Jane. *Vibrant Matter: A Political Ecology of Things*. Durham: Duke University Press, 2010.

Bingham, Robert. "In the Shadow of Crisis: Dance and Meaning in the Anthropocene." In *Dance and the Quality of Life*, edited by Karen Bond, 67–80. Cham: Springer, 2019.

Braidotti, Rosi. *The Posthuman*. Cambridge: Polity Press, 2013.

———. "Affirmative Ethics and Generative Life." *Deleuze and Guattari Studies* 13, no. 4 (2019): 463–491. https://doi.org/10.3366/dlgs.2019.0373.

Clark, Marianne. "Reimagining the Dancing Body through Barad." In *Sport, Physical Culture, and the Moving Body: Materialisms, Technologies, Ecologies*, edited by Joshua I. Newman, Holly Thorpe and David L. Andrews, 209–228. New Brunswick: Rutgers University Press, 2020.

Clark, Nigel. "Cultural Studies for Shaky Islands." In *Cultural Studies in Aotearoa New Zealand: Identity, Space, and Place*, edited by Claudia Bell and Steve Matthewman, 3–18. South Melbourne: Oxford University Press, 2004.

Clark, Nigel, and Bronislaw Szerszynski. *Planetary Social Thought: The Anthropocene Challenge to the Social Sciences*. Cambridge: Polity Press, 2020.

Crutzen, Paul J. "Geology of Mankind." *Nature* 415, no. 6867 (2002): 23. https://doi.org/10.1038/415023a.

East, Alison. "Dancing Aotearoa: Connections with Land, Identity and Ecology." *Dance Research Aotearoa* 2, no. 1 (2014): 101–124. https://doi.org/10.15663/dra.v2i1.24.

———. "Performing Body as Nature." In *Moving Consciously: Somatic Transformations through Dance, Yoga, and Touch*, edited by Sondra Fraleigh, 164–179. Urbana: University of Illinois Press, 2015.

Fraleigh, Sondra. *Land to Water Yoga: Shin Somatics Moving Way*. New York; Bloomington: iUniverse, Inc., 2009.

———, ed. *Moving Consciously: Somatic Transformations through Dance, Yoga, and Touch*. Urbana: University of Illinois Press, 2015.

———. *Back to the Dance Itself: Phenomenologies of the Body in Performance.* Urbana: University of Illinois Press, 2018.

———. "Canyon Consciousness." In *Dance and the Quality of Life*, edited by Karen Bond, 23–44. Cham: Springer, 2019.

———. "Body and Nature: Quest for Somatic Values, East and West." *Journal of Dance & Somatic Practices* 13, no. 1–2 (2021): 113–123. https://doi.org/10.1386/jdsp_00040_1.

Fraleigh, Sondra, and Robert Bingham. eds. "Performing Ecologies in a World in Crisis." Special Issue. *Choreographic Practices* 9, no. 1 (2018).

Graham, Ian J., ed. *A Continent on the Move: New Zealand Geoscience into the 21st Century.* Wellington: Geological Society of New Zealand, 2008.

Hanna, Christina, Iain White, and Bruce C. Glavovic. "Managed Retreats by Whom and How? Identifying and Delineating Governance Modalities." *Climate Risk Management* 31 (2021): 1–14. https://doi.org/10.1016/j.crm.2021.100278.

Hunter, Victoria. *Site, Dance and Body: Movement, Materials and Corporeal Engagement.* Switzerland: Palgrave Macmillan, 2021.

Jeffrey, Allison, Karen Barbour, and Holly Thorpe. "Entangled Yoga Bodies." *Somatechnics* 11, no. 3 (2021): 340–358. https://doi.org/10.3366/soma.2021.0364.

Jones, Brandon. "Mattering." In *Posthuman Glossary*, edited by Rosi Braidotti and Maria Hlavajova, 244–247. London; New York: Bloomsbury, 2018.

McSaveney, Eileen. "Story: Historic Earthquakes." *Te Ara: The Encyclopedia of New Zealand*. (June 12, 2006). https://teara.govt.nz/en/historic-earthquakes.

Mortimer, Nick, and James M. Scott. "Volcanoes of Zealandia and the Southwest Pacific." *New Zealand Journal of Geology and Geophysics* 63, no. 4 (2020): 371–377. https://doi.org/10.1080/00288306.2020.1713824.

Neimanis, Astrida. "Material Feminisms." In *Posthuman Glossary*, edited by Rosi Braidotti and Maria Hlavajova, 242–244. London; New York: Bloomsbury, 2018.

Oppermann, Serpil. "Storied Matter." In *Posthuman Glossary*, edited by Rosi Braidotti and Maria Hlavajova, 411–418. London; New York: Bloomsbury, 2018.

Pihkala, Panu. "Toward a Taxonomy of Climate Emotions." *Frontiers in Climate* 3 (2022): 1–22. https://doi.org/10.3389/fclim.2021.738154.

Ray, Sarah Jaquette. *A Field Guide to Climate Anxiety: How to Keep Your Cool on a Warming Planet.* Oakland: University of California Press, 2020.

Royal, Te Ahukaramū Charles. "Story: Māori Creation Traditions." *Te Ara: The Encyclopedia of New Zealand*. (February 8, 2005). Accessed November 29, 2022. https://teara.govt.nz/en/maori-creation-traditions.

———. "Story: Whenua—How the Land Was Shaped." *Te Ara: The Encyclopedia of New Zealand*. (June 12, 2006). Accessed November 29, 2022. https://teara.govt.nz/en/whenua-how-the-land-was-shaped.

Schulmeister, James, ed. *Landscape and Quaternary Environmental Change in New Zealand.* Amsterdam: Atlantis Press, 2017.

Sheridan, Mary P., Amélie Lemieux, Ashley Do Nascimento, and Hans Christian Arnseth. "Intra-Active Entanglements: What Posthuman and New Materialist Frameworks Can Offer the Learning Sciences." *British Journal of Educational Technology* 51, no. 4 (2020): 1277–1291. https://doi.org/10.1111/bjet.12928.

Stewart, Kathleen. *Ordinary Affects*. Durham: Duke University Press, 2007.

Te Kāwanatanga o Aotearoa New Zealand Government, and Te Hau Mārohi ki Anamata. n.d. "Towards Productive, Sustainable and Inclusive Economy: Aotearoa New Zealand's First Emissions Reduction Plan." *Te Kāwanatanga o Aotearoa New Zealand's First Emissions Reduction Plan*. (2022). Accessed November 29, 2022. https://environment.govt.nz/publicati ons/aotearoa-new-zealands-first-emissions-reduction-plan/.

The Intergovernmental Panel on Climate Change (IPCC). Climate Change 2021: The Physical Science Basis. Contribution of Working Group I to the Sixth Assessment Report of the Intergovernmental Panel on Climate Change. The Intergovernmental Panel on Climate Change (IPCC), (2021). www.ipcc.ch/rep ort/ar6/wg1/.

Williams, Paul W., ed. *New Zealand Landscape: Behind the Scene*. Amsterdam: Elsevier, 2017.

PART II

Horse, Lion, Queer Animal, Skin

7

CRITTERCAL SOMATICITY

Rewilding Our Horse Senses

Stephen Smith
Pitt Meadows, British Colombia, Canada: 49.3058° N, 122.6057° W

Sondra Fraleigh recalls a moment when, with her back to a paddock fence, she became aware of a horse standing behind her. She felt a "gentle weight" as the mare reached down and rested her head on her shoulder. There was, says Fraleigh, a "warmth" to the gesture when it seemed something of mutual significance was being communicated between horse and human. "The horse was saying *yes,* yes to life."[1] This larger and more mobile life form was drawing Fraleigh out of her self-containment and affirming a vital contact with forces of nature attenuated all the more by the lightness of the mare's touch. There was in that very moment a sense of life overflowing the margins of otherwise separate selves.

This awakening to interspecies life resonates with a passage in which Martin Buber recalls his inspiration for contemplating the "I-Thou" relationship. Buber describes being at his grandparents' farm as a child and stealing away to the stables to spend time "gently strok[ing] the neck of my darling, a broad dapple-grey horse." He tells of experiencing "not a casual delight but a great, certainly friendly, but also deeply stirring happening." This manner of making contact with the horse, feeling with caressing hands an intimate connection, brought with it an expansive sense of life which Buber explained as sensing "the Other, the immense otherness of the Other" which "let me draw near and touch it."[2] He continues,

> When I stroked the mighty mane, sometimes marvelously smooth-combed, at other times just as astonishingly wild, and felt the life beneath my hand, it was as though the element of vitality itself bordered on my skin, something that was not I, was certainly not akin to me,

DOI: 10.4324/9781003390985-10

> palpably the other, not just another, really the Other itself; and yet it
> let me approach, confided itself to me, placed itself elementally in the
> relation of THOU and THOU with me.[3]

Buber makes clear that the horse is another sentient being whose vitality
touches upon, yet exists beyond, one's own frame of reference. There needs
to be no possession of the physical forces and motional powers of this
creature. Even in repose, while standing in a barn stall, the horse exhibits
a potential wildness that is not so much tamed and domesticated as it is
held in reserve. The vitality which Buber senses in petting the horse is this
felt sense of Otherness gained through the very means available to us of
sharing the plenitude of life with other creatures.

Buber went on to write of a bodily awareness, and particularly of the
physical sensations with which he became enamored when being with
the stabled horse. He drew attention to what he was experiencing of the
caressing motions and to that which turned this interaction into a personal
amusement. The horse became, as it were, the young Buber's plaything.
In doing so, however, this self-indulging consciousness constitutes a
withdrawal from the very moment when engagement with another might
reveal a surplus of life. There intrudes a self-consciousness which finds
pleasure in a tactility folding back on itself, no longer reaching out and
connecting vitally with other animate life forms.

The issue in becoming self-indulgent is not just about incorporating
the life of another in one's own but about the sensory limitations to this
shift of attention that curtails more motionally and emotionally finessed
interactions between human beings and across interspecies lines. What if,
instead of indulging in the pleasurable feelings we are having, we were
to find ourselves "letting go of the self" as a prepossession, exploring
contact and connection as a permeable palpability, to "uncover warmth
and concern for others and connect wordlessly with them in nature"?[4]
What if the reflexive moment were sustained by a greater somaticity than
the tactile fun of caressing a horse's neck and running one's fingers through
the mane? What if, instead of fixating on one's own feelings of being
in the horse's presence, we had recourse to a more extensive repertoire of
postural, positional, gestural, expressive means of motional and emotional
resonance? Prior to any deflection and defective reflectivity, there might be
an upwelling, upsurging, overflowing sense of life as mutual and reciprocal
expansiveness and as symbiosis, symphysis even, and sympathetic,
kinesthetic communion.

Human *and* horse could be "saying *yes*, yes to life."[5] We could each
revel in that which animates our otherwise separable existences. For
human beings who have felt with Fraleigh the "gentle weight" of another

animate being and been touched by what Buber called "the Other, the immense otherness of the Other" in a particularly animated way, saying yes to life can become a matter, as I will describe in this chapter, of moving inspirationally, mimetically, and energetically with horses. Bringing up life in this very motile manner can become what I call a *critter*cal somaticity of rewilding our horse senses.

In what follows, I will explicate what this somatic sensibility entails in animated horse play with as few constraints as possible. First of all, *wildness* will tellingly conceptualize the interactions that become possible in open spaces and when drawing upon one's own kinetic and kinesthetic capacities to tap into horses' motional powers. Indeed, before any thought of harnessing this wildness, what is *critter*cal will be described as the somatic awareness of how we, horse and human, can play off of one another. Next, *kinning*, as the verbal form of kith and kin, will encapsulate a repertoire of unfettered motional attunements which I discuss in the kinning considerations that follow as specific practices of *critter*cal somaticity. All of which inform the project of *rewilding* our horse senses to better appreciate these and other critters as more-than-human life forms who each in their own motile ways have us say yes to life.

Wildness

Richard Louv contends that while "[d]eep communication with animals" may be attributed to "practical knowledge, empathy and intuition," it is also likely a function of "the use of senses we have forgotten we have or could better develop."[6] His collation of critter encounters draws attention to so many sensory modalities that might be cultivated and of which we tend to rely on only the most obviously external ones. Horses, by comparison with, say, octopuses, eagles, turtles, and bears, are more sensorially recognizable to us. Their hearing is far more acute than ours. We would be hard-pressed to match their sense of smell. And while our taste palette is more richly varied, one would be more inclined to trust horses than humans in knowing what is good for each of us to eat. Horses have electromagnetic sensitivities in multiples of our own protecting them from harm and enabling them to find their way home. Where we come sensorially together, although far from equally so, is with respect to tactility and especially inner touch. We can move together proprioceptively and connect kinesthetically. There is an inner unfolding tactility and outer enfolding palpability which levels the playing field. This haptic sensory connection is, after all, the very basis of the domestication of wild horses and the breeding of the creatures so many ride today.

Moving inspirationally, mimetically, and energetically with horses requires going deeper sensorially than stroking, patting, and feeling their warmth. We are each in our own ways highly mobile beings afforded a somatic sensibility that the afore-mentioned instances of feeling a connection intimate. Finding ourselves in the company of equine beings, we, too, can come to appreciate a "pathos-with" one another, which Michel Henry took phenomenological pains to describe.[7] Pathic resonance reverberates in the very lines, creases, and folds of interspecies life. Becoming more-than-human, becoming horse, holds no mystery for those whose lives are spent with these creatures since it is borne by the postures, positions, gestures, and expressions of motional meaning-making that become etched in one's corporeal being.[8] What I call *critter*cal somaticity— the breathing inspiration for connecting with others, the motional means of meeting them on common ground, and the kinesthetic depth to this critterly attunement—can be cultivated in practices of playing with horses.

We come to our horse senses where "wild and brute being" can be motionally felt as extending the range and reach of our sensory bodies.[9] Emmanuel Falque, who draws upon Merleau-Ponty's interrogation of "wild and brute being," admits that it is not about attempting to grasp the "wildness" of nature but of finding ourselves on common ground and in this "wild region" of being drawn out of our separate selves to make us even more "ourselves."[10] Falque asks: "What 'thought' or 'language' would be adequately suited to speak of that 'wild being' without immediately, so to speak, domesticating it?"[11] The thought most suited would be, I suggest, the very somatic sensibility arising from the ground up and carried in the sensations and feelings of wildness when up close and in motion with critters of an equine kind.

Paddocks, fields, and arenas can be places not of domesticated dullness but of liberatory possibilities for coming to our horse senses. Wildness is about giving freer rein to the potent powers of equine agency, unbridled exercise of these forces of nature, all the while moving inspirationally, mimetically, and energetically with them. There can be a tending to the vital powers of movement as rooted in what Maxine Sheets-Johnstone called "the sheer experience of aliveness, the sheer nonverbal kinetic experience of ourselves and others as animate forms."[12] In removing all equipment, all constraints, and playing with horses in the open and at liberty, an ecosomatic practice of wildness can be taken up. Likening it to "dancing with horses," this practice puts paid to training procedures of human exceptionalism that desensitize horses to environmental stimuli and thereby dull their and our horse senses.[13] It is a practice not simply of training horses to move through patterns and forms with modulated, cadenced gaits and transitions, but of playing tactilely, proprioceptively, and kinesthetically with this aesthetic.

Kinning

For more mutually and reciprocally satisfying interactions with horses, I take to heart what Maurice Merleau-Ponty called a "lateral" rather than "hierarchical relation [...] that does not abolish kinship" but strengthens it all the more.[14] Kinship is more precisely a verbal expression rather than a nominal one. It is, as various contributors to the *Kinship* series describe, a "kinning" with one another in the more-than-human world.[15] I propose that breathing together, moving mimetically, and attuning somatically are motional gerunds of kinning that can be described further as the very means of coming literally and not just figuratively to our horse senses in moments, passages, and durations of interactive playfulness.

My own practice of coming to my horse senses is informed through sustained phenomenological scholarship, being a long-time student of horse-person-ship and, most especially, by engaging in daily practices of sharing breath, moving mimetically with horses, and finding in this realm of motile responsiveness the practical means of staying energetically in tune with these critters. The reflexivity of cultivating this practice need not be about indulging in self-assuring feelings, although they are not to be denied from time to time, but become a *critter*cally transformative, motional, and emotional rewilding of our senses, sensitivities, and ecological sensibilities.

The following descriptions are of my interactions with Lucente and Spartacus who are Spanish-bred horses of considerable vim and vigor. We play together in an open arena where they are free to run away if they wish and where it is of the very nature of our interactions to tap into an inherent *wildness* we share. We do so through kinning postures, positions, gestures, and expressions. I refer specifically in the following accounts to how we breathe together, move mimetically, and attune somatically with one another in terms of key aspects of Shin Somatics that are extended to this *critter*cal somaticity. Attention to breathing, matching practices, and "contact unwinding" have distinctive applications to saying yes to life with horses.

Breathing Together

Lucente gallops off around the expanse of the outdoor arena as soon as he is freed of the lead line clasped to his halter. He has been in the barn stall overnight and now, in the glaring sunshine of a summer morning, he has pent-up energy to burn (*see* Figure 7.1). He rounds the arena a few

FIGURE 7.1   Lucente, Westwind Farm, Pitt Meadows, British Colombia, June 2018. Photograph courtesy of the photographer, Michele Black.

times, kicking his hind hooves in the air and whinnying to other horses in pastures nearby. His exuberance is still evident as he *piaffes* and *passages* back to me, these high-stepping motions exclaiming: "Look at me! Life is great!" We come face to face, him snorting and sucking air, while I hold a more tranquil pose to absorb his breathing. Lucente's inhalations lengthen and deepen; his exhalations moisten the air between us.

He settles into a slower, more rhythmical breathing pattern. The flaring nostrils of just moments ago soften as inhalations and exhalations become deeper and more sustained. We are now almost nose to nose. I can smell the alfalfa-enriched hay on his breath. My own breathing settles into an easy cadence much practiced in *Hatha Yoga* and through *Gi Gong* exercises. Breathe in, breathe out, sustain the exhalation just a little longer. Respiration becomes a kind of osculation. We breathe together as the basic physiological mechanism of entrainment and in sharing breath we, both human and horse, move from the singularity of our respective functions and forms to mutual feelings and shared flow motion. We, of different creaturely kinds, come face to face, nose to nose, with one another literally as kindred spirits.

Fraleigh's Shin Somatics emphasizes the felt physiology to this practice of breathing into shared exchange. Diaphragmatic breathing works at the level of "the complex third chakra" to energize, integrate, and coordinate the overall posture.[16] Deep diaphragmatic breathing is essential to centering, grounding, and finding therein the means of mutual respiration with horses. Breath becomes the respirational passage of shared aspiration between human and horse. So, if "[m]editational practices allow me to deconstruct myself, as any thoughts of 'self' or 'other' flow by on a river of breath," an active meditation with horses can incorporate breathing as the very means of flowing motionally and emotionally in unison.[17]

James French communicates with horses and has developed a way of breathing together for which he has coined the term the "trust technique."[18] With an extensive background in Reiki massage and yogic practices, French stands quietly beside horses, his head bowed, and through deep and measured breaths has them drop their necks and soften their bodies to the point of relaxation where they are quite willing to lay down. French uses breath to bring his own brain wave frequencies within the Alpha range and, in so doing, affects electromagnetically the horses' brain wave states to the extent of having them shift into the slow wave sleep frequencies. Linda Tellington-Jones uses a more evidently tactile approach to relax and soften horses. Her "TTouch method" involves circling finger motions against the horse's skin which, again in tandem with attention to breath, bring humans and horses into emotional resonance.[19] Other methods such as the percussive tapping pioneered by J. P. Giacomini and promoted by Paul Dufresne work equally well in connecting with horses.[20] But what the more somatic means of breath resonance afford is the capacity to feel into the motional life of the horse. Breathing together inspired by wildness is what Merleau-Ponty termed the "chiasmic" overlap, the crisscrossing play of forces, the mutuality and reciprocity sustained until the inevitable "dehiscence" of being out of breath (*see* Figure 7.2).[21]

On certain days, I lead Lucente into the outdoor arena and do not release him immediately. I feel his life through the longe line still clasped to his halter. He circles around me on a perimeter made possible by this rope connection. Would he run off otherwise? This seeming constraint consolidates the basic patterns of the more tangibly felt connection I am seeking. I release my hold on him after just a few revolutions, yet he does not run away. There is no lessening of energy. We remain in touch with one another as if the longe line were still attached. I raise my energy accordingly, not to compound what Lucente is offering, but to feel this rush, this burst and upsurge of motional activation. My breathing draws deeper still, calling up resources to do much the same thing as Fraleigh describes when entering an animated place.[22] I become both "spongelike"

FIGURE 7.2 Lucente and author, Westwind Farm, Pitt Meadows, British Colombia, June 2018. Photograph courtesy of the photographer, Michele Black.

in my inspirations and bellows-like in my expirations. How much I take in and how much I expel is the moment-to-moment decision-making of keeping Lucente on an invisible cord of contact as he circles around me. "Suddenly, I am less sure of who I am and more aware of being a part of something wildly beyond me."[23] We run and stop and then take off again, until we are both heaving and out of breath.

Sharing breath is the sense I have of not just an energy flowing back and forth between Lucente and me but of the waxing, swirling, eddying synergies of moving together. I feel in my chest and diaphragm an absorption of the freshness of this morning. I take in Lucente's deep inhalations as we move vigorously in this outdoor arena. I hear the exhalations he makes as he settles into cadenced movements and feel the qualitative shift in effort as we tire of this play. I pant forcefully and hear him doing likewise as if on cue. The inspiration to run around with Lucente is about divining together an Otherness that is nowise apart from either of us. It is as if we are moving beyond self-same familiarity into a wider affirmation of life.

Moving Mimetically

Spartacus has enjoyed liberty play for some years now. He became accustomed as a young stallion to the longe line and learned to move in controlled circles and make changes of gait on cue. Now, with this repertoire ready in hand, we can "match" our movements, posturally, gesturally, and expressively, in attuned responsiveness to one another.[24] We are able to create serpentines, figure-of-eights, and circles within circles, shaping into patterns and sequences what becomes possible as human and horse "move together in a flow of open responsiveness."[25]

I match the motions Spartacus offers up. I walk beside him, bringing up my energy through a more upright, chest-out posture, and longer, more emphatic strides, and then must keep pace with him as he breaks into a trot. He feels the freshness of this crisp morning, leans into a canter, and soon has me running shoulder-to-shoulder with him. This matching of motions "affords opportunities for finding unity and difference."[26] There is unison in step, direction, and motional intention (*see* Figure 7.3). There are differences in amplitude, weight, and power requiring "surrender to unknown variables that inevitably appear in the give and take."[27]

FIGURE 7.3   Spartacus and author, Westwind Farm, Pitt Meadows, British Colombia, May 2018. Photograph courtesy of the photographer, Michele Black.

I feel the emotions in Spartacus's motions. Running side-by-side through our paces is a kind of palpation or touch at one remove. A "draw" is built into these interactions such that a reservoir of affectivity makes it possible to "drive" Spartacus out wider and have him change direction, all the while holding the invisible cord of connection. There must be enough "draw" in the relationship so that any "driving" motions will not result in the horse breaking away to find pressure relief.[28] It is a palpably tactful connection which matching motions engender. "One cannot repattern movement without first finding what patterns are present through matching them."[29] With Spartacus there is motional matching to pick up the natural patterns of movement and embellish them as improvised sequences of circles, spirals, rollbacks, 180-degree turns, and changes of direction. Self-movement and moving with the horse become coterminous. One does not so much anticipate the movement gestures and expressions as embellish them on the spur of the moment.

This liberty play may at first be considered a kind of "contact improvisation" in which there is felt "a primary impulse, a life force, which animates the places inhabited."[30] I am inclined to say that "I connect elementally with life as a self-activating force and move in an animate world with animal consciousness [....] in concert with the motions of other creatures" and that this is especially so when playing with horses at liberty.[31] But the practice of "contact unwinding" comes even closer to what is happening when human senses attenuate a horse's motions. Contact unwinding is "a spontaneous unfolding of intuitive movement" in which a "skillful listening touch taps into the kinesthetic sensitivity of the mover and can enable a fuller exploration and expression of movement to take place."[32]

> Contact Unwinding differs from Contact Improvisation, where the roles of the partners are equal or parallel and where both partners in this process act as movers, giving and taking weight in a spontaneous and sometimes explosive dance. Both processes involve listening at their core. However, Contact Unwinding is a more subtle engagement using attuned listening, which employs specific touch techniques that are learned in somatic bodywork.[33]

There is a "deep listening" to what the horse offers by way of response to the cues given.[34] I listen to what Spartacus is saying through the vigor and amplitude of his movements and to what the space I hold allows him to express. There is a "deep sounding" to this interactional space to which his senses are already attuned and which I am trying to pick up in the very manner of our exchanges.[35] I do not lose myself in this interaction.

I am becoming more my animated self in the postures, positions, gestures, and expressions of moving together where I am still capable of "holding presence for the mover" while "interacting with them in the process."[36]

Attuning Somatically

Fraleigh writes that: "Affect attunement, matching, and flow repatterning are aspects of our work involving techniques of moving together and being moved."[37] "Becoming other," when it is possible to "clear away the clutter of everyday concerns through merging with nature in meditative attunement," can be an active and interactive process of "becoming horse in the duration of the moment."[38] Attuning somatically with horses is part and parcel of breathing together and moving mimetically with one another with particular attentiveness to the moment-to-moment bursts, surges, rushes, and gushes along with the fadings and wanings of energy. Synergy is premised on mutually feeling and expressing these "vitality affects" in keeping with our respective motional repertoires.[39]

Spartacus is a slower starter than Lucente. But his power and range of motion become readily apparent when the juices are flowing. I run beside him and am careful not to compound at first the amplitude of his canter because that may result in him bouncing on the spot (in what classical horse trainers call the *terre-a-terre*) and then going up in the air (in *levade*). What the schools of equestrian art call *airs above the ground* come easily to Spartacus due to his powerful morphology. I cue him with an arm-waving gesture to circle wider and he takes this reservoir of energy into a gallop around the expanse of the outdoor arena. I follow him round and round, not so much literally going after him as focusing on his frame and leaning forward with motional intentionality into his powerful hind quarters.

My smaller circle in the center of the arena becomes an axis of Spartacus' centripetal flight. The more vigorously I walk around this circle and extend my energy outside its relatively small circumference, the more animated Spartacus becomes. But there is a limit. This burst of energy abates a little a few circumferences later. I pull back a bit, relaxing my posture, and Spartacus transitions to a canter. His head turns to the inside as I soften more and create a recessive space to draw him back to me. He wheels around in an arc as I start stepping backward, gathering a little speed and keeping just enough distance that he does not lose the inclination to join up. Spartacus runs directly toward me. I maintain the draw, holding the invisible cord of connection and running backward as if reeling Spartacus in. And just as he is about to bump into me, I press forward, standing upright to deflect the momentum upward. Spartacus bounces on the spot, alternately on his hind and forequarters. I press and pull back, press

FIGURE 7.4 Spartacus and author, Westwind Farm, Pitt Meadows, British Colombia, July 2016. Photograph courtesy of the photographer, Michele Black.

forward again and then soften, each time seeing Spartacus sink deeper and deeper onto his hind legs. I lean backward for added effect and Spartacus rears up in exuberance (*see* Figure 7.4).

Nothing is planned. This is spontaneously relational play. Spartacus knows the patterns and sequences of liberty play. He and I have gone through these motions many times. Yet each time brings something new. Frédéric Pignon who founded the *Cirque du Soleil* equestrian show, *Cavalia*, points out that some of his most compelling performances are when he is not sure what his horses are going to do in "moments when I am not totally in control, when the horses on their own initiative do something spectacular either to please the audience or me–or for the sheer joy of doing something without compulsion."[40] Similarly and perhaps even more so when no performance is at stake, playing with horses at liberty can be an improvisational practice of attuning somatically to the rushes of feeling, the gushes of gladness, and the bursts of joy.

Jean-Louis Chrétien describes "spacious joy" as "the very excess that is happening to us."[41] Spartacus' rearing lifts both our spirits. He takes me up with him. It is as if we are both now poised between earth and sky. How long can he hold this position? For as long as this burst of joy allows.

Every joy is fueled by a pure yes, rising like a flame, without curling back on itself. One never says yes to oneself, which is why one is never truly oneself except in saying yes.[42]

Attuning somatically with horses reaches the zenith of expressivity in such moments of excessive joyfulness (*see* Figure 7.5). Life manifestly overflows the bounds of our separate selves.

The reversibility of human and horse reveals in this joy-in-excess a further reversibility. Backgrounds become foregrounded. Breathing together, moving mimetically, and attuning somatically give vent to not only an interspecies animation but also to a *critter*cal somaticity in which the landscape itself inspires and moves us. Joy sustains what Fraleigh describes as "a relational unity perspective that the body and environmental nature are intrinsically one."[43] The reversibility of foreground and background, which feels at first like the landscape enlivens our motions, becomes a joy in being alive in an inherently animated world. Perhaps East puts it best in saying we are "a kind of holographic representation" of this place. We are "nature performing nature–or nature performing herself."[44]

FIGURE 7.5 Spartacus and author, Westwind Farm, Pitt Meadows, British Colombia, May 2018. Photograph courtesy of the photographer, Michele Black.

Rewilding

There is in this liberty play with horses what Fraleigh refers to in Shin Somatics as "moving with and as nature." She writes that we "are not separate from nature's elements; rather, they dwell within, fluid as water and murky. Nature is not simply in surroundings. *Nature is in us.*"[45] East nuances this principle of nature's immanence, explaining that "we are already part of, and participating with, the elements and energies of the planet."[46] This very principle is what I recognize when coming to appreciate how "[e]lemental movements, emotions, inspirations, enthusiasms, and environmental engagements are somatically based."[47] The enjoyment of playing vigorously with horses at liberty, which may well be turned into amusement and public entertainment, shall remain a kinetic, kinesthetic, and energetic dynamic of connecting with life forces and the wildness of nature in the flesh. Such playfulness is essentially a practice of rewilding a movement sensibility otherwise dulled and domesticated to human affairs.

Rewilding was originally coined by Dave Foreman to emphasize how "to make a place wild again" through the reintroduction of native species.[48] It has been broadened to now include the human species and our place connectedness. Human rewilding and the deepening of this ecological movement to include bodily practices have been of particular interest to me in terms of the emergence of natural and paleo movement forms, nature parkour, earthing, forest bathing, and wild swimming. This interest becomes even more evidently ecosomatic when, as Devin Johnston puts it, "creatures are enlisted as emissaries of the senses."[49] He concludes that "[a]t high altitude, ocean depths, or close to the earth, they enact modes of attention. If anthropomorphism interprets the world in human terms, we can with patience arrive at its inversion: not humanizing but creaturely."[50]

We need not rewild our senses to the extent of Charles Foster's burrowing, floating, foraging experiments in *Being a Beast* or Chisa Hidaka's and Benjamin Harley's *Dolphin Dance Project* in which humans and critters encounter one another when moving together in the open ocean.[51] If "[t]he main aim of rewilding is to restore to the greatest extent possible ecology's dynamic interactions," then what better way of doing so than by, say, rewilding our horse senses through a *critter*cal somaticity of breath, motional responsivity, and kinesthetic connectivity that can be readily experienced when playing with horses?[52]

Equine-assisted therapy and learning have brought general awareness of the bodily resonances between humans and horses.[53] Yet fine-tuned somatic appreciation of horses' and humans' capacities to move synchronously and synergistically together exists mostly in classical horse-riding traditions and circus and performing arts.[54] Visual artist, choreographer, and somatic practitioner, Paula Josa-Jones, has created equestrian performances involving dancers and aerialists where, through a system of wires, harnesses, and pulleys, human beings can at last move with speed, amplitude, and distance coverage to match the horses' motions.[55] Frédéric Pignon and Sylvia Zerbini have taken their liberty play with horses to a much wider public.[56] What is more fundamentally at stake than performative virtuosity, however, is how we can readily breathe and move and match and merge with horses and, in so doing, come to experience the somatic means of connecting vitally with forces and powers beyond ourselves and our anthropocentric devices. Rewilding in this manner is not just a critical and contemplative exercise of coming to our ecological senses but also a *critter*cal one of cultivating the kinetic, kinesthetic, and energetic capacities to move with creatures as they and we harness the forces of life.

Conclusion

This chapter has explored the practice of playing with horses at liberty as a means of coming to our horse senses. It is a practice of drawing upon the motional repertoire available to us as animate beings to move together, matching what we are each capable of doing and contact unwinding in support of the fuller expression of life. While playing with horses is not for everyone, what can be learned from it is a way of breathing together, moving mimetically, and attuning somatically to one another "in the radical experience that each has of sharing with others the originary affectivity of a single life-force."[57]

Joy that is spacious and in excess of any feelings of fun or amusement becomes the expression of a "transcendental affectivity."[58] Joy is the plenitude to life that begets the feelings we have for one another. It is not a summary state but a moment-to-moment interplay of moving oneself and being moved by another. Joy is an upsurging, upwelling of life forces in liberty play with horses that can be experienced with other domesticated creatures. It can also be experienced, I suggest, in wider ecosomatic realms. These extensions of coming to our horse senses are indicated in earlier writings on the "vitality affects" of moving in and with natural geographies.[59] My purpose in the present essay has been to press more deeply, both phenomenologically and practically, into a *critter*cal somaticity that makes motional sense of these life-wide affirmations.

At stake is the recovery of an inner wildness that is at the root of receptivity to the more-than-human world. This wildness resonates kinetically, kinesthetically, and energetically with what might otherwise appear as an outer wildness in creatures even of a domesticated kind. The practice of rewilding our horse senses thus becomes not just a *critter*cal somaticity but also an ecological project of attuning to a more widely and inherently animated world. The affirmation of saying yes to life which Fraleigh felt when a horse rested her head gently on Sondra's shoulder can be sustained laterally across interspecies lines. This affirmation which Buber experienced as Otherness does in fact require a turn inward in a reflection that is not preoccupied with personal feelings but is the very somatic means of staying true to life as it presents itself in bursts and rushes and ebbs and flows. This somaticity to our horse senses is the gift of living more fully in the moment, with nature, and in keeping with a capacity for moving ever more vitally with others of a human and more-than-human kind.

Notes

1 Fraleigh told the story of her encounter with the horse in response to a presentation given with Rebecca J. Lloyd on motional ways of saying yes at the "Somatic Practices in Nature" conference organized by Eastwest Somatics Institute, Elmira, NY, October 16–17, 2021. *See also* Sondra Fraleigh and Karen Barbour, "What Saying 'Yes' Affirms—We Feel; We Move; We Do," *Theatre, Dance and Performance Training* 13, no. 2 (2022). https://doi.org/10.1080/19443927.2022.2066372.

2 Martin Buber, *Between Man and Man* (New York: Routledge & Kegan Paul, 1947), 27.

3 Ibid.

4 Sondra Fraleigh, "Body and Nature: Quest for Somatic Values, East and West," *Journal of Dance & Somatic Practices* 13, no. 1–2 (2021): 116. https://doi.org/10.1386/jdsp_00040_1.

5 Fraleigh and Barbour, "What Saying 'Yes' Affirms," 247. *See also* Rebecca Lloyd and Stephen Smith, "Leaning into Life with Somatic Sensitivity: Lessons Learned from World-Class Experts of Partnered Practices," *Journal of Dance & Somatic Practices* 14, no. 1 (2022). https://doi.org/10.1386/jdsp_00072_1.

6 Richard Louv, *Our Wild Calling: How Connecting with Animals Can Transform Our Lives; and Save Theirs* (Chapel Hill: Algonquin Books of Chapel Hill, 2019), 80–81.

7 Michel Henry, *Material Phenomenology*, 1st ed. (New York: Fordham University Press, 2008).

8 "More-than-human" is a term coined by David Abram. *See* David Abram, *The Spell of the Sensuous: Perception and Language in a More-than-Human World*, 2017 ed. (New York: Vintage Books, 1996).

9 Maurice Merleau-Ponty, *The Visible and the Invisible*, trans. Alphonso Lingis (Evanston: Northwestern University Press, 1968), 168.

10 Emmanuel Falque, *The Loving Struggle: Phenomenological and Theological Debates* (London; New York: Rowman & Littlefield International, 2018), 65.

11 Ibid., 66–67.

12 Maxine Sheets-Johnstone, *The Primacy of Movement* (Amsterdam; Philadelphia: John Benjamins Publishing, 1999), 225.

13 Klaus Ferdinand Hempfling, *Dancing with Horses: Collected Riding on a Loose Rein: Trusting Harmony from the Very Beginning* (North Pomfret: Trafalgar Square Publishing, 2001).

14 Maurice Merleau-Ponty, *Nature: Course Notes from the Collège de France*, trans. Robert Vallier and ed. Dominique Seglard (Evanston: Northwestern University Press, 2003), 268.

15 Gavin Van Horn, Robin Wall Kimmerer, and John Hausdoerffer, eds., *Kinship: Belonging in a World of Relations*, vol. 1–5 (Libertyville: Center for Humans and Nature, 2021).

16 Sondra Fraleigh, "Somatic Movement Arts," in *Moving Consciously: Somatic Transformations through Dance, Yoga, and Touch*, ed. Sondra Fraleigh (Urbana: University of Illinois Press, 2015), 6.

17 Ibid., 14.

18 "The Trust Technique," accessed July 5, 2022. https://trust-technique.com.

19 "TTouch Method," accessed July 5, 2022. www.ttouch.ca.

20 "Training for Courage," accessed July 5, 2022. www.pauldufresne.com.

21 Merleau-Ponty, *The Visible and the Invisible*, 130–155.

22 *See* Fraleigh, "Body and Nature."

23 Ibid., 121.

24 Sondra Fraleigh, *Dancing Identity: Metaphysics in Motion* (Pittsburgh: University of Pittsburgh Press, 2004), 268n.4. Elizabeth Behnke, "Matching," in *Bone, Breath and Gesture: Practices of Embodiment*, ed. Don Johnson (Berkeley: North Atlantic Books, 1995).

25 Karin Rugman, "Contact Unwinding," in *Moving Consciously: Somatic Transformations through Dance, Yoga, and Touch*, ed. Sondra Fraleigh (Urbana: University of Illinois Press, 2015), 202.

26 Fraleigh, "Somatic Movement Arts," 33.

27 Ibid.

28 Jonathan Field, *The Art of Liberty Training for Horses: Attain New Levels of Leadership, Unity, Feel, Engagement, and Purpose in All You Do with Your Horse* (North Pomfret: Trafalgar Square Books, 2014), 117–120.

29 Fraleigh, "Somatic Movement Arts," 35.

30 Stephen J. Smith, "A Pedagogy of Vital Contact," *Journal of Dance & Somatic Practices* 6, no. 2 (2014): 234. https://doi.org/10.1386/jdsp.6.2.233_1.

31 Ibid.

32 Rugman, "Contact Unwinding," 197, 200.

33 Ibid., 200.

34 Paula Josa-Jones, *Our Horses, Ourselves: Discovering the Common Body: Meditations and Strategies for Deeper Understanding and Enhanced Communication* (North Pomfret: Trafalgar Square Books, 2017), 120–122.

35 Ibid., 124.

36 Rugman, "Contact Unwinding," 203–204.

37 Fraleigh, "Somatic Movement Arts," 46.

38 Sondra Fraleigh, "We Are Not Solid Beings: Presence in Butoh, Buddhism, and Phenomenology," *Asian Theatre Journal* 37, no. 2 (2020): 476, https://doi.org/ 10.1353/atj.2020.0037; Stephen J. Smith, "Becoming Horse in the Duration of the Moment: The Trainer's Challenge," *Phenomenology & Practice* 5, no. 1

(2011), https://doi.org/10.29173/pandpr19833. Stephen J. Smith, "Riding in the Skin of the Movement: An Agogic Practice," *Phenomenology & Practice* 9, no. 1 (2015): 44. https://doi.org/10.29173/pandpr25361.

39 Daniel N. Stern, *Forms of Vitality: Exploring Dynamic Experience in Psychology, the Arts, Psychotherapy, and Development* (Oxford and New York: Oxford University Press, 2010), 41.

40 Frédéric Pignon, Magali Delgado, and David Walser, *Gallop to Freedom: Training Horses with the Founding Stars of Cavalia* (North Pomfret: Trafalgar Square Books, 2009), 138.

41 Jean-Louis Chrétien, *Spacious Joy: An Essay in Phenomenology and Literature*, trans. Anne A. Davenport (Lanham: Rowman and Littlefield, 2019), 1.

42 Jean-Louis Chrétien, *The Call and the Response*, trans. Anne A. Davenport (New York: Fordham University Press, 2004), 123.

43 Fraleigh, "Body and Nature," 118.

44 Alison East, "Performing Body as Nature," in *Moving Consciously: Somatic Transformations through Dance, Yoga, and Touch*, ed. Sondra Fraleigh (Urbana: University of Illinois Press, 2015), 166.

45 Sondra Fraleigh, "Canyon Consciousness," in *Dance and the Quality of Life*, ed. Karen Bond (Cham: Springer, 2019), 30.

46 East, "Performing Body as Nature," 167.

47 Stephen J. Smith and Rebecca J. Lloyd, "Promoting Vitality in Health and Physical Education," *Qualitative Health Research* 16, no. 2 (2006): 264. https://doi.org/10.1177/1049732305285069.

48 Qtd. in J. B. MacKinnon, *The Once and Future World: Finding Wilderness in the Nature We've Made* (Boston: Houghton Mifflin Harcourt, 2013), 64.

49 Devin Johnston, *Creaturely and Other Essays* (New York: Turtle Point Press, 2009), 12.

50 Ibid.

51 Charles Foster, *Being a Beast: Adventures across the Species Divide* (New York: Metropolitan Books, Henry Holt and Company, 2016). For a description of Hidaka's and Harley's project, *see* Una Chaudhuri, "Interspecies Diplomacy in Anthropocenic Waters: Performing an Ocean-Oriented Ontology," in *The Routledge Companion to the Environmental Humanities*, eds. Ursula K. Heise, Jon Christensen, and Michelle Niemann (New York and London: Routledge, 2017).

52 George Monbiot, *Feral: Rewilding the Land, the Sea, and Human Life* (Chicago: The University of Chicago Press, 2014), 83–84.

53 Stephen J. Smith, "Being with Horses as a Practice of the Self-with-Others: A Case of Getting a FEEL for Teaching," in *Catalyzing the Field: Second-person Approaches to Contemplative Learning and Inquiry*, eds. Olen Gunnlaugson et al. (New York: SUNY Press, 2019), 59.

54 Stephen J. Smith, "Human-Horse Partnerships: The Discipline of Dressage," in *Sport, Animals, and Society*, eds. James Gillett and Michelle Gilbert (New York: Routledge, 2014), 37–40; Stephen J. Smith, "Dancing with Horses: The Science and Artistry of Coenesthetic Connection," in *Domestic Animals and Leisure*, ed. Neil Carr (New York: Palgrave Macmillan, 2015), 225–228.

55 Paula Josa-Jones, "Flight," Video, *The Horse Dances* (2017). www.paulajosajo nes.org/the_horse_dances/flight.html.
56 Pignon, Delgado, and Walser, *Gallop*.
57 Raphaël Gély, "Towards a Radical Phenomenology of Social Life: Reflections from the Work of Michel Henry," in *Michel Henry: The Affects of Thought*, eds. Jeffrey Hanson and Michael R. Kelly (London: Bloomsbury, 2012), 162.
58 Henry, *Material Phenomenology*, 81.
59 Stephen J. Smith, "Gesture, Landscape and Embrace: A Phenomenological Analysis of Elemental Motions," *Indo-Pacific Journal of Phenomenology* 6, no. 1 (2006): 3, https://doi.org/10.1080/20797222.2006.11433914; Stephen J. Smith, "The First Rush of Movement: A Phenomenological Preface to Movement Education," *Phenomenology & Practice* 1, no. 1 (2007): 66. https://doi.org/10.29173/pandpr19805.

Bibliography

Abram, David. *The Spell of the Sensuous: Perception and Language in a More-than-Human World*. 2017 ed. New York: Vintage Books, 1996.
Behnke, Elizabeth. "Matching." In *Bone, Breath and Gesture: Practices of Embodiment*, edited by Don Johnson, 317–337. Berkeley: North Atlantic Books, 1995.
Buber, Martin. *Between Man and Man*. New York: Routledge & Kegan Paul, 1947.
Chaudhuri, Una. "Interspecies Diplomacy in Anthropocenic Waters: Performing an Ocean-Oriented Ontology." In *The Routledge Companion to the Environmental Humanities*, edited by Ursula K. Heise, Jon Christensen and Michelle Niemann, 144–152. New York; London: Routledge, 2017.
Chrétien, Jean-Louis. *The Call and the Response*. Translated by Anne A. Davenport. New York: Fordham University Press, 2004.
———. *Spacious Joy: An Essay in Phenomenology and Literature*. Translated by Anne A. Davenport. Lanham: Rowman and Littlefield, 2019.
East, Alison. "Performing Body as Nature." In *Moving Consciously: Somatic Transformations through Dance, Yoga, and Touch*, edited by Sondra Fraleigh, 164–179. Urbana: University of Illinois Press, 2015.
Falque, Emmanuel. *The Loving Struggle: Phenomenological and Theological Debates*. London; New York: Rowman & Littlefield International, 2018.
Field, Jonathan. *The Art of Liberty Training for Horses: Attain New Levels of Leadership, Unity, Feel, Engagement, and Purpose in All You Do with Your Horse*. North Pomfret: Trafalgar Square Books, 2014.
Foster, Charles. *Being a Beast: Adventures across the Species Divide*. New York: Metropolitan Books, Henry Holt and Company, 2016.
Fraleigh, Sondra. *Dancing Identity: Metaphysics in Motion*. Pittsburgh: University of Pittsburgh Press, 2004.
———. "Somatic Movement Arts." In *Moving Consciously: Somatic Transformations through Dance, Yoga, and Touch*, edited by Sondra Fraleigh, 24–49. Urbana: University of Illinois Press, 2015.
———. "Canyon Consciousness." In *Dance and the Quality of Life*, edited by Karen Bond, 23–44. Cham: Springer, 2019.

———. "We Are Not Solid Beings: Presence in Butoh, Buddhism, and Phenomenology." *Asian Theatre Journal* 37, no. 2 (2020): 464–489. https://doi.org/10.1353/atj.2020.0037.

———. "Body and Nature: Quest for Somatic Values, East and West." *Journal of Dance & Somatic Practices* 13, no. 1–2 (2021): 113–123. https://doi.org/10.1386/jdsp_00040_1.

Fraleigh, Sondra, and Karen Barbour. "What Saying 'Yes' Affirms—We Feel; We Move; We Do." *Theatre, Dance and Performance Training* 13, no. 2 (2022): 246–247. https://doi.org/10.1080/19443927.2022.2066372.

Gély, Raphaël. "Towards a Radical Phenomenology of Social Life: Reflections from the Work of Michel Henry." In *Michel Henry: The Affects of Thought*, edited by Jeffrey Hanson and Michael R. Kelly. London: Bloomsbury, 2012.

Hempfling, Klaus Ferdinand. *Dancing with Horses: Collected Riding on a Loose Rein: Trusting Harmony from the Very Beginning.* North Pomfret: Trafalgar Square Publishing, 2001.

Henry, Michel. *Material Phenomenology.* 1st ed. New York: Fordham University Press, 2008.

Johnston, Devin. *Creaturely and Other Essays.* New York: Turtle Point Press, 2009.

Josa-Jones, Paula. "Flight." Video, The Horse Dances. (2017). Accessed December 10, 2018. www.paulajosajones.org/the_horse_dances/flight.html.

———. *Our Horses, Ourselves: Discovering the Common Body: Meditations and Strategies for Deeper Understanding and Enhanced Communication.* North Pomfret: Trafalgar Square Books, 2017.

Lloyd, Rebecca, and Stephen Smith. "Leaning into Life with Somatic Sensitivity: Lessons Learned from World-Class Experts of Partnered Practices." *Journal of Dance & Somatic Practices* 14, no. 1 (2022): 91–108. https://doi.org/10.1386/jdsp_00072_1.

Louv, Richard. *Our Wild Calling: How Connecting with Animals Can Transform Our Lives; and Save Theirs.* Chapel Hill: Algonquin Books of Chapel Hill, 2019.

MacKinnon, J. B. *The Once and Future World: Finding Wilderness in the Nature We've Made.* Boston: Houghton Mifflin Harcourt, 2013.

Merleau-Ponty, Maurice. *The Visible and the Invisible.* Translated by Alphonso Lingis. Evanston: Northwestern University Press, 1968.

———. *Nature: Course Notes from the Collège de France.* Translated by Robert Vallier and edited by Dominique Seglard. Evanston: Northwestern University Press, 2003.

Monbiot, George. *Feral: Rewilding the Land, the Sea, and Human Life.* Chicago: The University of Chicago Press, 2014.

Pignon, Frédéric, Magali Delgado, and David Walser. *Gallop to Freedom: Training Horses with the Founding Stars of Cavalia.* North Pomfret: Trafalgar Square Books, 2009.

Rugman, Karin. "Contact Unwinding." In *Moving Consciously: Somatic Transformations through Dance, Yoga, and Touch*, edited by Sondra Fraleigh, 195–211. Urbana: University of Illinois Press, 2015.

Sheets-Johnstone, Maxine. *The Primacy of Movement.* Amsterdam; Philadelphia: John Benjamins Publishing, 1999.

Smith, Stephen J. "Dancing with Horses: The Science and Artistry of Coenesthetic Connection." In *Domestic Animals and Leisure*, edited by Neil Carr, 216–240. New York: Palgrave Macmillan, 2015.

———. "Gesture, Landscape and Embrace: A Phenomenological Analysis of Elemental Motions." *Indo-Pacific Journal of Phenomenology* 6, no. 1 (2006): 1–10. https://doi.org/10.1080/20797222.2006.11433914.

———. "The First Rush of Movement: A Phenomenological Preface to Movement Education." *Phenomenology & Practice* 1, no. 1 (2007): 47–72. https://doi.org/10.29173/pandpr19805.

———. "Becoming Horse in the Duration of the Moment: The Trainer's Challenge." *Phenomenology & Practice* 5, no. 1 (2011): 7–26. https://doi.org/10.29173/pandpr19833.

———. "Human-Horse Partnerships: The Discipline of Dressage." In *Sport, Animals, and Society*, edited by James Gillett and Michelle Gilbert, 35–51. New York: Routledge, 2014.

———. "A Pedagogy of Vital Contact." *Journal of Dance & Somatic Practices* 6, no. 2 (2014): 233–246. https://doi.org/10.1386/jdsp.6.2.233_1.

———. "Riding in the Skin of the Movement: An Agogic Practice." *Phenomenology & Practice* 9, no. 1 (2015): 41–54. https://doi.org/10.29173/pandpr25361.

———. "Being with Horses as a Practice of the Self-with-Others: A Case of Getting a FEEL for Teaching." In *Catalyzing the Field: Second-Person Approaches to Contemplative Learning and Inquiry*, edited by Olen Gunnlaugson, Charles Scott, Heesoon Bai and Edward W. Sarath, 59–71. New York: SUNY Press, 2019.

Smith, Stephen J., and Rebecca J. Lloyd. "Promoting Vitality in Health and Physical Education." *Qualitative Health Research* 16, no. 2 (2006): 249–267. https://doi.org/10.1177/1049732305285069.

Stern, Daniel N. *Forms of Vitality: Exploring Dynamic Experience in Psychology, the Arts, Psychotherapy, and Development*. Oxford; New York: Oxford University Press, 2010.

Van Horn, Gavin, Robin Wall Kimmerer, and John Hausdoerffer, eds. *Kinship: Belonging in a World of Relations* Vol. 1–5. Libertyville: Center for Humans and Nature, 2021.

8

MOVING WITH CATS

Shannon Rose Riley

The Berlin Zoological Garden, Berlin, Germany: 52.5079° N, 13.3378° E

Grimmuseum, Berlin, Germany: 52.4911° N, 13.4127° E

Private multispecies dwelling, Fremont, California, U.S.A.: 37.5483° N, 121.9886° W

David Wood's classic essay, "Thinking with Cats," opens out from Derrida's reflections on an encounter with his own cat in order to raise the question of what it means to address and to be addressed by the "animal" Other.[1] Wood traces this phenomenology of the look and of being looked at through Sartre and Lévinas and queries human-animal relations *vis-à-vis* both Wittgenstein's lion and Derrida's cat. Wood makes clear that two of Derrida's many contributions to "the question of the animal" are to affirm that there is "an intimate connection between our thinking about animals and our self-understanding" and that "our carnivorous and other exploitative practices need to be called out for what they are: violence and genocide."[2] Wood concludes with the provocation that we humans need to "recognize that our inter-*est*, our *inter-esse*, our being-connected, being-related, is in need of enlightenment even for the sake of our own survival."[3]

In what follows, I take a "performance as research" (PAR) approach that engages phenomenology and ecosomatics in order to ask what else might be learned or learned differently—by moving with cats.[4] I offer two situated responses to this question: one in the context of a performance art piece in Berlin and the other in the context of a somatic movement session in a private multispecies dwelling in Northern California. I conclude with some reflections on what is at stake in this work as well as with suggestions on how you may safely and respectfully begin to incorporate or invite moving with cats, or perhaps other critters, into your ecosomatic practice.

DOI: 10.4324/9781003390985-11

Performing *Lie-in (Lion)*

On May 18, 2013, I performed Joanne Bristol's *Lie-in (Lion)* as part of Ilya Noé's *Erogate/Surrogate (E/S) Performance Series* at Month of Performance Art-Berlin.[5] The person who was originally scheduled to perform the piece was unable to participate, so Noé asked if I would step in. I agreed.[6]

Bristol's instructions were simple. The performer would go to the zoo and observe a lion for 2 hours on the day of the scheduled performance. The performer would then go to Grimmuseum and "attempt to mimetically reenact the gestures and spatial usage of the lion" over a period of 2 hours in the front gallery space. I was not to take any notes or photographs; Bristol feels anyone should be able to perform the piece based only on observation.[7]

Carnivore House, Berlin Zoological Garden

It was a cold, rainy day in gray Berlin and I was unprepared for the weather. Ilya Noé lent me a coat, but flipflops exposed my feet to the elements as we made our way by train and on foot to the Berlin Zoological Garden, or Berlin Zoo, which was established in 1844. The Zoo featured many of the hallmarks of colonialism and anthropocentrism that you might expect to find, from the names of its features, like the "Carnivore House" and its antiquated display philosophy to the orientalist gates that mark its two entrances in a heraldic manner: the Elephant Gate on *Budapester Straße* and the Lion Gate on *Hardenbergplatz*.

We entered through the Lion Gate and made our way toward the building that held the large cats. It was a sad zoo—you could feel it in the air. The philosopher, Edward S. Casey, uses the term "peri-phenomenological" to emphasize the peripheral dimensions of emotions that are "displayed in scenes *around (peri-)* us and thereby deliver [the emotions] *to* us."[8] What I felt that day as we passed under the Lion Gate and spent time in the "Carnivore House" was reaffirmed later when I learned that the Berlin Zoo has been described as one of "Europe's Cruellest Destinations."[9]

The display pathway inside the big cat exhibit was designed to move people in a counterclockwise direction in front of several glassed-in enclosures that circled the outer edge of the space. In, around, and out again. I had to consciously resist the ready-made scopophilia to which this kind of zoo architecture caters. I did so initially by not moving through the space as intended but by stopping at the first large cage, which contained an old male and female lion. Folks came in behind me and continued past on their quick tour of the wild cats; most people stayed with these lions for only a moment or two. While the rest of the viewers talked loudly, gesticulated wildly, and moved on quickly, I remained quiet, attentive, rooted in place. I remained for the next 2 hours. During that time, some people noticed my odd zoo performativity—one of Noé's photos shows a little boy staring quizzically at me as I focus my attention and intention on the lioness.

What must be said about these cat enclosures? They were little more than concrete cells with turquoise tiled walls and blue metal doors—none of the greens, tans, or browns of a more natural habitat, save for a few beige ropes hanging from the wall like misplaced toys from a domestic cat condo. One or two tree stumps were piled on the concrete floor, while a larger tree branch was propped against a wall—a couple of clinical looking benches were attached there like prison cots or examination tables.

The old lioness was sleeping when I first arrived but it was almost feeding time, so I got to witness her awaken, stretch, eat, pace, urinate, yawn, and so on. It would be clear to anyone who took the time to consider that she would move very differently if she were free. At several points, she and I made and held eye contact with one another (*see* Figure 8.1). I was reminded of Emmanuel Lévinas, who theorized that we come into being through the face-to-face encounter with the Other.[10] But Lévinas could not extend the ethics of this relation to the animal-human encounter.[11]

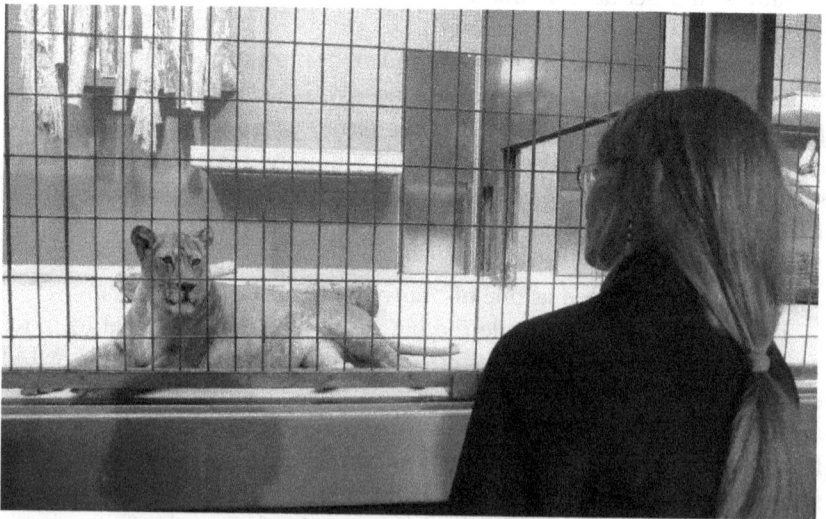

FIGURE 8.1   Riley and the female lion, Berlin Zoo, May 2013. Photo by Ilya Noé, courtesy of the photographer, 2013.

The first time the lioness and I held each other's gaze, I remained as still as possible and focused on matching her breath and muzzle, about which I say more below. What struck me, almost immediately, was a feeling of profound connection with this being, this lion-*esse*. In that moment of seeing and feeling myself seen by her, of breathing with her, I was affected with a barrage of different kinds of emotion from awe and joy to shame and something else along the edge of a great silent sadness shared between us—an embodiment of the "abyssal ruptures" that open up in human-animal encounters.[12] More superficially, I realized that while I had purposefully paid no attention to my clothing choices that day, there was an uncanny parallel between us—her fur, my hair, and jeans all happened to be a similar color; my ponytail perhaps akin to her tail. Even the frames of my glasses appeared similar in color to the tiled walls of her cell. I had not anticipated any of these connections.

Given my background in somatic approaches to performance, I was concerned that basing my performance primarily on visual observation would replicate some of the problematic components of zoo performativity, namely, its brutal colonizing gaze. Such a strategy also reifies us—the lion and me—as separate subjects or more properly as subject/object rather than exploring what nuclear physicist and feminist theorist Karen Barad describes as our "mutual constitution," our "deeply entangled agencies."[13]

I allowed myself to be fully present to the unfolding "intra-action" between us rather than only trying to mimic her gestures or movement. Developed by Karen Barad, "intra-action" signifies such mutual constitution and entanglement and offers a theoretical challenge to the notion of interaction, which conceives of *a priori* separate agencies.[14] For Barad, intra-action highlights that "distinct agencies do not precede, but rather emerge through, their intra-action."[15] The notion of mutual emergence may seem similar to Lévinas' articulation of the ethical subject emerging through human interaction, namely, the encounter with the human face. However, Barad's formulation is much more nuanced in its understanding of agencies and mutual constitution; it is flexible enough to extend the ethics of the face-to-face relation to the encounter between the human and more-than-human. Thinking Lévinas alongside Barad, we can approach moving with cats as a phenomenology of intra-subjective responsibility.

Foregrounding this kind of relation, I used a somatic approach to *being-with* the lion as well as to my subsequent performance in the gallery, which I discuss below.[16] Most simply, I drew from my movement practice and what I have described elsewhere as "embodied perceptual practices," in order to discern and pay close attention to the various streams of perception that converge simultaneously in the lived body: proprioception, "the perception of touch and the movement of muscles and joints"; interoception, "the perception of the state of the organs and viscera"; and exteroception, "perception of one's relationship to their environment".[17] For neurologist Antonio Damasio, these streams are part of what constitute the process of mind, which he understands to be embrained, embodied, and environed.[18]

I then extended my own somatic awareness to the lion's embodiment, trying to become sensitive to her streams of perception. Said again, I attempted to attune myself, as much as possible, to a sense of her body in space and the movement of her limbs; to her sense of breath, of hunger, of needing to urinate; and to her sense of confinement, her very clear awareness of humans in the space, of being on display, of us looking at her, and so on. I did this at first by simply trying to match my breath with hers and to match her muzzle with my own while holding on to the energetic connection that was building between us (*see* Figure 8.2). I also paid attention to any feelings, images, or sensations that emerged while I focused on matching and perception. In this manner, I intended to match my state of awareness with that of the lion and to try to map her perceptions and awareness onto my own.

FIGURE 8.2 Riley and the female lion, matching breath and energy in the muzzle, Berlin Zoo, May 2013. Photo by Ilya Noé, courtesy of the photographer, 2013.

I tried on her movement and her breath while I paid attention to my emotional responses to these positions. I observed the energetic places in her body and was quickly able to locate and trace a vibrant line in her being, from her breath, muzzle, and tongue through her spine and tail. I continued trying to match her energy, breath, and postures, while focusing especially on this muzzle-spine-tail connection.[19] In Figure 8.3, my hands match her front paws and my crossed feet match the position of her back legs. I am trying to feel the weight and heft of those big front paws in my own hand.

What is going on with this technique—this encounter? For one thing, somatic strategies like embodied perceptual practices and matching offer generative possibilities for being-with-others, including the more-than-human. This manner of being-with-others utterly rejects anything resembling the Cartesian subject with its imagined apartness and its mind/body split. Matching, originally described by phenomenologist and somatic educator, Elizabeth Behnke, consists of three parts, which can flow together in practice: "(1) awareness of something in one's own body; (2) an inner act of matching or aligning oneself with this; and (3) allowing something to change."[20] For Behnke, matching describes an attunement with and through one's embodied self-awareness.[21] In Fraleigh's critique of Heidegger, however, she speaks further of matching *wesen*—whatness— or "the sway of beings," rather than being concerned with mastery over the Other.[22] Indeed, as Fraleigh pointed out to me, the "lion's dilemma" stems from her being "the object of mastery and masters."[23] Matching, instead, is relational—it offers a way of practicing noninterference while being-with the Other. Matching is a form of attunement that extends itself toward and receives the Other. Casey notes that bodily interaction, or "interembodiment," is a prerequisite to "affective attunement."[24] Said another way, "the lived body is the sine qua non" for this kind of attunement.[25]

On one level, to attune simply means to bring into harmony, but the generative concept has been developed in numerous areas, including psychology and phenomenology, for thinking about human interactions in the clinical setting as well as in society, and as a method for self-cultivation and attending to the lived body through somatic practice (the bodily attunement of Nagatomo and Yuasa).[26] Casey's discussion of attunement focuses especially on the attunement between human beings and the "outlying environment."[27] In order to build his argument, he draws from Cynthia Willett's *Interspecies Ethics*, which "extends the range of socially affective attunement to interspecies interactions of many kinds," and from Marjolein Oele's theory of "e-co-affectivity,"

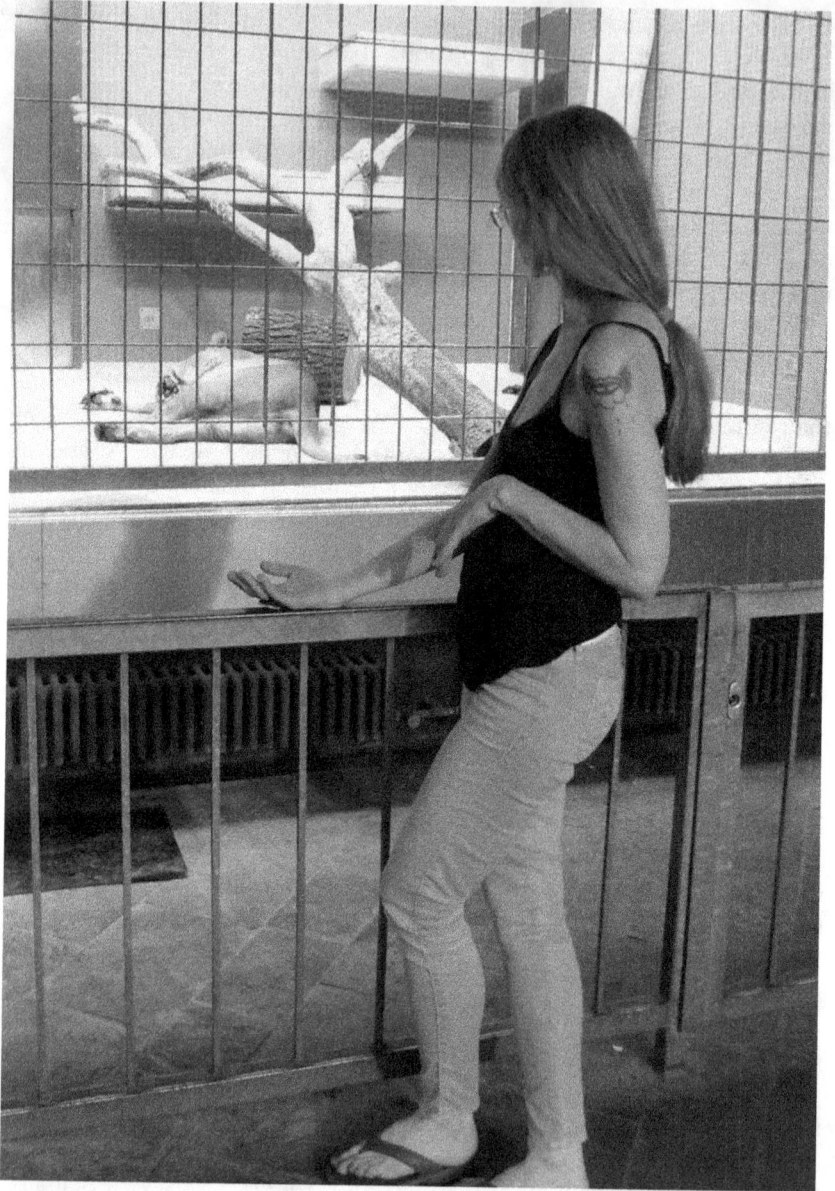

FIGURE 8.3 Riley matching the female lion, Berlin Zoo, May 2013. Photo by Ilya Noé, courtesy of the photographer, 2013.

which postulates that the "inherent biomorphic materiality" of all living beings fosters a "fundamental shared affectivity" in and between us.[28] Casey's work opens up the notion of attunement to place and the peri-phenomenology of emotions. If we take Barad, Casey, Willett, and Oele seriously, we are always already interembodied with other species, spaces, and shared emotions whether we intend to be or not. We are always already involved in intra-action. Through matching, we can begin to shift the energy of mastery that has been foundational to the human-animal encounter so that we can speak instead of shared affect and attunement.[29] I feel confident claiming that the interembodiment that unfolded between the lion and me—moving together vis-à-vis matching—constituted a shared "affective attunement." There can be little doubt, for me, that the lion felt us humans in the space at the level of her very lived body—perhaps she felt me as a distinct being. I accept Barad's theorizations of our "deeply entangled agencies," and thus, I believe that in some manner, the lioness did.

Some scientists have attempted to locate the capacity for affective and cognitive attunement in the body at the level of biology. One study in the 1990s found "mirror neurons" functioning in macaque monkeys when they observed humans engaging in activity.[30] They theorized that mirror neurons are a functional mechanism of "embodied simulation" that consists of the "automatic, unconscious, and noninferential simulation in the observer of actions, emotions, and sensations carried out and experienced by the observed."[31] In this view, "mirror neurons" are the "neural underpinning" of "interpersonal relations."[32] Other scientists problematize this view; Gregory Hickok argues that

despite its widespread acceptance, the proposal [the mirror neuron theory of action understanding] has never been adequately tested in monkeys, and in humans there is strong empirical evidence, in the form of physiological and neuropsychological (double) dissociations, against the claim.[33]

Rather than looking "inside" for a biological source of the capacity for intra-action, let's follow Casey and Barad and consider first, that attunement, "interemotionality," has peri-phenomenological aspects, and second, that it may emerge at the scene of, or *vis-à-vis*, intra-action. "There is no single proper place for emotions–least of all inside the human subject."[34]

Place Matters

In 2018, floods destroyed the Carnivore House enclosures and all five wild cats escaped, including my lion—what else can I call her but that? All five were captured and given permanently to other zoos, the outdated and cruel enclosures finally gutted. The Berlin Zoo, which had been criticized for its unethical practices, installed more "natural" outdoor habitats, including heated caves, that might allow for increased natural movement and behavior.[35] However, even in improved habitat, movement and behavior will be limited and decidedly un-natural. While discussing the Berlin Zoo in particular, the People for the Ethical Treatment of Animals (PETA) Foundation makes clear that "All zoos are morbid prisons for animals. Enclosures are small, natural needs are suppressed and 'zoochosis'— abnormal, self-destructive behaviour such as pacing around, biting cage bars and head-bobbing—often results."[36] Let us reflect upon our own morbidity.[37]

Grimmuseum, Berlin

I felt sad to leave my lion, but mostly, I felt exhausted and numb, even angry. I wondered if that was how she felt on a daily basis, month after month, year after year. As we humans made our way again by train and on foot to the gallery where I would perform, I kept picturing her, feeling her breath in my body. I did not feel like talking much before the performance—I felt isolated and strange and did not want to dissipate any of the energy that remained in my body from our encounter. En route, I pondered my performance strategy. I later described to Bristol and Noé that I was hesitant to rely on notions of "becoming the other" and that I chose to work with the same somatic signals that I used at the Zoo, such as the breath and tongue work and connecting my tailbone/tail to the tip of my nose/muzzle as opposed to making "an attempt to internalize lion-ness."[38]

Once on site, I stayed in the gallery's office until it was time to begin. Other than removing my glasses and shoes, I made no other changes to my appearance. I made sure to give myself a bit of time to be in the performance space alone because I wanted to feel on display from the very start and to be fully engaged in my somatic experience well before the audience entered. I climbed up onto the platform and began to focus on matching my breath to what I recalled of hers. Before and during the performance, I worked on locating and activating my own muzzle (*see* Figures 8.4 and 8.6) and trying to match her dorsal power (*see* Figure 8.7). Throughout, I worked improvisationally with my embodied re-collections

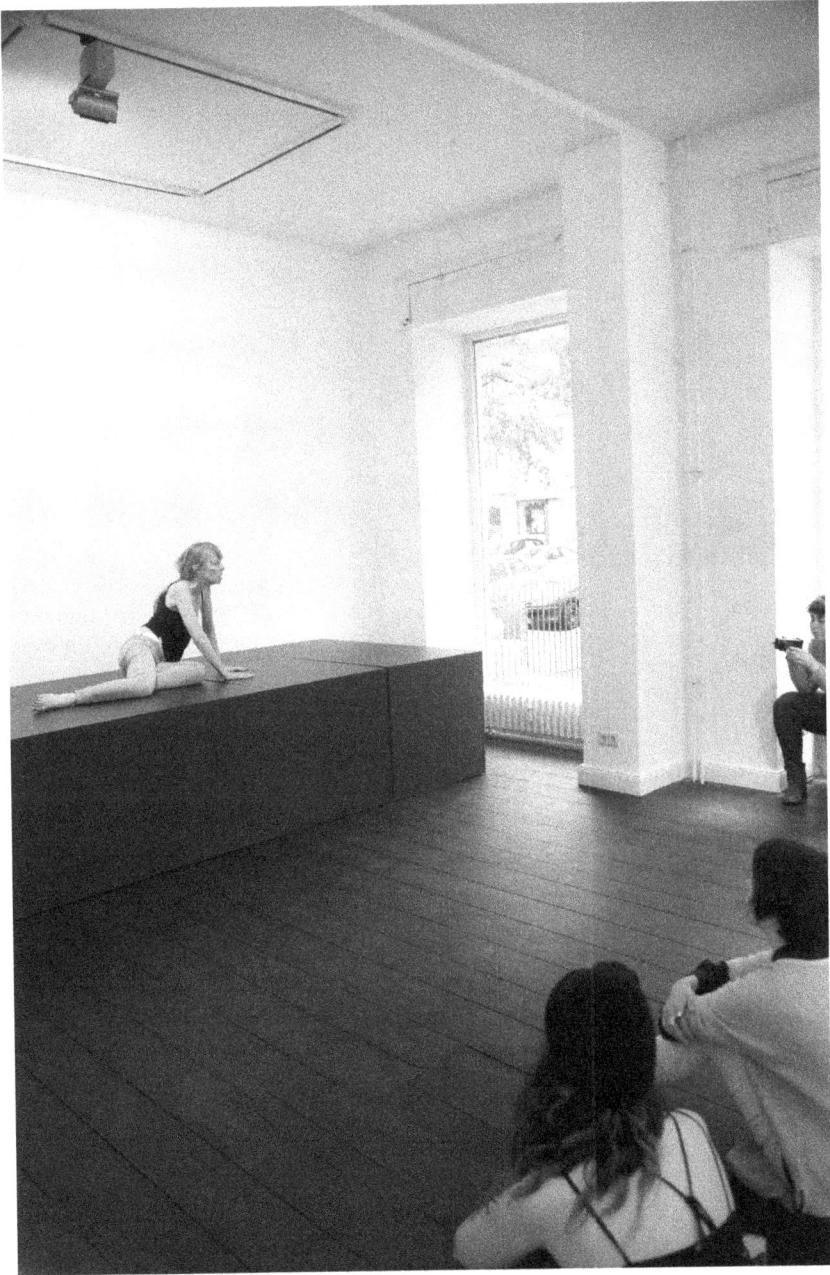

FIGURE 8.4   Riley matching energy in the muzzle in performance at Grimmuseum, Berlin, May 2013. Photo by Amanda Ribas Tugwell, courtesy of the photographer.

FIGURE 8.5   Riley focusing on matching lion energy while performing at Grimmuseum, Berlin, May 2013. Photo by Amanda Ribas Tugwell, courtesy of the photographer.

FIGURE 8.6   Riley paying attention to the breath and spine while performing at Grimmuseum, Berlin, May 2013. Photo by Amanda Ribas Tugwell, courtesy of the photographer.

FIGURE 8.7   Riley matching the lion's stretch in performance at Grimmuseum, Berlin, May 2013. Photo by Amanda Ribas Tugwell, courtesy of the photographer.

of that encounter without trying to replicate anything in particular or in any order. When the audience was finally let in, I was well-established in the space—restless, bored, pacing back and forth on an edge between angst and anguish, then tired, despondent, sitting, taking them all in (*see* Figures 8.4, 8.5, 8.6, and 8.7). My tormenters.

No one in the audience addressed me. They watched; they took photographs. They whispered to themselves. They pointed. They stayed. They left. Being on display in that manner is degrading and very stressful. I knew this when I was at the Zoo, through some combination of attunement and peri-phenomenological affect, but I felt it more intensely at Grimmuseum because of taking on the position of the observed object. As I performed, the parallel places of exhibition—zoo and gallery—were also mapped onto one another in my emotional landscape. The bars outside of the gallery window confirmed my impression (*see* Figure 8.4).

In matching, you make space for allowing something to change. Taking time to myself while traveling to the gallery and before the audience entered gave me that space. To be sure, the embodied simulation and matching at the Zoo had a strong impact on me, but the overall impact became more fully apparent during my performance at Grimmuseum. In a post-performance email, I wrote about becoming "deeply sad and melancholic."[39] One of the benefits of a PAR methodology, in which doing or performing something—embodied research—produces knowledge, is that it takes embodied reflection and response seriously.[40] One of Bristol's goals with the work is to "transform the ways humans hold and behold nonhuman others" and I believe that occurs most fully when the observer at the Zoo becomes the observed at the gallery.[41]

After performing, I wrote

the two-hour performance was both compressed, fast, and very drawn out. [...] It was exhausting. I became lethargic at moments, my eyelids felt heavy—my breath driving the movement of my spine, the flicker of my tongue. It was through performance that the scopophilic drives of the zoo-space/time receded and I began to experience situated knowledge about feeling lion-sadness as a result of containment and display.[42]

It may be clear to the astute reader that PAR can be a phenomenological methodology.

Movement in Depth with a Housecat, Fremont, California, United States

For many years, I have practiced what is called "movement in depth" or "authentic movement," which I describe elsewhere as "a form of relational movement therapy based on a psychoanalytic model of mind."[43] In the most basic form—the dyad—there is a mover and a witness. The mover moves with eyes closed, starting in silence and stillness; "the mover's eyes are closed in order to bring the mover's attention to her other senses and into a more perceptive state of awareness."[44] The witness holds space for the mover and observes the movement that is unfolding. There are different ways to begin and end this basic process but at the conclusion of the movement period, there is often time for processing individually as well as together, usually beginning with the mover's own experience first.[45]

Typically, my witness and I meet twice a month in a shared physical space in Berkeley, California; however, since the COVID-19 pandemic, we have had sessions via Zoom technology. This has allowed for some interesting developments—namely, moving with cats. I am lucky enough to cohabitate with two cats and a dog. The oldest cat is a 14-year-old once-feral Maine Coon named Squeak; the newly rescued feral Tabby, Katniss, is around 8 months old; and Cairo, the mixed breed Min Pin, is six. Of the three, Squeak is the most independent. He rarely jumps into my lap to be pet, dislikes cuddling, and is usually on his own somewhere.

Even the still-skittish kitty will occasionally grace my lap for a minute or two. However, Squeak is the only one of the three who joins me in this movement practice. Cairo and Katniss usually vacate the living room when they see what is about to unfold. They know the signals when I start to set up my laptop in a certain place, prepare the rug, and get down on the floor. By the time the witness rings her bell three times, Squeak is already on the floor or sitting on the couch in the role of second witness.

My eyes are closed when I move, but I can usually hear his purr before I have a sense of his location in space. I do not seek him out; I do not try to play or move with him. I just continue about my business, which is to follow the impulse to move in the present. That said, there are moments when we make contact by accident. I have been in a forward hanging position and felt the long fur of his tail caress my chin as he has walked under me as if I were a bridge. In these moments, I have the opportunity to engage or to let the interaction pass. The first few times he joined me in movement, I let the interactions pass.

At a session in December 2021, I knew we were moving together even with eyes closed; I heard him nearby and felt his fur. More than once, he initiated direct contact. Suddenly, in response to him, I felt something uncanny emerge from my own body: the elongated lion stretch that I had experienced at Grimmuseum (*see* Figure 8.7). I was surprised to find the lion's vocabulary and sensation in my body after more than eight years—I was especially surprised that a movement experience with my old cat triggered the memory. I was so moved—full of e-motion. After completing that tongue-to-toe stretch, I simply lay down on my right side, cat-like, content; I could still hear him purring nearby. One rarely opens one's eyes in this practice, but I felt compelled to do so. When I did, Squeak was lying on his right side, kitty corner from me, his head aligned with my feet. His eyes were open and he was staring at me, too. I could see his ribs rise and fall with each purr. We kept this shared connection for several minutes before the bell rang three times, signaling the end of the movement period.

My witness, Susan Bauer, and I have discussed the kinds of things Squeak does and what it is like to incorporate a cat into this practice. She has described a sense of having two movers to witness—me and a cat. At other times, she has described moments where I, the mover, have two witnesses—she and the cat.[46] It has also been interesting to feel Squeak drop more deeply into this time together. He seems to be as excited for the sessions as I am. After movement, I usually draw or write for a few moments in order to process the experience in another medium. I now have several drawings that include the figure of the cat-witness or cat-mover. In Figure 8.8, I have depicted myself in a blue and orange forward

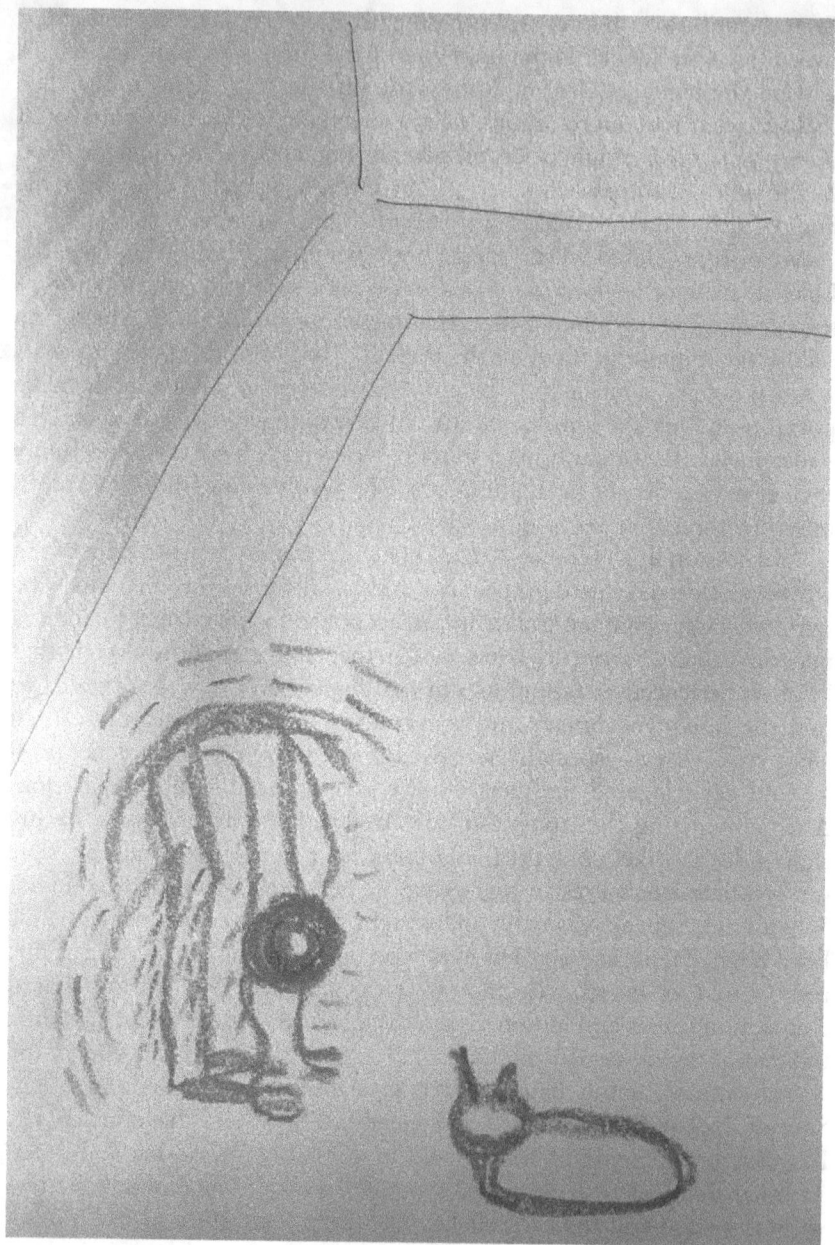

FIGURE 8.8 Shannon Rose Riley, Authentic Movement with Squeak, 2021, pastel and ink on paper, courtesy of the author. Drawing by Shannon Rose Riley.

fold with many energetic lines emanating from my limbs and head. A gray cat sits on the floor nearby and simply bears witness.

How to Move with Cats

How to Be-with a Cat Who Is in Captivity, Whether in a Cage or "Natural" Habitat Enclosure

1 Hopefully you will no longer be able to encounter a large cat in a cage or cell. But if you do, please be mindful of how you enter the space. Zoo space is designed for your gaze to colonize the lion again and again. Don't do it—find another way.
2 With any luck, the wild cats will be in their heated caves and won't have the time of day for you if you come upon them in an outdoor enclosure. Whether they are visible or not, please be quiet and respectful. Do not attempt to enter the habitat—this almost always ends in death for the animal and sometimes for the offending human.
3 Approach the limits of the space respectfully with your head slightly bowed. Observe silently without making large gestures, laughing, or talking to the cat or to anyone else. Focus all of your intention on being present for/with one cat.[47]
4 Stand still for a while and be-with.
5 From stillness, begin matching your physical position and breath to that of the cat that you have chosen or that has chosen you. Move slowly and with respect.
6 Meet them at the level of muzzle, tongue, breathe, spine, and tail. Practice matching the energies at these sites and processes with-in your own body.
7 Stay with them for at least an hour, alternating between periods of focused matching in which you focus on the breath, muzzle, and tail, and periods of focused stillness in which you rest and simply be-with. Two hours is better. If you get distracted, simply begin again by matching breath.
8 If you find yourself in a face-to-face encounter, remember that for Lévinas, this relation is foundational; it defines us as ethically responsible to the Other. He says that the human face "orders and ordains" us; it calls us into "giving and serving" the Other.[48] What happens if we extend this ethical relation to this lion in this moment? To the more-than-human—even in just this one instance? It might mean that you have an obligation to return again to see this particular cat. Perhaps you should learn their name. You may have briefly broken through their

tedium and anxiety—or your own. They may have felt something—you may have felt something. They will look for you. What else can you do? What does it mean to carry this responsibility for and toward the Other forward?

9 When you go home, give yourself some time to allow something to change. Find a safe place where you can move or remain still for a few minutes—by yourself is best. It is important that you try to re-locate and take on the lion's breathing for yourself; try to feel the muzzle-spine-tail connection that you may have witnessed. Be-with this different physicality for a few moments; invite any emotions that may arise. Perhaps you would like to draw or journal about your experience afterward. What did you learn, if anything?

How to Move with a Housecat

1 Avoid instrumentality; do not approach the cat *as* a pet—especially if they are *your* pet. Do not pet them, baby talk to them, cuddle them as you normally would (just for now). You want to circumvent the normal vocabulary of movement that you have established between you. Perhaps start with someone else's cat.

2 Either way, you must begin with yourself.

3 Close your eyes and find a place you feel comfortable sitting or lying down on the floor, if at all possible. The floor is likely your pet's domain; you enter their space when you get on the floor. If you cannot easily access the floor, do not worry; you can begin wherever you are. Close your eyes.

4 Prepare to meet your cat (or other domestic critter) by focusing on your own breath.

5 On your next breath in, bring your attention to the feeling of your body against the floor, chair, or other surface. Exhale.

6 On your next breath in, bring your attention to the feeling of your lungs filling up—and then emptying out. Maybe you feel your belly as you breathe.

7 On your next breath in, bring your attention to the environment around you. Listen for the cat—is it nearby? Exhale.

8 Bring your attention to the energetic centers of the muzzle, tongue, spine, and tail. Rather, try to find and feel those places in your own body. Extend your muzzle forward as the epicenter of knowing. Practice matching the cat's energies at these sites by trying to feel them with-in your own body. If the cat is not around, focus on finding these centers in your own body. Explore them; play with them, perhaps like a cat.

9 Allow space for something to change.

10 From this calm and focused position, open your eyes and make eye contact with the cat. If you can.

11 If the cat won't make eye contact with you—if they won't address you—do not worry. Stay focused on your breath. Eventually they will become curious about your atypical energy and affect. Even better, they will continue to ignore you and you will have to become more like the cat yourself—aloof, independent, resistant, and deeply embodied. Just sit here, now, breathing deeply, perhaps locating some sunlight on the front of your muzzle. Perhaps you feel a little purr emerging in your chest or throat. It's best to be like the cat—do this for yourself.

Practice Matters

Developing an ecosomatic practice in relation to the more-than-human matters. It matters, as Wood notes, because of the "intimate connection between our thinking about animals and our self-understanding."[49] If we want to know ourselves better, we must reflect on and change our relations to the more-than-human Other. It also matters because "our being-connected, being-related, is in need of enlightenment even for the sake of our own survival."[50] Indeed. The U.N.'s Intergovernmental Panel on Climate Change published a report in April 2022 that identifies human consumption of meat to be one of the major negative factors impacting climate change.[51] Engaging in regular ecosomatic practice can be a useful tool for making space for such personal changes.

So let me conclude—first, by asking what moving with cats has contributed to my own self-understanding, and second, by reflecting on ecosomatic practice as a critical tool. At a time in my life when I have been pacing in my own cage—under COVID-19 lockdown, over 60, with some other personal stresses—I can still feel the sadness of that lion's enclosure and with no way to escape. I have learned, even in writing this chapter, that I still carry that old lioness—and that I can and need to do more in the face of such cruelty.

Casey notes that "Lived bodies put emotions into motion—precipitating literal *e-motions*—by breaking free from enclosure in the individual conscious subject where emotions have been held in lockdown for far too long."[52] While Casey's goal is to liberate emotions from their false prison "inside" the human subject and instead to locate them more properly in peri-phenomenological place and intra-relation, I like to think that my lived experience with the lion, and my embodied recollections of her—both onstage in Berlin and in that stretch on my living room floor with my old cat nearby—have afforded something to break free from its enclosure, wherever it may be located. To be sure, my encounter with the old lioness and embodying her in a somatic way was one of the most profound experiences I have had in performance. I have been powerfully surprised to discover that years later, it continues to be meaningful.

I have also learned of the unexpected generosity of an old cat whom I thought I already knew: Squeak. After many years, we have found a different way of "holding and beholding" each other, to use Bristol's language. We have a closer relationship, even after so long together. I have found myself wondering if I am attempting a kind of rewilding of self even as I question what that might even mean. I do know that this kind of ecosomatic practice has allowed me to build empathy for and connection with the Other at a deeply embodied level and I feel my own life is enriched as a result.

But beyond the self-understanding and empathy that can be accessed *vis-à-vis* this kind of ecosomatic practice, a critical edge can be developed and put into play, perhaps even in one's daily life.[53] Moving with cats—indeed, engaging in any ecosomatic practice—is a way to "let the real cat, the cat that can deconstruct our pigeonholes through play, out of the bag."[54] It is a way to develop "phenomenological attentiveness," a technique of attending-to, of giving a particular kind of attention to the more-than-human, which is part of the "tripartite practice" that Wood believes is necessary to the task of "jamming the anthropological machine."[55] We need to jam the anthropological machine in order to approach the cat "as such," in order to be able to decenter the human when inquiring into the "truth about animals," and in order to root out the gluttony, violence, and genocide that blind us to "the possibilities of mutually enhancing engagement."[56]

Wood suggests that through "careful attention" we might not romanticize animals but instead "be receptive to the exquisite detail of their behavior, and especially to the ways in which they challenge our assumptions about what animals are and can do." He calls for connecting with a "range of creatures in their natural settings," but here, he specifically excludes zoos.[57] I realize he does this because he believes that the zoo animal cannot teach us about the nature of the animal *per se*—but is this really true? How do we know? For Wood, only creatures "in their natural settings" can teach us such things. It begs the question—is that not another form of romanticization? And where is this "natural setting" outside of what Anna Tsing calls "feral ecologies?"[58] Very few such places exist—indeed, if we humans are in any particular place to observe anything at all, it is likely to be full of, or nearby to, human-built infrastructure. At the very least, careful attention to the wild animal kept in a human-built zoo can teach us something about the human animal and our own carnivore houses.

For me, there is an ethical responsibility to be with the living beings that we keep in our zoos—and perhaps this is another takeaway from the old lioness in Berlin. If we humans are the reason for their isolation or oppression in a zoo—or for the need for conservation—then we cannot look away. We are ethically bound to bear witness to our own morbidity and to develop a form of attunement to the more-than-human, especially in the contexts of oppression that we create. Allow me to extend Lévinas to this scenario and claim that we must "not let the Other alone, even in the face of the inexorable."[59] So, I say yes to giving careful attention to and to building relationships with wild animals in captivity, even if only one. This Saturday, I go to the Oakland Zoo.

Notes

1 Originally published in 2004, *see* David Wood, "Thinking with Cats," in *Thinking Plant Animal Human: Encounters with Communities of Difference* (Minneapolis: University of Minnesota Press, 2020), 117–135.
2 Ibid., 117. Wood discusses Derrida's analysis of the "abyssal ruptures" that open up in human-animal encounters and how the logic(s) of sacrifice, parricide, and carnophallogocentrism are key to understanding our behavior toward and treatment of animal Others. *See also* Jacques Derrida, "The Animal That Therefore I Am (More to Follow)," *Critical Inquiry* 28, no. Winter (2002).
3 Wood, "Thinking," 135.
4 This chapter is dedicated to Sondra Fraleigh, who encouraged me to take up the subject. On performance as research, *see* Shannon Rose Riley and Lynette Hunter, *Mapping Landscapes for Performance as Research: Scholarly Acts and Creative Cartographies* (New York: Palgrave Macmillan, 2009).
5 Month of Performance Art-Berlin (MPA-B) was "a 31-day, city-wide, non-profit and non-funded collaborative platform devoted to supporting and advancing independent performance art practices in Berlin, and from around the world." It occurred in May, from 2011 through 2015. https://mpa-blog.tumblr.com/ABOUT.
6 Bristol's work was originally to be performed by "surrogate" performer, Sarah Lüdeman. While I am still sorry that Sarah could not perform as planned, I feel very fortunate that I was able to step into this amazing work. *See* Ilya Noé, "E/S 3: lie in (lion)," *Ilya Noé*, December 18, 2013. www.ilyanoe.com/es3-lie-in-lion/.
7 Joanne Bristol in Shannon Rose Riley, Joanne Bristol, and Ilya Noé, May 23, June 2, June 3, 2013. Ilya took photographs at the Zoo but I did not review the images until after the performance.
8 Edward S. Casey, *Turning Emotion Inside Out: Affective Life beyond the Subject* (Evanston: Northwestern University Press, 2022), 7.
9 PETA, "Europe's Cruellest Destinations." www.peta.org.uk/living/euro pes-cruellest-destinations/. It has also been described as "a prison." Janna Dotschkal, "Life in Captivity: Portraits of Germany's Zoo Animals," *National Geographic* (2014). www.nationalgeographic.com/photography/article/musi ngs-elias-hassos-zoos.
10 Emmanuel Lévinas, "Ethics and Infinity," *CrossCurrents* 34, no. 2 (1984).

11 Lévinas makes clear that he is concerned with the "inter-human relationship." Ibid., 195. *See also* David Wood, *Thinking Plant Animal Human: Encounters with Communities of Difference* (Minneapolis: University of Minnesota Press, 2020), 124.

12 Wood, "Thinking," 125. *See also* Derrida, "The Animal That Therefore I Am" 399.

13 Karen Michelle Barad, *Meeting the Universe Halfway: Quantum Physics and the Entanglement of Matter and Meaning* (Durham: Duke University Press, 2007), 33.

14 Ibid. Barad cautions us to remember that agencies are distinct only in a relational sense and strictly speaking do not exist as "individual elements."

15 Ibid.

16 Here, I borrow Heidegger's notion of *Mitsein*, "being-with" or "being-with-others." Martin Heidegger, *Being and Time* (1927), trans. John Macquarrie and Edward Robinson (New York: Harper & Row, 1962).

17 Shannon Rose Riley, "Embodied Perceptual Practices: Towards an Embrained and Embodied Model of Mind for Use in Actor Training and Rehearsal," *Theatre Topics* 14, no. 2 (2004): 453.

18 As I have argued elsewhere,

> Neurologist Antonio Damasio postulates that what we mistakenly call "the mind," as if it were an object located in the brain, is more accurately an interactive relational process between brain, body, and environment. Specifically, the process of mind is constituted by what he calls "multiple, parallel, converging streams," coded as images, which flow throughout body and brain in response to the environment, forming multiple temporary "feed-forward and feed-back projections."
>
> (ibid., 451)

19 While I did not know it at the time of the performance, the muzzle-spine-tail connection that I perceived is akin to the "head-tail" pattern of total body connectivity outlined in Bartenieff Fundamentals. Irmgard Bartenieff and Dori Lewis, *Body Movement: Coping with the Environment* (New York: Routledge, 2002), 229–262.

20 Sondra Fraleigh, *Dancing Identity: Metaphysics in Motion* (Pittsburgh: University of Pittsburgh Press, 2004), 268n.4. *See also* Elizabeth Behnke, "Matching," in *Bone, Breath and Gesture: Practices of Embodiment*, ed. Don Johnson (Berkeley: North Atlantic Books, 1995).

21 Fraleigh, *Dancing Identity*, 122–126.

22 Ibid., 145.

23 In a personal communication, February, 2022.

24 Casey, *Turning Emotion Inside Out*, 128.

25 Ibid.

26 Shigenori Nagatomo, *Attunement through the Body* (Albany: State University of New York Press, 1992); Yasuo Yuasa, *The Body: Toward an Eastern Mind-Body Theory* (Albany: State University of New York Press, 1987); Richard G. Erskine, "Attunement and Involvement: Therapeutic Responses to Relational Needs," *International Journal of Psychotherapy* 3, no.3 (1998).

27 Casey, *Turning Emotion Inside Out*, 177–178.

28 Ibid. *See also* Cynthia Willett, "Affect Attunement: Discourse Ethics across Species," in *Interspecies Ethics* (New York: Columbia University Press, 2014). Casey, *Turning Emotion Inside Out*, 178. *See also* Marjolein Oele, *E-Co-Affectivity: Exploring Pathos at Life's Material Interfaces* (Albany: State University of New York Press, 2020).

29 For Fraleigh's full development of the theme of "matching not mastery," *see* Fraleigh, *Dancing Identity*.

30 *See* Vittorio Gallese, Morris N. Eagle, and Paolo Migone, "Intentional Attunement: Mirror Neurons and the Neural Underpinnings of Interpersonal Relations," *Journal of the American Psychoanalytic Association* 55, no. 1 (2007): 132, https://journals.sagepub.com/doi/10.1177/00030651070550010601; Katy A. Cross et al., "Controlling Automatic Imitative Tendencies: Interactions between Mirror Neuron and Cognitive Control Systems," *Neuroimage* 83 (December, 2013) https://doi.org/10.1016/j.neuroimage.2013.06.060. For critiques, *see* Gregory Hickok, "Eight Problems for the Mirror Neuron Theory of Action Understanding in Monkeys and Humans," *Journal of Cognitive Neuroscience* 21, no. 7 (July, 2009), https://doi.org/10.1162/jocn.2009.21189, www.ncbi.nlm.nih.gov/pubmed/19199415; Gregory Hickok, *The Myth of Mirror Neurons: The Real Neuroscience of Communication and Cognition* (New York: W. W. Norton & Company, 2014).

31 Gallese, Eagle, and Migone, "Intentional Attunement."

32 Ibid.

33 Hickok, "Eight Problems."

34 Casey, *Turning Emotion Inside Out*, 167.

35 PETA notes that "The Berlin Zoo has a reputation for flaunting its animals as entertainment without caring for their welfare." PETA, "Europe's Cruellest Destinations."

36 Ibid. In 2020, the Berlin Zoo announced its acquisition of three young African lion siblings, who will live outside in new heated caves. Imanuel Marcus, "Berlin: King of the Jungle Returns to Zoo," *The Berlin Spectator* (2020). https://berlinspectator.com/2020/11/20/berlin-king-of-the-jungle-returns-to-zoo-1/.

37 There are compelling arguments in favor of the conservation and rescue work performed by many zoos today. The Association of Zoos and Aquariums (AZA) notes that "research and conservation projects conducted by AZA-accredited zoos and aquariums improve biodiversity. AZA members are contributing to scientific discoveries about how animals in their care are vulnerable to climate change, habitat loss, and other threats." www.aza.org/connect-stories/stories/benefits-of-zoos?locale=en. Nonetheless, it is predominantly human activity that causes the need for conservation in the first place.

38 Riley, qtd. in a personal email between Riley, Bristol, and Noé.

39 Riley, qtd. in ibid.

40 Riley and Hunter, *Mapping Landscapes*.

41 Bristol, qtd. in a personal email between Riley, Bristol, and Noé.

42 Riley, qtd. in ibid.

43 Riley, "Embodied Perceptual Practices," 449. Mary Starks Whitehouse, who developed the form, used the term Movement in Depth; Sondra Fraleigh uses Depth-Movement Dance. Janet Adler and Susan Bauer use the term Authentic Movement. My current practice has been facilitated by Susan Bauer since 2017.
44 Ibid.
45 Ibid.
46 Bauer and I in various Zoom conference calls, 2020 and 2021.
47 At the School of the Museum of Fine Arts, Boston, performance artist Marilyn Arsem would instruct students to perform any activity using "to/for/at/with/in spite of" as ways to approach the audience. We would repeat the activity five times, exploring a different relation in each.
48 Lévinas, "Ethics," 195, 202.
49 Wood, "Thinking," 117.
50 Ibid., 135.
51 The UN's Intergovernmental Panel on Climate Change (IPCC) published a report in April titled *Climate Change 2022: Mitigation of Climate Change*. It argues we must immediately reduce emissions of greenhouse gases and one of the major mitigation actions the report suggests is to reduce the demand for meat. *See* www.ipcc.ch/report/ar6/wg3/.
52 Casey, *Turning Emotion Inside Out*, 164.
53 Wood, "The Truth about Animals I: Jamming the Anthropological Machine," 142–147.
54 Ibid., 141.
55 Wood's three-part system of critique consists of (1) "semiotic suspicion", (2) "critical hermeneutics", and (3) "phenomenological attentiveness." Ibid., 142–148.
56 Ibid., 142.
57 Ibid., 146.
58 Anna L. Tsing, *Feral Atlas: The More-than-Human Anthropocene* (Stanford University Press, 2021). http://feralatlas.org/.
59 Lévinas, "Ethics," 195, 202.

Bibliography

Barad, Karen Michelle. *Meeting the Universe Halfway: Quantum Physics and the Entanglement of Matter and Meaning*. Durham: Duke University Press, 2007.
Bartenieff, Irmgard, and Dori Lewis. *Body Movement: Coping with the Environment*. New York: Routledge, 2002.
Behnke, Elizabeth. "Matching." In *Bone, Breath and Gesture: Practices of Embodiment*, edited by Don Johnson, 317–337. Berkeley: North Atlantic Books, 1995.
Casey, Edward S. *Turning Emotion Inside Out: Affective Life beyond the Subject*. Evanston: Northwestern University Press, 2022.
Cross, Katy A., Salvatore Torrisi, Elizabeth A. Reynolds Losin, and Marco Iacoboni. "Controlling Automatic Imitative Tendencies: Interactions between Mirror Neuron and Cognitive Control Systems." *Neuroimage* 83 (December, 2013): 493–504. https://doi.org/10.1016/j.neuroimage.2013.06.060.

Derrida, Jacques. "The Animal That Therefore I Am (More to Follow)." *Critical Inquiry* 28, no. Winter (2002): 369–418.

Dotschkal, Janna. "Life in Captivity: Portraits of Germany's Zoo Animals." *National Geographic*. (2014). www.nationalgeographic.com/photography/article/musings-elias-hassos-zoos.

Erskine, Richard G. "Attunement and Involvement: Therapeutic Responses to Relational Needs." *International Journal of Psychotherapy* 3 no. 3 (1998): 235–244.

Fraleigh, Sondra. *Dancing Identity: Metaphysics in Motion*. Pittsburgh: University of Pittsburgh Press, 2004.

Gallese, Vittorio, Morris N. Eagle, and Paolo Migone. "Intentional Attunement: Mirror Neurons and the Neural Underpinnings of Interpersonal Relations." *Journal of the American Psychoanalytic Association* 55, no. 1 (2007): 131–175. https://journals.sagepub.com/doi/10.1177/00030651070550010601.

Heidegger, Martin. *Being and Time (1927)*. Translated by John Macquarrie and Edward Robinson. New York: Harper & Row, 1962.

Hickok, Gregory "Eight Problems for the Mirror Neuron Theory of Action Understanding in Monkeys and Humans." *Journal of Cognitive Neuroscience* 21, no. 7 (July, 2009): 1229–1243. https://doi.org/10.1162/jocn.2009.21189, www.ncbi.nlm.nih.gov/pubmed/19199415.

Hickok, Gregory. *The Myth of Mirror Neurons: The Real Neuroscience of Communication and Cognition*. New York: W. W. Norton & Company, 2014.

Lévinas, Emmanuel. "Ethics and Infinity." *CrossCurrents* 34, no. 2 (1984): 191–203.

Marcus, Imanuel. "Berlin: King of the Jungle Returns to Zoo." *The Berlin Spectator*. (2020). https://berlinspectator.com/2020/11/20/berlin-king-of-the-jungle-returns-to-zoo-1/.

Nagatomo, Shigenori. *Attunement through the Body*. Albany: State University of New York Press, 1992.

Noé, Ilya. "E/S 3: lie in (lion)," *Ilya Noé*, December 18, 2013. www.ilyanoe.com/es3-lie-in-lion/.

Oele, Marjolein. *E-Co-Affectivity: Exploring Pathos at Life's Material Interfaces*. Albany: State University of New York Press, 2020.

PETA. "Europe's Cruellest Destinations." www.peta.org.uk/living/europes-cruellest-destinations/.

Riley, Shannon Rose. "Embodied Perceptual Practices: Towards an Embrained and Embodied Model of Mind for Use in Actor Training and Rehearsal." *Theatre Topics* 14, no. 2 (2004): 445–471.

Riley, Shannon Rose, and Lynette Hunter. *Mapping Landscapes for Performance as Research: Scholarly Acts and Creative Cartographies*. New York: Palgrave Macmillan, 2009.

Tsing, Anna L. *Feral Atlas: The More-than-Human Anthropocene*. Stanford University Press, 2021. http://feralatlas.org/.

Willett, Cynthia. "Affect Attunement: Discourse Ethics across Species." In *Interspecies Ethics*, 80–99. New York: Columbia University Press, 2014.

Wood, David. *Thinking Plant Animal Human: Encounters with Communities of Difference*. Minneapolis: University of Minnesota Press, 2020.

———. "Thinking with Cats." Chap. Nine. In *Thinking Plant Animal Human: Encounters with Communities of Difference*, 117–135. Minneapolis: University of Minnesota Press, 2020.

———. "The Truth about Animals I: Jamming the Anthropological Machine." Chap. Ten. In *Thinking Plant Animal Human: Encounters with Communities of Difference*, 137–151. Minneapolis: University of Minnesota Press, 2020.

Yuasa, Yasuo. *The Body: Toward an Eastern Mind-Body Theory*. Albany: State University of New York Press, 1987.

9

EMBODYING ISLANDS

Ecosomatics and the Transnational Queer

Fei Shi

Nex̱wlélex̱m *(Bowen Island), Canada: 49.3768° N, 123.3702° W*
Chongming Island, China: 31.6813° N, 121.4820° E

Here I am, as I move and write on what is now called Bowen Island off the coast of Vancouver in British Columbia, Canada. A troubling yet reverent sense of place and belonging emerges from my heart through my fingertips onto my notepads and keyboards. I am gratefully moving and writing on this beautiful island—*Nex̱wlélex̱m* "beat a fast rhythm" or *Xwlíl'xhwm* "fast drumming ground"—lush and moist from the typical Pacific Northwest winter morning rain, the traditional unceded territory of the Coast Salish peoples, including the Squamish and Tsleil-Waututh nations.[1] Yet what does this gratitude entail? What does it mean for me to relate to this place where I live and love? How do I belong here?

As one encounters the question of the geography of self and the "geographies of us"—the self-mappings of collective belonging—one inevitably confronts the truth of the place where one resides and the place one comes from as a migrant and a settler. Considering the hundreds of years of lived experiences that happened here on this island, I am merely a sojourner, an impermanent resident, an unsettled settler, a passer-by, a traveler, a stranger, and the other. Yet ironically, through this travel and the encounter with the other, and recognizing being the other, I gained life-changing knowledge about myself. I was born and grew up in a time and place that in Marguerite Yourcenar's words, "for each who comes out, there are ten who do not, and a hundred who have never even admitted to themselves that they are."[2] I was one of the 100, on another island—Chongming Island—off the coast of Shanghai, where the Yangzi River meets the Pacific Ocean; it is now purportedly the "eco-island" of China.[3]

DOI: 10.4324/9781003390985-12

Settling into the privileged position of a university educator in Squamish via the public graduate education in Northern California, I started to question and reveal to myself what I concealed. Not only my desires, but more importantly my own sense of self-worth. I learned the importance of life itself—what kind of life is considered intelligible, visible, and ultimately livable and what is not according to social and cultural norms.[4] Being queer, for me, became a political choice, a life-affirming choice. Entering a different cultural and geographical context offered me opportunities for profound transformations. Yet as every traveler, migrant, and immigrant knows intimately, one holds their personal history and cultural legacy wherever they go. It is the intimate knowledge of what we renounce and what remains deep in the heart when we cross linguistic, cultural, social, and geo-political borders. Such renunciations, adjustments, and changes indispensably involve our bodies, emotions, and memories, similar to what Raymond Williams calls the "structure of feelings" and "meanings and values as they are actively lived and felt."[5] Our yearning for being who we are and being accepted as who we are in its most brutal, intimate, and cherished nakedness necessitates an embodied rework of the complicated and layered personal politics of here and there or as Trinh T. Minh-ha poetically puts it, "elsewhere, within here."[6]

This critical and emotional self-location of embodying "the embeddedness of identity within local and global political, economic, and social systems that structure experiences of oppression, discrimination, power, and privilege" becomes my entry way into somatics as the movement framework to structure my lived experiences as a queer migrant; and ecosomatics—the body work deeply situated in place, land, and environment and their meanings, histories, and knowledges—to account for the queer existences of the other, including the socially marginalized, the racially and ethnically minoritized, the Indigenous, the displaced, the courageous border crossers, and cultural straddlers, in relationship with the more-than-human world in the age of environmental crisis.[7] This embodied self-location also makes apparent the privileges I currently have as an employed, middle-class, young, able settler, and partnered working parent with flexible mobility between affluent regions of some powerful nations across the Pacific.

How do I situate myself geographically as a sociopolitical being between the United States (more specifically the agricultural part of Northern California), China (the city of Shanghai and the Island of Chongming), and west-coast Canada (Vancouver, West Vancouver, and *Nexwlélexm*)? My Californian queer coming of age in graduate school challenged and decolonized my years of heteronormative and masculinist PRC nationalist education during its post-reform economic boom and its status as the factory of the world. My collaborations with the very few fellow BIPOC faculty members in our institution greatly enriched and empowered my research orientations and teaching pedagogies. My awakening to the histories of the land of the First Nations where I work and live initiated me into a humbling journey of continual learning and allyship. During this profound process of continuous transformations, I wonder how my Chineseness endures and manifests as borders are being crossed and destinations are being departed from. During a Chinese New Year celebration gala in West Vancouver Community Center, I found myself attempting to disentangle layers of my national identity and multiple borders inscribed on my body. Various waves and generations of "Chinese" immigrants of different historical moments, from disparate locations, such as mainland China, Hong Kong, Taiwan, Southeast Asia, and Europe, gather in one of the wealthiest neighborhoods of "flexible citizenship" in Canada, celebrating almost anachronistically the Republic of China, with dancers in Gaoshan clothing, that of the Indigenous peoples of Taiwan.[8] One could not help wondering how all this happened so political-incorrectly, yet nonetheless seemingly so harmonious with missed, mixed, and crosscut imaginations and yearnings. This precise uneasy feeling of being out of place, ironically during a cultural moment of emplacement, becomes my energy source and critical impetus for mapping my identity as a transnational Chinese queer migrant self-ethnographically and ecologically. The necessity of examining and reawakening the body in its layered histories leads to my ecosomatic project of reconciliation and becoming—a critically informed and practice-based somatic framework to account for the "natureculture" body capable of remembering, healing, as well as open explorations and experimentations.[9]

What are the intersections of queer migration and queer indigeneity? What are their affinities that we can embody somatically to formulate new understandings of gender expressions in the era of environmental crisis? How do senses of place and non-anthropocentric belonging reorientate us toward new gender becomings? How can somatic practices and exercises help to inspire new ecological existences? How does this new ecological imagination awaken our memories and stories as queer migrants? How does ecosomatics—embodied ways of seeing, feeling, exploring, performing, and relating to the world—give flesh to the theoretical ponderings of queer existences in continuum with their places? With these critical questions hovering above, I embark on the explorations of my ecosomatic body and the "reorganization of [its] multiplicity and heterogeneity."[10]

Inspired by Sondra Fraleigh's provocative questions—"What kind of culture will we make of our nature?" and "what kind of nature will we make of our nature?"—my ecosomatic project is determinedly autobiographical and self-ethnographical in recognizing particular conditions of my embodied-ness, while providing an open framework for collective explorations of shared threads of identities and experiences.[11] Through embracing my "ecological body, cycling with environmental phenomena, polluted or pure, imploring our feelings and testing our liberating and not-so-liberating choices," I traverse with memories and movements between two different islands: Chongming Island (the island where I was born and grew up) and Nex̱wlélex̱m (where I now call home and where my two baby daughters were born—a space I move through while carrying them on me).[12] They represent together the memories of quotidian movements lodged within my tangible body as well as the memories of my movements across multiple borders, be it physical, geographical, national, sociopolitical, cultural, or metaphorical. In alignment with Christine Bellerose's calling for "depatterning somatic amnesia and repatterning ecosomatic senses [...] to access [our] own sense of enchanted kinship, [our] own fundamental relationship to the land," I write and move in the same process, developing movement/memory/storywork to remember and reembody my experiences of Chongming on Bowen, along with reflections of the island-bound "straight" child and the migratory "queer" parent, linking eco-body through queer animal explorations and intergenerational recalling.[13] While exploring my own storied body as a mode of critical inquiry in Part I, I have devised simple movement exercises adaptable for others who want to awaken their displaced body in connection with past memory through movement and storywork. I share these in the italicized sections of Part II.

Part I

Chongming/Nex̱wlélex̱m (Island Memories as Movement)

How does my individual memory relate to the national and global discourses of environmental crisis and ecological yearning? For many in China, Chongming Island evokes the pristine image of an eco-island of wetland and migratory sea birds. Situated some 30 miles away from the metropolitan area of Shanghai separated by the Yangzi River, it is one of the largest alluvial islands in the world—gradually living and expanding ever since the Tang dynasty (618–907 AD) with depositing sand and soil from the tributaries and upper streams of China's longest river. After some failed international eco-development projects on the island since the Changjiang Tunnel-Bridge Expressway linked the island to the mainland in 2009, Chongming Island provokes sociologist Julie Sze's transnational critique of neoliberal reframing of eco-futures through capitalistic productions of eco-desire and eco-development. For her, "Chongming as a model 'ecological' site for Shanghai's development reflects a metaphor, an imaginary 'island' of ecological virtue in a city and a world defined by narratives (and the reality) of environmental chaos and decline."[14] Chongming has always been considered the backwater rural and agricultural area of underdevelopment (not even worthy of being a remote suburb because of its unique island identity, population, and dialect) in the shadow of Shanghai and a bucolic and boorish "fantasy island" for the sophisticated city visitors. In a somewhat similar way, because of its proximity to Vancouver, Bowen Island's modern history in the postcolonial era is dominated by resort tourism operated by the Terminal Steamship Company (1900–1920) and the Union Steamship Company (1920–1962).[15] Yet more recently, because of the real estate pressure from the Vancouver area and lower mainland, Bowen emerges as a desirable suburban location for retirees, young families, and professionals of flexible work schedules, while continuing its status as a "within reach, beyond compare" eco-tourism destination a short ferry ride away.[16]

So, what do the "eco-islands" of Chongming and Bowen mean to me? How does my ecological yearning play out in my movement between here and there? How to account for this intimacy of growing up on Chongming and living on Bowen? Does "island" represent for me a nostalgia for a lost home, a defined and enhanced identity of enclosure, or yearning for the outside world by ferrying away? What does my movement (both as journeys of transnational migration and ecosomatic practice) mean? What is the alternative story of Chongming as an intimate expression of ecological existence—a personal one that entangles and reveals? I realize the potential of this eco-yearning as a practice and turn to the power of art as a conduit of transformation to discover the extraordinary in the ordinary of my personal intimacy with Chongming Island. It is a deeply personal method of connecting one's history with the environment and one's natureculture being. Through sharing my identity, my movement, and my story with fellow artists, learners, explorers, and readers of this essay, I want to find a way to gain wisdom by somatically mapping out my body and its history and story. Through weaving the threads of queer diaspora (exploring belonging and the sense of home-making while acknowledging bodies of intersectional experiences and expressions), storywork (foundation to evoke one's storied memory and identity in fleshed forms), queer animal (trans-corporeal and metamorphic movements in the forest), and writing (capturing the evanescent body in motion and words), I move ecosomatically with island memories and migrate back to Chongming like a sea bird.

Storywork (Where Queer Diaspora and Indigeneity Meets)

As Gayatri Gopinath argues, "if 'diaspora' needs 'queerness' in order to rescue it from its genealogical implications, 'queerness' also needs 'diaspora' in order to make it more supple in relation to questions of race, colonialism, migration, and globalization."[17] Queer diaspora not only finds it difficult to trace its entangled roots to its national origin (and sometimes problematic so because of homophobia and internalized oppression) but also struggles in home-making because of rejections and refutations of homonormativity and other racist, classist, colonialist, etc. discriminations. In order to decolonize both "queerness" and "diaspora," I turn to the work of Q'um Q'um Xiiem, Jo-Ann Archibald, and the wisdom of the land where I live. In *Indigenous Storywork: Educating the Heart, Mind, Body, and Spirit*, Archibald develops the idea of storywork as an Indigenous storytelling research methodology with First Nations elders and pedagogy for future generations. She proposes seven principles "respect, responsibility, reciprocity, reverence, holism, interrelatedness, and synergy" as the foundation for First Nations storytelling as "intergenerational pedagogy of learning."[18] These principles clarify for me the moral compass of storytelling. They also provide me with grounding energy and flying inspiration to reach out to my own memory, history, and ancestry. In making affinity through shared oppressions while acknowledging the differences between queer diaspora and indigeneity, I find ways to bring back my own history of displacement by reliving and reimaging my story. Storywork, for me, is an intentional way to *story* one's body through movement and memory—an ecosomatic exploration to relive one's memory in place(s). It follows the process of moving-evoking-remembering-storying. How do I notice my body in motion? How do I sense my body in connection? How do I remember my body in place? How do I language my body in story?

How can somatics possibly respond to the intersections of migration, diaspora, racial and ethnic minority, sexual minority, multiple gender expressions and identifications, transnationalism, and settler colonialism? Our soma and sensations, indeed, are already entangled with our culture, history, and experiences. Body remembers. Evoking them through memory and storywork provides a renewed sense of reconciliation, healing, being, and belonging as Dr. Archibald argues, "[t]his concept of memory entails the development of a storied memory, the living of storied lives, the disruption of memory stories, and the awakening and resurgence of storied memories."[19] Storywork, for me, as a form of self-ethnography and cultural reconciliation through ecosomatics, invites conscious efforts of paying attention to the cultural multiplicities of our bodies when memories emerge. Centering on particularity and multiplicity, whether settler colonial layering, transnational imagination, or home-making and meaning-making as queer migrants, individual healing and revelation shed light on the communities where we come from and where we belong, composed by those who share our experiences, close and far.

Queer Animal (Experimenting and Becoming)

"To be one is always to become with many."[20] With this, Donna Haraway incisively suggests this becoming a multiplicity requires being *with*—as in intimate conjunction with our bodies and/as material environment throughout. This "with" is both a "provocation of involvement" and an "excitation to fold materiality over itself to become more and other."[21] Eco—environment in its fundamental sense of relationality and materiality—plays a key role in somatic remembering and becoming. It provokes, excites, and facilitates bodily and psychological transformation of history into story. To explore and to experiment in breaking away from the repetitions of history and reaching out to deeper senses of meaning-making, I embrace the fluidity of queer animal bodies, as ways of visualization and movement in the forest, to envision and embody trans-corporeal and metamorphic figures of eco-existence. This visualization and movement exercise enables my efforts in, what Cary Wolfe calls, "excavating and examining our assumptions about who the knowing subject can be; and in embodying that confrontation."[22] If *queer* denotes nonnormative amorous intimacies, *animal* symbolizes the affective relations between human and the more-than-human world. Queer animal, for me, is both a natureculture expression and an ecosomatic method—an experimental and explorative figure of leaning, extension, connection, nonetheless capable of aches and pleasure in the process of metamorphosis.

As Susan Stryker puts it, "our life is always a trans-life, that simultaneously occupies the ranked and contested intersections of Man, the human, the species of homo sapiens, and the animate nonhuman," queer animal represents to me the *trans* figure of the murky contingency of environmental existence.[23] It suggests transition, transgression, transformation, and in my case, translation and transnationality that I need to thoughtfully and courageously embody the intricate relationship between body, knowledge, memory, desire, and environment. This figure is materialized such that it takes on an embodied form of changing without a teleological purpose in the real flux of co-existence and alliance with other intimate material realities around and abound. Through the exercise of this figure of becoming, intimately (in a bodily sense) and ultimately (in an existential sense) as a queer bending and blending, I hope to awaken my own natureculture existence where "flesh and signifier, bodies and words, stories and worlds" join together.[24]

Writing (Body in/to Words)

How to capture the evasive, the ephemeral, and the unapproachable of my own ecosomatic movements? How do words flow through my limbs and fingertips, attempting to capture the fleeting presence on this page? What makes movements and performances, in this case my own, aesthetically and politically open via my own interpretations in this art of "disappearance"?[25] Peggy Phelan argues that

> in moving from the grammar of words to the grammar of the body, one moves from the realm of metaphor to the realm of metonymy. [...] Metaphor secures a vertical hierarchy of value and is reproductive, it works by erasing dissimilarity and negating difference; it turns two to one. Metonymy is additive and associative; it works to secure a horizontal axis of contiguity and displacement.[26]

Thus, writing compromises and contends simultaneously with its own limit. It approaches the evasive and captures the elusive while having to stand on its own. It reveals in itself the moments of slippage, displacement, disappearance, and loss in its storytelling.

As I examine the relationship among movement, performance, and writing, I envision writing as an equally important component in the framework of ecosomatics as one's reconciliation and healing process with history. Writing straddles gracefully between solemn introspection and earnest sharing. If Phelan privileges metonymy for being additive and associative, I also return to metaphor in ecosomatic writing as equally powerful. Metonymy suggests and metaphor actualizes. Metaphor brings body into the writing and gives the writing the confidence and necessity to both account for its inadequacy and at the same time fulfill its own embodiment as stories. To translate our movements on the page, one has to take notice of the motions, feelings, and memories that are aroused by the materiality of the body and to replace the inevitable displacement of the bodily sensations when performance ends and writing begins. Instead of recording what the body-in-motion presents itself, writing contemplates the perceptional and emotional magic it arouses while conjuring stories through memories. Writing of/as movement explores the possibility of language through its physical, artful manifestation and contestation as narrative and poetry. Exploration of words starts from an embodied practice of resignification and metaphorization.

Thus, it is my strategy of storywork with ecosomatic movement—to install a rhythm of the body in the process of remembering and writing as poetic moments, as breaths, as unfolding in places, as stories. Each exercise follows various steps of moving, journaling, recalling, awareness, storying with space and time to reflect and imagine. The practice anchors upon the self-ethnographical particular to access my own migratory and queer identity and ecological history. It is also an open conversation and exploration with you—my readers and fellow movers—who are keen to explore with freedom your own stories and episodes of personal history while sensing the places of your movements within, away, and between. Nomads and migrants, cultural shifters, and border crossers, let's root ourselves again and again as ways of return.

Part II

Breathing the Waves

Stand firmly, but not forcefully, by the shoreline, on a river's edge, close to the water, and face the waves (colossal or minute). Breathe. Slightly bend your arms, either up above your head or modestly stretch them to the sides. Your torso should be in a relaxed position, perhaps like a bow, charged with strength in the lower belly, but so at ease that there is no need for it to fully unbind or straighten itself. Your knees flexibly help to ground your lower body; and your legs stretch out or refrain in co-ordination with your arms.[27]

Give yourself time and space to breathe.

Listen to the rhythm of the waves, the nuances of them.

Breathe with the waves and bring the rhythm of the waves into your body; notice the patterns of the movement throughout your body.

Let this slow and rhythmic pattern of breathing coincide with the continuous ebb and flow of the waves. Each inhalation reaches the edges of the lower back and almost forces the torso to bend; on each exhale, your body rebalances itself. Let it shift into a slightly different position that continues the energy flow without negating the former position.

Breathe.

Body on the border between land and water: keep breathing and moving while on this border.

Let the first sound emerge from your body through your throat. Let this sound follow the rhythmic pattern of the breath and wave.

Give time and space to this sound.

Let the sound and your body movement merge to create a continuous flowing and receding pattern.

Let the first simple melody and words appear out of this sound. Let the metrical patterns of the vocalizing lines correlate with your movement: repetitive, continuous, almost monotonous, but charged with nuanced variations because of the breath.

Repeat until this sound/singing becomes you. Repeat until the sound of the wave become you. Repeat until the waves of sound become you.

"路遇大姐得音讯，九里桑园访兰英...."[28] As I breathe and let the first sound emerge and take shape, the opening line of the Yue opera scene *He Wenxiu—Paying Visit to the Wife in Mulberry Garden* follows the rhythmic movement of the folding and unfolding of my shoulder and chest. It is the theatre of a classical southern Chinese local opera, depicting the secret visit of a Confucian scholar-official to his wife for reunion after years of separation because of injustice. Off guard, I follow the line and repeat in tonal variations of this invigorating prospective of secret homecoming, "...*obtaining news from an older lady through chance encounter, I visit Lanying of nine-mile mulberry garden.*"[29]

I listen to my own voice—the music and performance I grew up with and obsessed over as a child and see the reflection of myself in the water as consumed and crossed-dressed Sheng and Dan.[30] The tenderness of that voice; and the beautiful mind that was forming in this reflection where my own sense of aesthetics and self-beauty developed. Memorizing long stretches of arias, I performed in front of family, friends, and sometimes in public gathering, unabashed in my own theatricality. Seeing images of myself in elaborate Yue opera costumes during rare and exciting photography studio visits arranged by my mother, I remember luxuriating in gender play and that I found a sense of belonging in shifting my Chongming dialect to a musical Yue opera accent.

I gaze into the waves and afar. The misty landscape gradually formulates a Chinese mountain—water ink painting with lines blurring and seeping into empty stretches of sky, traced open by the ferry crossing and departure.

Off from home island, into the city, and away to another country; having traveled from China and settled in at the University of California, Davis, I remember all the courage I mustered to attend my first LGBTQ Student Association meeting—some handsome man looked at me. I turned away, crumbled inside. Inadequacy, silence, mixture of unspeakable feelings.

I remember learning from the grace and audacity of my dance professor, Della Davidson, about art as a means to survive and thrive. I started to formulate my go-to questions for my own arts experiments and practice with new-found freedom and personal truth:

What are the connections?
Who are you as an artist?
What are the frameworks?
What are the practices?
What do you do with your practices? (Your ethics? your values? Your
    risks? Your defiance? Your resistances? Your identities?)
Individual or community based? How do you and who do you
    collaborate with?
How to account for time/place?
How to move, feel, and write?
How to be free?

Queer Animal in the Forest

Bare your hands and feet and stand among the trees. Breathe gently and quietly and close your eyes. Make the first step on the ground and feel the touch point. Start to move slowly and initiate contact with the soil. Switch different contact points between your body and the earth. How about your forehead, shoulder, cheek, your back, your limbs, the tip of one finger, your inner thigh, the back of your neck, your lips, etc.? Slowly and gently. Feel the touch.

How about the accidental touch points? Little rocks, fallen twigs, branches, foliage, soft leaf mold or moss. How about being touched by the unidentified, the unclassified, the unknown? How does your body stretch—and being stretched, move—and being moved, roll—and being rolled by the subtle changes of textures around? How do texture and temperature cajole or disturb your body?

Visualize internally an animal in this environment and start to move and embody it. Awaken the senses of this becoming-animal of your body and in steps, focus on the sense of smell and move with the smells; focus on the sense of listening and move with the sounds; focus on touching and move with the touches, focus on morphing and move with this metamorphosis.

Suspend any judgement as "the becoming-animal of the human being is real, even if the animal the human being becomes is not; and the becoming-other of the animal is real, even if that something other it becomes is not."[31]

Open your eyes and keep moving as this embodied animal. How is each view framed with each movement? How do shades of color represent different textures? How does your animal body balance and rebalance as it merges with textural nuances of the forest?

Gradually soften your gaze and pause the movement. Take a deep breath and walk easily back to where you started.

Slug moves. Awakened senses—seeing, smelling, tasting, touching, listening, emoting, morphing on a single branch, caressing decayed leaves and fungi underbelly, toward the seduction of fresh greens and scent of wet moss (*see* Figure 9.1). This materialized becoming-slug—multiplicity and processes of sensing *with* and responding to, forgetting coherence and originality. A technique of fabricating the soft and oozing body, a unique cyborg, a queer practice. My figure of becoming-slug twists, stretches, contracts, and drenches softly, capable of bodily intercourses, exchanges, permutations, gendering, and engendering.

Sword fern stem, broken between teeth.
Modes of desires and beings.
Swoon…

Asleep in a wheelbarrow in Dongping Forest Park on Chongming Island, formerly a socialist forest farm enclosed from adjacent farmlands and established in 1959. I was lost in the woods until a forest worker put me on top of tree trunks and branches in the cart and wheeled me back to an office building. Through frequent visits, my childhood accompanied and followed the transition time of this forest farm into a national public forest park in 1989. As the biggest man-made forest in Eastern China, predominantly planted with 水杉 dawn redwood (*Metasequoia glyptostroboides*), it represents the yearning of the Chongmingnese to labor upon fast-growing trees for public resource while making special the scenery of a place: an almost long-lost tree species of fossil status, propagated into a monocultural spectacle, and later dotted with faux Spanish-style buildings for tourist attractions in the 1990s. Hence this is the site where Chongming's eco-island reputation, and my people's ecological yearning coupled with practical economic incentives, started.

The unique scent and shape of dawn redwood, a non-native of the island, occupied my memories of nature-bound home.

FIGURE 9.1 The author's hand and foot in the forest with a slug, Mount Gardner, 2023. Photography by Fei Shi.

Circling as a Ritual

Recognize a personal place of nature that you frequent in everyday life. Identify a circle or a loop of this place, a manageable size that you can complete in a reasonable time. Walk the loop once a day and develop the ritual of circling this place over a week's period. When you walk, pay attention to the following:

1. *Notice the texture of the air on your skin and through your breath into your body.*
2. *Notice the texture under your feet with every step.*
3. *Notice the smells of the surrounding vegetation.*
4. *Notice all of the sounds around you, near and far.*
5. *Notice the patterns of how you "tie up" the loop while negotiating its topography.*
6. *Notice the locations of aches or discomfort in your body.*
7. *Notice the relationship you develop with this circle.*

The daily ritual of circling Killarney Lake with Iris, and later Céleste on my chest and on my back, defines me as a parent. Relaxation and labor at the same time, I learn the sensations of weight sinking into my body and the gentle twisting of life forms stretching into my stomach muscle with little undulations of breaths, punctuated by gentle crescendos or sudden bursts of crying. Shifting my body, humming in Chongmingnese, eager to contain, pacify, and release, I manage to balance and rebalance positions of holding and sustaining (*see* Figure 9.2).

The quotidian place of this circle becomes the privileged site of my naturecultural being, extending and connecting my mammal life into water on one side and forest on the other. Each circling repeats with almost unnoticeable variations, until the senses are guided and focused into a dominant dimension of awareness.

Gnarly tree roots surface and tangle with rocks. Narrow streams flow down along edges of pebbles and stones until soaking into the dense carpet of damp leaf mold. Dappled light, further veiled by hazy drizzle, accentuates the trunks of the Douglas firs. Crisp air, scented with ferns and mosses. Movement as burden, movement as careful negotiation of every step, movement as completion of a circle—a beginning with an end, and an end as beginning.

I remember the continuous patches of rice fields in the country on Chongming and the exhilarating pride with which I dragged a bag of corn upon seeing the grin of my grandmother. I remember walking seemingly endless stretches of raised riverbank of Yangzi, questioning and yearning

FIGURE 9.2 Circling Killarney Lake with the children: Céleste's pinky up as a little dancer and Iris, on the author's back, surveying the lake. Killarney Lake, 2023, Photography by Alix Melchy.

for the other side. I remember the open decks of the ferry and the wind blowing with salty moisture. I remember the clasped hands of love along the rocky beaches and driftwoods. I remember my father's sorrowful eyes and raspy voice in his years of adjustment and preparation to eventually find his balance in expressing love for his child. And that parental love, though sometimes challenged by difficult circumstances, gives the profound meaning to parenting as self-sacrifice and self-acceptance. Gratefully moving and moved by the offering of this circle, I remember.

"If the notion of departure is linked to so many difficulties on the one hand, to so many misunderstandings on the other hand, is it worth leaving home?" Marguerite Yourcenar asks, and yet "we feel that despite everything, our travels like our readings, like our encounters with others, are means of self-enrichment that we cannot refuse."[32] Moving and remembering ecosomatically, I trace back to my migratory travels while embracing two islands and the enormity of each single moment. The places of ecological anchoring give us the personalized opportunities for self-reflection, self-questioning, and self-location. Ecosomatic movements allow me to reach beyond, to confront, to be challenged, to be unsettled, to explore, to travel, and eventually to hit home.

As neo-expressionist choreographer Pina Bausch tells us, "when you work on a piece, something comes from the side, something important to you."[33] Ecosomatic movement and storywork help me to recognize and embrace the freedom and courage of sharing. Through sharing in writing and practice, the individual body connects to bodies of communities. An ethnographical particular and literary singular translates to a vision of empowerment via understanding, affinity, and alliance—different experiences, histories, values, cultures, desires, belongings, yet never too remote—the geographies of us. Natureculture identity is an embodied history—a prism through which one articulates being and belonging in the world—and art is the flesh of this history revisited, represented, and reimagined.

Notes

1 See www.bowenheritage.org/a-short-history-of-bowen-island.html and https://bowenislandmuseum.ca/first-nations-on-bowen/.
2 Marguerite Yourcenar, *Le Tour de la Prison* (Paris: Gallimard, 1991), 44. My translation.
3 Julie Sze, *Fantasy Islands: Chinese Dreams and Ecological Fears in an Age of Climate Crisis* (Oakland: University of California Press, 2015), 5.
4 Judith Butler, *Gender Trouble: Feminism and the Subversion of Identity* (New York: Routledge, 1999), xxii.
5 Raymond Williams, *Marxism and Literature* (Oxford: Oxford University Press, 1977), 132.
6 T. Minh-Ha Trinh, *Elsewhere, Within Here: Immigration, Refugeeism and the Boundary Event* (London: Routledge, 2011), 2.
7 Scott Kouri, "Settler Education," *International Journal of Child, Youth & Family Studies* 11, no. 3 (2020): 62. https://doi.org/10.18357/ijcyfs113202019700.
8 Aihwa Ong, *Flexible Citizenship: The Cultural Logics of Transnationality* (Durham: Duke University Press, 2004), 18.
9 Donna J. Haraway, *The Companion Species Manifesto: Dogs, People, and Significant Otherness* (Chicago: Prickly Paradigm Press, 2003), 20.
10 Isabelle Ginot, "From Shusterman's Somaesthetics to a Radical Epistemology of Somatics," *Dance Research Journal* 42, no. 1 (2010): 25. https://doi.org/10.1017/S0149767700000802.
11 Sondra Fraleigh, *Dancing Identity: Metaphysics in Motion* (Pittsburgh: University of Pittsburgh Press, 2004), 117.
12 Ibid., 118.
13 Christine Bellerose, "I Dance Land: An Apprenticeship with Wind and Water: Depatterning Somatic Amnesia, Repatterning Ecosomatic Senses" (Ph.D. York University, 2021), 1. https://yorkspace.library.yorku.ca/xmlui/handle/10315/38698.

14 Sze, *Fantasy Islands*, 18.
15 *See* https://bowenislandmuseum.ca/union-steamship-company-the-resort-era/.
16 *See* www.tourismbowenisland.com.
17 Gayatri Gopinath, *Impossible Desires: Queer Diasporas and South Asian Public Cultures* (Durham: Duke University Press, 2005), 11.
18 Jo-Ann Archibald, *Indigenous Storywork: Educating the Heart, Mind, Body, and Spirit* (Vancouver: UBC Press, 2008), ix; Jo-Ann Archibald and Q'um-Q'um Xiiem, "Indigenous Storytelling," in *Memory: Essays on How, Why, and When We Remember*, eds. Philippe Tortell, Mark Turin, and Margot Young (Vancouver: Peter Wall Institute for Advanced Studies, The University of British Columbia, 2018), 237.
19 Archibald and Xiiem, "Indigenous Storytelling," 233.
20 Donna J. Haraway, *When Species Meet* (Minneapolis: University of Minnesota Press, 2008), 4.
21 Eva Hayward and Jami Weinstein, "Introduction: Tranimalities in the Age of Trans Life," *Transgender Studies Quarterly* 2, no. 2 (2015): 197. https://doi.org/10.1215/23289252-2867446.
22 Cary Wolfe, "Human, All Too Human: 'Animal Studies' and the Humanities," *PMLA: Publications of the Modern Language Association of America* 124, no. 2 (2009): 571. https://doi.org/10.1632/pmla.2009.124.2.564.
23 From Susan Stryker's foreword to Douglas A. Vakoch, *Transecology: Transgender Perspectives on Environment and Nature* (Abingdon: Routledge, 2020), xvii.
24 Haraway, *Companion Species Manifesto*, 20.
25 Peggy Phelan, *Unmarked: The Politics of Performance* (New York: Routledge, 1993), 27.
26 Ibid., 150.
27 One often finds such relaxed and ordinary positions with grounded inner strength in Taiji movement with lower body grounding and upper body softness and flexibility.
28 This is the first line of the Yue Opera scene-play (zhezixi 折子戏) "He Wenxiu – Paying Visit to the Wife in Mulberry Garden" (何文秀 桑园访妻).
29 My translation of the first line that opens this paragraph. Yue opera is probably the second most popular Chinese opera after Beijing opera. On Yue Opera, *see* Jin Jiang, *Women Playing Men: Yue Opera and Social Change in Twentieth-Century Shanghai* (Seattle: University of Washington Press, 2011); Zhenzhong Mu, "Yue Opera in Queer Gaze: Chinese Gay Fans of Yue Opera in Negotiation with Heteronormative Mainstream Society and Identity Politics" (ProQuest Dissertations Publishing, 2022).
30 In Yue opera, and many other local Chinese operas, Sheng is the character type of a young man, usually a young Confucian scholar-official; Dan is the character of a young woman, often demure and household-bound, yearning for romantic encounter and marriage beyond patriarchal confinement. In Yue opera, most character types, including Sheng and Dan are performed predominantly by female actors.
31 Gilles Deleuze and Félix Guattari, "Becoming-Animal," in *Animal Philosophy: Essential Readings in Continental Thought*, eds. Matthew Calarco and Peter Atterton (London and New York: Continuum, 2004), 87.

32 Yourcenar, *Le Tour*, 176. My translation.
33 Eva Hoffman, "Pina Bausch: Catching Institutions on the Wing," *New York Times*, September 11, 1994. www.nytimes.com/1994/09/11/arts/the-new-sea son-dance-pina-bausch-catching-intuitions-on-the-wing.html.

Bibliography

Archibald, Jo-Ann. *Indigenous Storywork: Educating the Heart, Mind, Body, and Spirit*. Vancouver: UBC Press, 2008.

Archibald, Jo-Ann, and Q'um-Q'um Xiiem. "Indigenous Storytelling." In *Memory: Essays on How, Why, and When We Remember*, edited by Philippe Tortell, Mark Turin and Margot Young, 233–242. Vancouver: Peter Wall Institute for Advanced Studies, The University of British Columbia, 2018.

Bellerose, Christine. *I Dance Land: An Apprenticeship with Wind and Water: Depatterning Somatic Amnesia, Repatterning Ecosomatic Senses*. Ph.D., York University, 2021. https://yorkspace.library.yorku.ca/xmlui/handle/10315/38698.

Butler, Judith. *Gender Trouble: Feminism and the Subversion of Identity*. New York: Routledge, 1999.

Deleuze, Gilles, and Félix Guattari. "Becoming-Animal." In *Animal Philosophy: Essential Readings in Continental Thought*, edited by Matthew Calarco and Peter Atterton, 87–100. London; New York: Continuum, 2004.

Fraleigh, Sondra. *Dancing Identity: Metaphysics in Motion*. Pittsburgh: University of Pittsburgh Press, 2004.

Ginot, Isabelle. "From Shusterman's Somaesthetics to a Radical Epistemology of Somatics." *Dance Research Journal* 42, no. 1 (2010): 12–29. https://doi.org/10.1017/S0149767700000802.

Gopinath, Gayatri. *Impossible Desires: Queer Diasporas and South Asian Public Cultures*. Durham: Duke University Press, 2005.

Haraway, Donna J. *The Companion Species Manifesto: Dogs, People, and Significant Otherness*. Chicago: Prickly Paradigm Press, 2003.

———. *When Species Meet*. Minneapolis: University of Minnesota Press, 2008.

Hayward, Eva, and Jami Weinstein. "Introduction: Tranimalities in the Age of Trans Life." *Transgender Studies Quarterly* 2, no. 2 (2015): 195. https://doi.org/10.1215/23289252-2867446.

Hoffman, Eva "Pina Bausch: Catching Institutions on the Wing." *New York Times*, September 11, 1994. www.nytimes.com/1994/09/11/arts/the-new-sea son-dance-pina-bausch-catching-intuitions-on-the-wing.html.

Jiang, Jin. *Women Playing Men: Yue Opera and Social Change in Twentieth-Century Shanghai*. Seattle: University of Washington Press, 2011.

Kouri, Scott. "Settler Education." *International Journal of Child, Youth & Family Studies* 11, no. 3 (2020): 56–79. https://doi.org/10.18357/ijcyfs113202019700.

Mu, Zhenzhong. *Yue Opera in Queer Gaze: Chinese Gay Fans of Yue Opera in Negotiation with Heteronormative Mainstream Society and Identity Politics*. ProQuest Dissertations Publishing, 2022.

Ong, Aihwa. *Flexible Citizenship: The Cultural Logics of Transnationality*. Durham: Duke University Press, 2004.

Phelan, Peggy. *Unmarked: The Politics of Performance*. New York: Routledge, 1993.

Sze, Julie. *Fantasy Islands: Chinese Dreams and Ecological Fears in an Age of Climate Crisis*. Oakland: University of California Press, 2015.

Trinh, T. Minh-Ha. *Elsewhere, within Here: Immigration, Refugeeism and the Boundary Event*. London: Routledge, 2011.

Vakoch, Douglas A. *Transecology: Transgender Perspectives on Environment and Nature*. Abingdon: Routledge, 2020. doi:10.4324/9780429023811.

Williams, Raymond. *Marxism and Literature*. Oxford: Oxford University Press, 1977.

Wolfe, Cary. "Human, All Too Human: 'Animal Studies' and the Humanities." *PMLA: Publications of the Modern Language Association of America* 124, no. 2 (2009): 564–575. https://doi.org/10.1632/pmla.2009.124.2.564.

Yourcenar, Marguerite. *Le Tour de la Prison*. Paris: Gallimard, 1991.

10

SKINBODY AND THE SKIN OF THE EARTH

Alison (Ali) East

Otepoti (Dunedin), Aotearoa (New Zealand): 45.8795° S, 170.5006° E

This chapter discusses my dance work in terms of the relationship between what I call the "skinbody" and what is considered to be the living "skin of the earth."[1] Quantum theorist, Karen Barad, uses the term "intra-action" to think about agency as a dynamism of forces rather than as something one has. In this sense, I understand that the skinbody—my skinbody—emerges within and as part of a dynamism of forces with the skin of planet Earth.[2] The relationship is co-constitutive and relational. In my movement workshops, I use a process that I refer to as ecosomatic intuitive movement practice (IMP) that draws from the environmental sciences, somatic approaches, and artistic performance.[3] This work prioritizes such a direct engagement between the human body and the Earth; one that is biologically driven and eco-politically situated. Its mode is both imaginal and real, containing elements of meditation and ritual.

I am interested in *both* the physical presence and the imaginal idea of the skinbody as a living organism in its own right—one that undergoes its own transformation—as well as a site of tangible meeting through knowledge of and with the other. As we merge with land, forest flora, and other bodies, we can experience a transformation of our consciousness and perhaps achieve a more egoless dissolving into the other. In keeping with this book's emphasis on poetics and performance, this chapter interweaves personal poetic reflection with the theoretical and philosophical. Indeed, Sara Ahmed and Jackie Stacey liken writing to skin, suggesting that both involve "materiality and signification, limits and possibilities [...] difference and identity."[4]

DOI: 10.4324/9781003390985-13

Skin Knowledges

The notions of skinbody and the skin of the Earth offer ways for us to consider the place of the human in relation to others. There are vastly different bodies of knowledge that we can draw on to support the notion of entanglement or oneness. From a Western perspective, Guy Claxton theorizes human skin not as a container but as the place where we are "joined to the world"; David Howes refers to "skinscapes" and "body as landscape," as part of the contours and textures of earth.[5] In a responsive dialogue, Howes describes the environment as "tattooing" our skin as it registers weather and changes of temperature. Meanwhile, the decayed skin and bones of all animals and organisms in turn help to form biocrusts, or the living skin of our planet.[6]

Writing in the 1950s, Suzanne Langer described the tattooed face of an old Maori Chief as reflecting the valleys and ridges of the land/*whenua* itself. The identifying term for the Indigenous Māori of Aotearoa, New Zealand is *Tangata whenua*—*Tangata* (people) and *whenua* (land)—expressing the material and spiritual oneness of people and land. *Whenua* is the word for both earth and placenta, the latter being ritually buried on tribal land following a baby's birth to mark its irrefutable belonging to that place. In similar sentiment, Native American medicine man, John (Fire) Lame Deer, commented that "We are part of the nature around us, and the older we get the more we come to look like it."[7] In "Inscribing Identity: Skin as Country in the Central Desert," Jennifer Biddle explains how the term, *kuruwarri*, from the Warlpiri Aboriginal people of the Australian northwest, refers both to marks or maps of the land with its "embodied traces and imprints" and "ancestral presences" and also to skin-marks: freckles, birthmarks, and ceremonial body painting. She ultimately argues that "skin is literally materially, the same substance as country in that it is equally a medium in which ancestral traces reside."[8]

The Cashinahua people of eastern Peru believe that it is the skin, rather than the brain that holds knowledge; *ichi una* means "skin knowledge" and it is possessed by those learned elders of the tribe. There is also hand knowledge, *meken una*; eye knowledge, *bedu una*; and ear knowledge, or *pabinka una*.[9] For Maxine Sheets-Johnstone, the whole body in dynamic action is intelligence. For her, "thinking in movement is the natural expression of this elemental biological character of life."[10] Perhaps the Cashinahua have always been right—our skin with its sensory organs does much of our thinking.

I slide the back of my hand and arm along the grassy surface
My nose, so close I can smell the musty earth.

This morning a man on the radio explained how
The top soil is a living habitat of a billion organisms
That communicate with plants, absorb carbon from the air.
And within my skin's crevices and follicles
A million tiny creatures
Also do their work—inhabit
This skinbody.

As dance artists and somatic practitioners, we recognize that our movement investigation is shaped by particular landscapes, by geography, sounds, climate, and other species. If we dwell there long enough, we assume the qualities and rhythms of a place, assimilate and re-interpret its properties via our artistic expression or spontaneous response. Dancers from the ancient and worn land of Australia move across the space in wide languid graceful steps, hops, turns, and pauses (often akin to desert animals—such as snakes or kangaroos), while those of us from neighboring Aotearoa leap and fall, roll, pause, and pivot erratically and dynamically— perhaps in response to our steep hills and valleys, our wild oceans, chilly weather, and melodic yet asymmetrical bird calls.

The dance within Aotearoa New Zealand is often dark, more gothic in tone—the kind that produces goose-bumps on the skin—a possible reaction to our dark dense forests where our ancient tree-beings were cut down. Earth's skin in Aotearoa is very thin, thrust upward by the volcanic fault line whose eruptions formed this place. Undercurrents of pent-up anger and frustration hide just beneath our characteristic easy-going personalities, evident in shameful gang and domestic violence. These are the same undercurrents found in explosive and vibrant artistic creativity. A damp mustiness seeds the earthskin here in Aotearoa, just the opposite of the dry sandiness of many parts of Australia. The skin's sensitivity to surface texture can influence a dancers' movement vocabulary, shape their dance and character: living forms with which we are in a process of intra-action afford or inhibit our gestural language and rhythm.

Skin and World

Skin is not simply a fleshy container or a marker of racialized identity, it is our means of connection with others and the world—where the body becomes etched, marked, or stimulated by the touch, sensation of the other. Here skin rubs, slips, and leans against the surface of another (as in Contact Improvisation and somatic matching) or traces the shapes and form of the landscape of another body or of the Earth's surface as in site-based ecosomatic improvisation (*see* Figure 10.1). The knowledge derived from

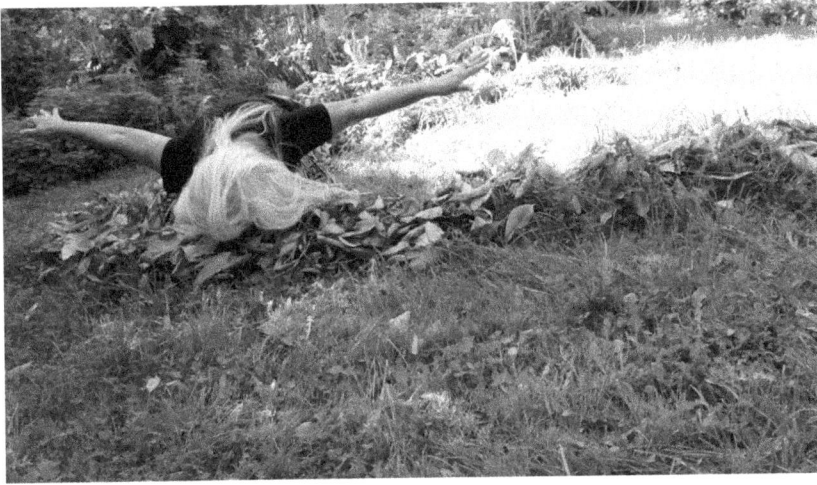

FIGURE 10.1   Ali East in *COVID Dance 11*. May 2020. Video still by Ali East.

touch is immediate. Sheets-Johnstone explains that "[s]urface sensitivity is the way creatures have made their way from the very beginning: knowing the world through touch and movement—through proprioception."[11] Aligning with these ideas, the relatively new area of ecosomatics has been defined as "a field which encourages direct sensory perception of one's body both *in* the natural environment, and *as* the natural environment."[12] As dancers we engage in an "ecography" of place—a phenomenological dialogue between the real (immediate present) and an imagined past via a "tactile synesthesia."[13]

Intuitive Movement Practice (IMP)

IMP is an approach developed over many years of teaching comparative somatics and choreographic methods. Its aim is to facilitate a searching beneath students' learned codified dance training, or prescribed (enculturated) behavior in order to access often unconscious somatic creativity and self-expression. IMP can also act as a metaphorical bridge between individual and collective creativity, between self-involvement and awareness of others or, by extension, between self and world. IMP has its origins in my artistic practice, my earlier incorporation of improvised moments into choreographed works, and my practice of interdisciplinary performance improvisation. During these spontaneous intuitive movement events, one may gain access to primal or psycho-choreographic impulses that appear to come from some prior evolutionary stage and yet remain embedded in our cellular memory. For example, in my dance, I have often performed as another species—as insect or bird—an ungendered moving abstraction or choreographic transfiguration. As I perform these other entities, I submit my total consciousness to a place that Maxine Sheets-Johnstone calls "creaturely knowing" or what Antonio Damasio refers to as "core consciousness."[14] My skin, while unchanged in texture and appearance, reshapes around the bones of my insect or bird-like gesture.

> I hover, sway in my skin
> Don't look at me as if you know me
> I am another, yet contained
> In my old skin
> I am both self
> and other.

As I perform and subsequently erase each former version of myself, I reshape my skinbody as I dance and open myself to being read—as data, as metaphor or fact, presence or resonance. In the liminal performance space, I can become other than myself. I am hyperreal. I am anima—the physical expression of the female soul of the world. I am small, like a tiny insect—a mere irritation on the skin of the planet scratching and burrowing in a desperate bid for survival. I am the dancer at home in my world of skin, a chameleon of rapidly evolving and transfiguring form. I am both self and other. However, as the following section suggests, I may also simply open myself to being inscribed by the Earth herself.

The Sensuous Dance of Skin: Ecosomatic Improvisation as Research Method and Practice

The goal of improvisational ecosomatic movement exploration, rather than that of artistic presentation, is to understand sensuously and tactilely the shape and form and intuitive intention of another's skinbody or the subtle sensuous secrets of particular place. In the process, each skinbody discovers something about itself and is changed by the interactive encounter. Braude and Schulman describe the fluid relationship between the moving dancer's body and the environment suggesting that we are always responding to an "internal" urge, motivation, and the affordances of the "external" world.[15] Albeit, in intra-action, boundaries of internal and external are more porous, like skin.

While viewed in certain contexts as artistic performance, these intuitive ecosomatic endeavors are also a form of research as we seek to understand both the quality of engagement with another, and also that understanding which may arise from this form of intuitive exploration of place/space, its imaginal sensory history and "psychogeography."[16] I have outlined a research methodology that I term "Intuitive Ethnography," whose data regarding the psycho-geographic nature of a place is largely gleaned through contact by the skin along with that of the other senses—sight, smell, hearing.[17] In these investigations, we acknowledge that each person, each skinbody, encounters the world differently and maps their own specific path, responding with the affordances of place in their own way. The ecosomatic dancer/researcher simply grazes through the landscape, practicing what James Gibson has referred to as ambulatory perception— a continuous moving across landscape absorbing sensation through the skin's surface—each of which stimulates some deep cellular memory as reflective image.[18] An ecosomatic sensory exploration of place offers radical information of a "geo-poetic" nature that acknowledges the body and its innate memory and imagination as a site and instrument of inter- and intra-relational knowledge.[19] Reflective writing and video documentation provide additional data that may supplement other forms of scientific and historical research.

As we merge with landscapes, flora, and others, we experience a transformation of our conscious being and an egoless dissolving into and merging with the other (*see* Figure 10.2). Just as trees cannot be considered separate from their nutritional source, we become part of the greater living energies around us. It is as if our skin has dissolved and sent fine filaments

FIGURE 10.2   Ali East during a solo improvisation, Doctor's Point, Aotearoa (New Zealand), 2019. Video still by Ali East.

downward into the soil and outward through wavering searching space, seeking connection.

> I search the air as bean tendril
> active yet passive
> letting the subtle wind guide
> my outreached arm around
> another body—to wrap, cling, climb
> our surfaces dissolving skin
> on skin.
> No leader, no particular direction.
> We move as one towards the light.

My need to bring my skin into contact with the surface of the Earth is also politically motivated. It is a form of subtle activism and might fit with what Ellen Dissanayake calls "an aesthetics of empathy."[20] She reminds us that every occurrence, object, or place will be experienced empathically and aesthetically differently by each individual. Sensory anthropologist David Howes warns of the consequences of a "tactile disengagement" from the world in modern-day cities and surmises that "[b]y distancing ourselves from the ground we lose not only the tactile experience of its surfaces but also other sensations that may be perceived when one has one's nose or ear to the ground."[21] Howes' warning is one that I wish to heed. Having one's "ear to the ground"—listening through our skin, paying attention to the state of the planet (pollution, global warming, water shortage, plague) may be a matter of our survival. When we consciously take our focus downward, we become aware of what is happening to the Earth. Dancer and somatic practitioner, Miriam Marler, speaks of "listening to the world through the skin" as a way of "countering our ocular-centric lifestyle."[22]

Performing Skinbody: Offerings toward Meaningful Connection

In what follows, I share a version of experiences from my movement workshops, offering a series of prompts that we might use so that you can explore the relationship between the skinbody and the skin of the Earth on your own. Some of the prompts are for a single mover and others are for small groups. Teachers might use them easily in classrooms or outdoors. If you are lucky enough to be able to include musicians, ask them to project or pitch the sound to different places in the room or environment and to different places in the body. Musicians and dancers create an intra-action of sound and movement that accumulates spontaneously through deep watching and listening.

In the studio or outdoors, aided by our imagination and rich verbal metaphor, we practice "dancing from our skin." Whether in the studio or outside environment, there are generally musicians or vocalists present who are also responding to the spoken images or prompts while simultaneously reacting to the spontaneous movement exploration happening around them as it evolves, along with the visual topography of the place. The spontaneous sound and movement languages inform each other as they track their own discreet pathway through time/space/place. Verbal prompts are inserted as part of the soundscape without interrupting the improvisational flow of the moment, providing additional information and guiding the research focus. Participants can choose to ignore or respond as they wish:

Prompt: Begin by standing, envisioning your *Tūrangawaewae*, or the place of tribal belonging where your *waewae*/feet are literally "woven" into the soil, connected to the earthskin of a particular place.

Prompt: From this grounding, allow a tree-like reaching and spiraling outward into the spaces between and toward another body. Allow this to become a surface-sensing and spiraling along the ground, walls, nearby trees, another's skin, etc.

Practice: "Tree listening"—a kind of listening and physical response that is activated by all the senses and which happens throughout your body. You might ask, how are the soles of my feet, the center of my back, listening, responding to surface sensations and sounds?

Prompt: How is my skin's surface intuitively responding to light and shade? How are my feet, fingertips, tailbone, etc. sensing, seeing/directing/coloring, reorienting my dance?

Prompt: Plant-breathing: Can you breathe through your whole skin's surface simultaneously—expanding, orientating your skinbody to receive maximum energy activation?

Action: Can you move with another body in a supportive "tactile, kinaesthetic matching," or "move with (match) the bodily organization of [your] partner [...] follow[ing] the paths of least resistance through patience and not fixing"?[23]

Action: In groups of three practice "tree reaching," supporting one person in reaching beyond their ability to maintain balance—*in "tree language" the analysis of a constant striving for light and nourishment—in human* *terms, a fearless seeking toward fulfillment and recognition of the communal body.*

Action: Can you bring every inch of your skin into contact with the floor, ground?

Note: *The un-sanitized intimacy of our skinbodies with place has suddenly taken on greater significance—become a transgressive act, a bid for connection and belonging.*

Prompt: Can you allow your appearance (skinscape) to morph and change shape as new meaning and metaphor?

Prompt: Can you dance from your exterior skin? Then from your internal fleshy contents? First as container then as content?[24]

Action: Can you find a *still* place (in the studio, forest, or beach) where you can practice being *present*—an egoless, thoughtless merging with place?

Prompt: Imagine tiny hair-like roots and vessels extending from your body into the ground, connecting, drawing sustenance upward—body and Earth intra-active.

Returning to Skinbody and Skin of the Earth

In this ecosomatic IMP, the human skinbody is considered a membrane of continuity and connection continuous with the skin of the Earth—a corporeo-geography of intra-activity and belonging—both a living and breathing, protecting and absorbing habitat for us and the multitude of microscopic species that dwell on and in our skin. Because, as human beings and dancers, we recognize our own surface contours, shapes, and habitual moving form, we are naturally attuned to the aesthetic shapes, sounds, and forms around us.

As the skin defines our form, shapes our containment boundary, acts as an organ of regulation monitoring temperature, displays the body's condition, protects sensitive organs, separates us from the rest of the world and from others, it remains our organ of negotiation, activism, mediation, and creative expression. As we move from vertical to horizontal, from bipedal to quadrupedal, adopt a reptilian slither and roll, or simply lie or sit very still, we begin to take on the perspective of the others who also dwell among us, often unseen and endangered. Our actions become political, engaged, participatory, posthuman, consciously performative, expressing connection and belonging. Monty Lyman suggests that "Our skin is not only a physical presence; it is an idea."[25] He elaborates by likening the body to a stage on which is performed the theatre of transformation.

The somatic practice that I am proposing here is immensely simple. As we engage in an ecosomatic intuitive exploration of self, others, and particular place through our skin and other sensory organs, we practice the art of being totally present to each moment and each stimulus. It is a form of Deep Seeing that, in ecological terms, can change both the seer and the thing seen. We become highly conscious, enactive, participatory, and engaged—yet nothing is planned or pre-thought. We are simply here "now" in this phenomenological moment.

When we are able to perceive our dancing skinbodies as part of nature, part of the environment, our surfaces akin to those of the planet—a fleshy geography of moist valleys, sensuous curved rolling plains, parched dry and scaley surfaces, flowing streams and toxic eruptions, we will have adopted a true ecological identity of "self plus other or self plus environment."[26] When we confer the same care, concern, and spiritual reverence for both our human skinbodies and the streams and mountains and deserts and oceans, then might we begin a holistic healing of both self and Earth. When we return to the notions of kinship with all species, animal and vegetable, held by many Indigenous peoples, then perhaps there will be a chance for us as human beings of the twenty-first century—a sustainable relationship and positive future on and for planet Earth. As dance artists and somatic practitioners, these research practices offer one contribution toward an interrelational empathic consciousness. Following a solo "COVID-19

FIGURE 10.3 Ali East in *COVID Garden Dance 1*, Otepoti Dunedin, NZ, May 2020. Video still by Ali East.

lockdown" ecosomatic surface-sensing exploration in my garden (*see* Figures 10.1 and 10.3), I wrote,

> In this time of silence, when the world has stopped its frenzy,
> Tune your ear to the sounds of birds, reach downwards,
> then open arms—wide,
> Sink into grass, lie, roll, recover,
> walk—nothing else to do—just walk
> back to where you began then
> start again.

Notes

1 In a 1950s geography textbook titled *The Skin of the Earth*, Miller uses a direct analogy of the anatomy and physiology of the Earth's living skin: "Meteorological charts and maps show the physiological state of the skin of the earth, its temperature, its moisture, and the circulation of air and water." In fact there are real similarities of structure and function between the human skin and the skin of the Earth. Both exist as layers through which circulate vessels, nerves, roots which transport nourishment and record sensation via hair or vegetal structures on the surface. A. Austin Miller, *The Skin of the Earth*, 2nd ed. (London: Methuen, 1964), 5. *See also* Matthew A. Bowker et al., "Biocrusts: the Living Skin of the Earth," *Plant and Soil* 429, no. 1 (2018), https://doi.org/10.1007/s11104-018-3735-1.

2 Karen Michelle Barad, *Meeting the Universe Halfway: Quantum Physics and the Entanglement of Matter and Meaning* (Durham: Duke University Press, 2007), 141.

3 Alison East, *Teaching Dance as If the World Matters: Eco-Choreography: A Design for Teaching Dance-Making for the 21st Century* (Saarbrücken: LAP Lambert Academic Publishing, 2011), 119.

4 Sara Ahmed and Jackie Stacey, *Thinking through the Skin* (London and New York: Routledge, 2001), 15.

5 Guy Claxton, *Intelligence in the Flesh: Why Your Mind Needs Your Body Much More than it Thinks* (New Haven: Yale University Press, 2015), 193–194. David Howes, "Skinscapes: Embodiment, Culture, and Environment," in *The Book of Touch*, ed. Constance Classen (Oxford and New York: Berg, 2005), 33.

6 Bowker et al., "Biocrusts: The Living Skin of the Earth."

7 John (Fire) Lame Deer and Richard Erdoes, *Lame Deer, Seeker of Visions* (New York: Simon and Schuster, 1972), 128–129.

8 Jennifer Biddle, "Inscribing Identity: Skin as Country in the Central Desert," in *Thinking through the Skin*, eds. Sara Ahmed and Jackie Stacey (London and New York: Routledge, 2001), 180.

9 Howes, "Skinscapes," 27–28.

10 Maxine Sheets-Johnstone, *The Corporeal Turn: An Interdisciplinary Reader* (Exeter; Charlottesville: Imprint Academic, 2009), 55.

11 Ibid., 140.

12 Nala Walla, "Ecosomatics at Work and Play: In the Landscape," *playGROUND* 1.0 (2009). www.bcollective.org/ESSAYS/playGROUND.v1.0.pdf.

13 "Ecography" is a term I have devised to describe ecosomatic improvised performances in the natural environment. *See* Alison East, "The Vegetal Body," in *Body and Awareness*, ed. Sandra Reeve (Axminster: Triarchy Press, 2021). *Re* "tactile synaesthesia," *see* Jennifer Fisher, "Tangible Acts: Touch Performances," in *The Senses in Performance*, eds. Sally Banes and André Lepecki (New York: Routledge, 2007), 167.

14 Sheets-Johnstone, *The Corporeal Turn*, 182. Antonio Damasio, *The Feeling of What Happens: Body and Emotion in the Making of Consciousness* (San Diego, New York, and London: Harcourt, Inc., 1999), 184.

15 Hillel D. Braude and Ami Schulman, "'What If …': A Question of Transcendence," in *Back to the Dance Itself: Phenomenologies of the Body in Performance*, ed. Sondra Fraleigh (Urbana: University of Illinois Press, 2018), 190.

16 Guy Debord coined this last term in a 1955 essay. Guy-Ernest Debord, "Introduction to a Critique of Urban Geography," in *Situationist International Anthology*, revised and expanded. Edited and translated by Ken Knabb. (Berkeley: Bureau of Public Secrets, 1981). Available online: https://www.bop secrets.org/SI/urbgeog.htm.

17 *See* Alison East, "Intuitive Ethnography: Ecosomatic Improvisation as Radical Pedagogy and Body-Centred Research Method," in *Creative Activism: Research, Pedagogy, and Practice*, ed. Elspeth Tilley (London: Cambridge Scholars, 2022).

18 James Gibson describes a number of modes of perception, including that which happens from one stationary point and that which happens as the perceiver moves through a place taking in cues as an animal might while grazing or hunting. *See* James J. Gibson, *The Ecological Approach to Visual Perception* (Boston: Houghton Mifflin, 1979).

19 Melinda Buckwalter explains how she uses Andrée Grau's term from the field of dance anthropology: "By geo-, I mean relating to the earth, and by *poetics*, an active artistic practice of knowing through making […]." *See* Melinda Buckwalter, "Dancing the Land: An Emerging Geopoetics," in *The Oxford Handbook of Improvisation in Dance*, ed. Vida Midgelow (New York: Oxford University Press, 2019), 611.

20 Ellen Dissanayake, *Homo Aestheticus: Where Art Comes from and Why* (New York; Toronto: Free Press; Maxwell Macmillan Canada; Maxwell Macmillan International, 1992), 150.

21 Howes, "Skinscapes," 29.

22 Marler argues that "listening to the world through the skin is part of our daily living, focusing deeply on this sense independently can counter our ocular-centric lifestyle" in Miriam Marler, "Stillness, Touch and Cultivating Intimacy with 'Vibrant' Landscapes," *Performance of the Real* 2 (2021): 75. https://doi.org/10.21428/b54437e2.b2522cdc.

23 Sondra Fraleigh, *Moving Consciously: Somatic Transformations through Dance, Yoga, and Touch* (Urbana: University of Illinois Press, 2015), 35.

24 Here, I am indebted to imagery from Bonnie Bainbridge-Cohen's Body-Mind Centering®.

214 Alison (Ali) East

25 Monty Lyman, *The Remarkable Life of the Skin: An Intimate Journey across our Largest Organ* (New York: Atlantic Monthly Press, 2020), 224.
26 Suzi Gablik describes an ecological identity as "self plus other plus environment." *See* Suzi Gablik, "Towards an Ecological Self," in *Theories of Contemporary Art*, ed. Richard Hertz (Upper Saddle River: Prentice Hall, 1993), 307.

Bibliography

Ahmed, Sara, and Jackie Stacey. *Thinking through the Skin.* London and New York: Routledge, 2001.

Barad, Karen Michelle. *Meeting the Universe Halfway: Quantum Physics and the Entanglement of Matter and Meaning.* Durham: Duke University Press, 2007.

Biddle, Jennifer. "Inscribing Identity: Skin as Country in the Central Desert." In *Thinking through the Skin*, edited by Sara Ahmed and Jackie Stacey, 177–193. London; New York: Routledge, 2001.

Bowker, Matthew A., Sasha C. Reed, Fernando T. Maestre, and David J. Eldridge. "Biocrusts: The Living Skin of the Earth." *Plant and Soil* 429, no. 1 (2018): 1–7. https://doi.org/10.1007/s11104-018-3735-1.

Braude, Hillel D., and Ami Schulman. "'What if ...': A Question of Transcendence." In *Back to the Dance Itself: Phenomenologies of the Body in Performance*, edited by Sondra Fraleigh, 183–204. Urbana: University of Illinois Press, 2018.

Buckwalter, Melinda. "Dancing the Land: An Emerging Geopoetics." In *The Oxford Handbook of Improvisation in Dance*, edited by Vida Midgelow, 611–633. New York: Oxford University Press, 2019.

Claxton, Guy. *Intelligence in the Flesh: Why Your Mind Needs Your Body Much More Than It Thinks.* New Haven: Yale University Press, 2015.

Damasio, Antonio. *The Feeling of What Happens: Body and Emotion in the Making of Consciousness.* San Diego; New York & London: Harcourt, Inc., 1999.

Debord, Guy-Ernest. "Introduction to a Critique of Urban Geography" in *Situationist International Anthology*, revised and expanded. Edited and translated by Ken Knabb. (Berkeley: Bureau of Public Secrets, 1981). Available online: https://www.bopsecrets.org/SI/urbgeog.htm

Dissanayake, Ellen. *Homo Aestheticus: Where Art Comes from and Why.* New York; Toronto: Free Press; Maxwell Macmillan Canada; Maxwell Macmillan International, 1992.

East, Alison. *Teaching Dance as If the World Matters: Eco-Choreography: A Design for Teaching Dance-Making for the 21st Century.* Saarbrücken: LAP Lambert Academic Publishing, 2011.

———. "The Vegetal Body." In *Body and Awareness*, edited by Sandra Reeve, 249–261. Axminster: Triarchy Press, 2021.

———. "Intuitive Ethnography: Ecosomatic Improvisation as Radical Pedagogy and Body-Centred Research Method." In *Creative Activism: Research, Pedagogy, and Practice*, edited by Elspeth Tilley, 407–433. London: Cambridge Scholars, 2022.

Fisher, Jennifer. "Tangible Acts: Touch Performances." In *The Senses in Performance*, edited by Sally Banes and André Lepecki, 166–178. New York: Routledge, 2007.

Fraleigh, Sondra. *Moving Consciously: Somatic Transformations through Dance, Yoga, and Touch.* Urbana: University of Illinois Press, 2015.

Gablik, Suzi. "Towards an Ecological Self." In *Theories of Contemporary Art*, edited by Richard Hertz, 301–309. Upper Saddle River: Prentice Hall, 1993.

Gibson, James J. *The Ecological Approach to Visual Perception.* Boston: Houghton Mifflin, 1979.

Howes, David. "Skinscapes: Embodiment, Culture, and Environment." In *The Book of Touch*, edited by Constance Classen, 27–39. Oxford; New York: Berg, 2005.

Lame Deer, John (Fire), and Richard Erdoes. *Lame Deer, Seeker of Visions.* New York: Simon and Schuster, 1972.

Lyman, Monty. *The Remarkable Life of the Skin: An Intimate Journey across Our Largest Organ.* New York: Atlantic Monthly Press, 2020.

Marler, Miriam. "Stillness, Touch and Cultivating Intimacy with 'Vibrant' Landscapes." *Performance of the Real* 2. (2021): 71–79. https://doi.org/10.21428/b54437e2.b2522cdc.

Miller, A. Austin. *The Skin of the Earth.* 2nd ed. London: Methuen, 1964.

Sheets-Johnstone, Maxine. *The Corporeal Turn: An Interdisciplinary Reader.* Exeter; Charlottesville: Imprint Academic, 2009.

Walla, Nala. "Ecosomatics at Work and Play: In the Landscape." *playGROUND* 1.0 (2009). www.bcollective.org/ESSAYS/playGROUND.v1.0.pdf.

PART III

Tree, River, Carbon, Stone

11

PRACTICING WITH TREES

Annette Arlander

Galway Road, Johannesburg, South Africa: 26.1658°S, 28.0223°E
David Bagares Street, Stockholm, Sweden: 59.3373° N, 18.0687°E
Kaivopuisto Park, Helsinki, Finland: 60.1557° N, 24.9557° E

Trees are respected and admired in many cultures, but also exploited in numerous ways, today often as monocultural plantations in commercial forestry. In Finland, the traditional cautionary tale, not to see the forest for the trees, should perhaps be turned around; we do not see the individual trees for the forest. As part of my project, *Meetings with Remarkable and Unremarkable Trees*, I have tried to encounter and attend to particular trees.[1] I have experimented with various techniques of posing together with specific trees for a video camera, either repeatedly for time-lapse videos, or only once for extended moments of stillness. I have used the yoga asana, two-legged tree pose, essentially a simple balancing exercise, next to and together with specific trees, often daily for a month. In the following text I will first describe the development of the practice briefly and then invite you, dear reader, to try the same, or similar, practice with a tree in your vicinity.

The title of my project, *Meetings with Remarkable and Unremarkable Trees*, alludes to the photography book by Thomas Pakenham, although my medium is performance for video, and the project is actually questioning what gets considered remarkable and what is deemed unremarkable.[2] Focusing on individual trees does not mean to suggest that trees are indivisible or separable from their surroundings, nor to deny that trees form networks and ecosystems or symbiotic relationships with other trees, with fungi, bacteria, and all kinds of micro-organisms; they are in a constant exchange with their environment, as humans are as well. Emphasizing individualism is a risky strategy in a time when focus on community is

DOI: 10.4324/9781003390985-15

needed more than ever. Attending to individual trees can nevertheless be an important step toward decolonizing our relationship with "nature." As late ecofeminist, Val Plumwood, pointed out, colonial thinking tends to emphasize a strong difference between "us" and "them", and to see "them" as all alike, stereotypical, non-individualized.[3] Thus, attending to particular trees as special and remarkable individuals might help us see trees as life forms that we have much in common with, despite our undeniable differences. If we consider trees as our teachers there is much to learn from them, like balance, rootedness, perseverance, and resilience, to begin with.

The first tree I balanced together with was an old oak tree on Galway Road in Johannesburg, South Africa, in February 2020, where I was part of an Arts Research Africa residency (*see* Figure 11.1).[4] The impulse for the experiment was a short notice in a Finnish yoga magazine "You, too, can become a tree," and a description of the two-legged tree pose, which I decided to explore with a suitable tree during the residency. The oak tree in the yard of the bed and breakfast where I stayed was the best tree partner to begin with, because I could leave the camera on a tripod behind me, unlike in public parks where an assistant looked after the camera while I was with a tree.

Trees that I have befriended and practiced the tree pose with since then include a maple tree in my home yard in Helsinki, the officially remarkable Spruce of Independence in Kaivopuisto Park in Helsinki in June, a small pine on the shore of Harakka Island in July, a tall Birch in Mustarinda residency in northeastern Finland in September, and an old pine on Örö Island in November 2020. After some alternative experiments I returned to the practice with a pine on Skifferholmen in Helsinki in February 2021, continued with a pine in Hailuoto Island during a residency in the northwest of Finland in April, with a maple tree and a crab apple on Eckerö in the Åland Islands in July, and a young ginkgo tree (*see* Figure 11.2) on David Bagares Street in the center of Stockholm at the end of the year 2021. The small ginkgo tree was supposed to be the last one. But while posing for a camera on tripod with an old pine tree (*see* Figure 11.3) as part of a follow-up project called *Pondering with Pines*, I nevertheless returned to the balancing practice.[5] I found that trying to become a tree with a tree was rewarding as a practice, perhaps even addictive in some sense.

The daily sessions are documented with two still-images on the project archive online; one image is of the tree alone, and the other includes the human practicing with them. The video material, the rough time-lapse videos consisting of sessions of a few minutes each, is edited into video works, usually added to the same pages as small files.

FIGURE 11.1   *The Oak on Galway Road*, Johannesburg, South Africa, video still, 2020. Courtesy of the author. Video still by Annette Arlander.

FIGURE 11.2   *Becoming Ginkgo*, David Bagares Street, Stockholm, Sweden, video still 2021. Courtesy of the author. Video still by Annette Arlander.

The video works vary in character depending on the trees and their environment, but the practice remains much the same. The pose does not look like a yoga asana, one can practice it as a simple stretching and balancing exercise. Although I practice in front of a video camera,

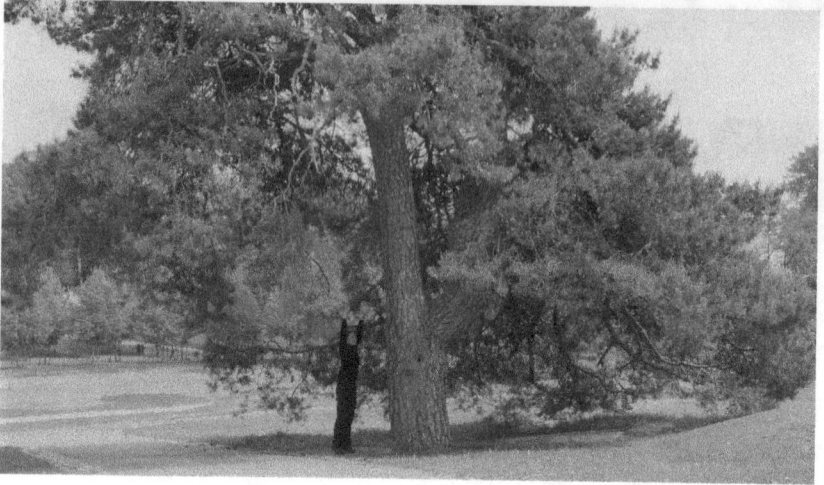

FIGURE 11.3   *With the Kaivopuisto Pine,* Helsinki, Finland, video still 2022. Courtesy of the author. Video still by Annette Arlander.

because my main objective is to create video works, the practice could well be undertaken without a camera, for the sake of the practice itself, and for the sake of the tree, in order to attend to the tree, befriend the tree, or learn from the tree.

The two-legged tree pose is a simple yoga asana, and you cannot really hurt yourself while practicing it. The instructions for executing the pose are simple: stand with the feet hip-width apart, lift your heels, raise your arms, focus your eyes on one point, and try to remain there. Some further advice is to keep the weight on your big toes and gently tighten your core muscles. The main idea, however, is to relax, to extend the out-breath, and to stay calm.[6] The only thing I add to this is the tree, to practice together with a tree. I soon discovered, however, that standing next to a tree, looking at a fixed point in the environment in front of me, often meant that the tree itself was not the center of my attention. Before and after the practice I usually made contact with the tree, though, often even touching it. After a while I noticed that it is a good idea to stand close to the tree, focusing on the trunk right in front of your eyes, letting the tree assist you in the balancing act. The tension between focusing on the tree and its environment and on one's body in the act of balancing can be resolved to some extent this way, by standing close to the tree and looking at the tree trunk. It might even happen that you feel like the tree is assisting you in reaching upward.

The most unremarkable tree becomes truly remarkable when attended to repeatedly, as a friend.

So, where to begin? Go out and look for a tree in your vicinity. Choose a tree that feels inviting to you, as a potential mentor or teacher, a tree that is easy to return to, daily or often. Greet the tree and ask permission to practice together with them in a manner that you find appropriate. Don't worry if you cannot feel a clear answer, try to remain sensitive and respectful. And then begin. If possible, stand in front of the tree, feet hip wide, raise your arms, focus your eyes, rise up on your toes, and breathe. If you cannot stand, try reaching upward in some other manner. When you cannot hold the balance any longer, lower your heels, lower your arms, and relax. Thank the tree in a manner you feel is appropriate and leave. Repeat the following day, and the next, and the next. That is all. And you can of course add your own variation, by writing a note or speaking a note, taking a photo, or singing a song....

Notes

1 *See* the project blog, *Meetings with Trees* https://meetingswithtrees.com and the project archive on Research Catalogue https://www.researchcatalogue.net/view/761326/761327.
2 Thomas Pakenham, *Meetings with Remarkable Trees* (New York: Random House, 1996).
3 Val Plumwood, "Decolonizing Relationships with Nature," in *Decolonizing Nature: Strategies for Conservation in a Post-Colonial Era*, eds. William M. Adams and Martin Mulligan (London: Earthscan Publications, 2003), 51–78.
4 The ARA residency resulted in an open access online publication. *See* Annette Arlander, *Meetings with Remarkable and Unremarkable Trees in Johannesburg with Environs* (Johannesburg: Arts Research Africa, 2020). https://wiredspace.wits.ac.za/items/5ee580a4-db3d-4de4-818e-7c6b65a16605.
5 *See* the project blog, *Pondering with Pines* https://ponderingwithpines.wordpress.com/ and the project archive on Research Catalogue www.researchcatalogue.net/view/1323410/1589526.
6 Jussi Kontala, "Kahden Jalan Puu [Tree with Two Legs]," *Ananda* 4, no. Winter (2019): 17.

Bibliography

Arlander, Annette. *Meetings with Remarkable and Unremarkable Trees in Johannesburg with Environs*. Johannesburg: Arts Research Africa, 2020. https://wiredspace.wits.ac.za/items/5ee580a4-db3d-4de4-818e-7c6b65a16605.
Kontala, Jussi. "Kahden Jalan Puu [Tree with Two Legs]." *Ananda* 4, no. Winter (2019): 15–19.

Pakenham, Thomas. *Meetings with Remarkable Trees.* New York: Random House, 1996.

Plumwood, Val. "Decolonizing Relationships with Nature." In *Decolonizing Nature: Strategies for Conservation in a Post-Colonial Era*, edited by William M. Adams and Martin Mulligan, 51–78. London: Earthscan Publications, 2003.

12

FEARLESS BELONGING AND RIVER-ME

Adesola Akinleye

Thames River, London, U.K.: 51.4925° N, 0.0288° W

Mystic River, Boston, Massachusetts: U.S.A.: 42.3979° N, 71.0797° W

Denton, Texas, U.S.A.: 33.2302° N, 97.1213° W

Us-Ness of Dance

It is in traditional Lakota and Yoruba dances that I first experienced dancing as tentacular, as being a part of everything and all, dancing with an enormous heart, lovingly pounding in every part of the world, every part of my sensing self. This is what I call, "becoming into us"—the sensation and awareness of the ongoing configuration and re-configuration of a web-like, networked, relational ontology, which the Lakota language confirms in the saying, *Mitákuye Oyás'iŋ*: we are all related/all my relations.[1]

Traditional dance ceremonies are meticulously handed down and methodically prepared to help dancers find such moments of us-ness. I am aware that the sacred nature of these traditional dances is to be lived more than written about. Traditional dances reveal the possibility of us-ness, which is a perspective that is lived beyond formal ceremony, spilling into everyday life experiences and into my love of dance in general. I am interested in how dance practices, in general, become an apparatus that reveals the us-ness of the assemblage of the moment.[2] For instance, when I take my contemporary dance practice to the riverbank, river-me is revealed in the foot-and-pebble negotiation of each step.

I am trained as a ballet and Graham dancer, growing up in the inner-city of London and later New York. Similar to so many artists, I have been homeless in both of these cities, easily lost in the rush of what it means to belong to the city. And I have also been on the stages of these cities' prestigious theatres, becoming the very product people visit the city to be a part of. In the Western dance studio, like so many other Black dancers,

DOI: 10.4324/9781003390985-16

I have experienced a similar spectrum, from being told I do not "belong" to becoming the diverse face of dance. Riding the tides of dancing and the city seems to require a kind of fearless belonging.

In Cambridge Massachusetts, I walk toward a mural of a woman's upper body, with painted birds around her, and a map of the area with the Charles River threading itself into the words "fearless belonging" below her. The mural was created by the Fearless Collective. Fearless is a South Asia-based public arts project that creates space to move from fear to love using participative public art. Out of the giant Wall-woman's mouth, reminiscent of the tourist maps around the area, a speech bubble reads, "you are here (with me)." I stop in my tracks as the mural woman reflects my presence "here" back at me in paint and brick. Wall-woman and I stand together for a moment in her fearless belonging, in the persistence of feeling "here" and the rawness of being a part of it all. As I start to walk again, Wall-woman leaves me with a sense of magnitude: the magnitude of responding to the environment, and of not allowing oneself to be cut off from the us-ness of "here."

It is not a question of if we belong to us-ness; we cannot step out of the assemblage of it all, it is more a question of how we enact our belonging. As Chief Seattle taught, we effect/cause and are effected/caused by all our relations with human, non-human, and environmental interactions.[3] The question is how fearlessly we are willing to claim and feel our belonging. How to be "here," belonging to the environment, the assemblage, when we are socially and politically conditioned to equate belonging with regulation and permission? Through the ecosomatic dances I describe in this essay, I suggest the sensitivities of dance reveal the impossibility of not belonging and the environmental and political importance to fearlessly claim belonging with care.

Fluidity

This chapter presents an account of lived experiences while dancing in the city—of dance experience that illustrates a fluid and co-created perception of self and environment. I relate this to the term "transaction" from philosopher John Dewey's lexicon, as he describes a collaborative, emergent reality of organism-environment.[4] For instance, perceived separations of self from world around ("organism" and "environment") have fluidity in how they differ across social, cultural, and political systems. Similarly, distinctions between subject and object are contested by scale. For example, I (my perceived body) am in the city's environment (I am an organism in my environment). At the same time, my perceived body is host to millions of bacteria and microbes (I am an environment for my organisms). Across this extension of my sense of presence, all that can be perceived as organism and all that can be perceived as environment are shaped and reshaped by their intra-action, the throb of relationships that my moving within the city spawns.

I use the term "intra-action" from feminist quantum theorist, Karen Barad, which they devised to describe the shifting relationships within this transactional ontology. That is not action between complete "things" coming together (inter-action), but action across the accumulation of the moment, emergent and co-created by the very action itself (intra-action).[5] This assemblage is multidimensional and in continuum, historied, anticipated, and simultaneously of the now—the moment.

We are all related, *Mitákuye Oyás'iŋ*, is to be understood by how it is lived, how it is danced in ceremony: "I" am the integrated experience of "we," moving with the ground, music, breath, and muscle of dance. Dance becomes a vehicle for engaging through and across the assemblage of the moment of my perception of "Self." Dance provides a distinct situation, a Deweyan "experience" that pulls the felt into conciseness as a memorable event.[6]

Yet dance remains slippery, in and of the situation, an example of the lived porosity of intra-action. Dance slithers across and between the would be subject and object, exemplifying the fabrication of their separation. When dancing, there is porosity between my perception of my body and my environment. Extensions and contractions, which Tonga philosopher 'Okusitino Māhina describes in the phrase "separations and connections" are marked by how the knowability of sensation is cradled in environmental dynamics.[7]

I feel mySelf dance, but when I dance, my sense of Self is not a static form. I use the terms *Self* or *mySelf* to indicate the accumulation of what I recognize as my identity. Through the situation of dancing, Self expands, connects, balances, becomes with, bends, touches, becomes without, and shifts within the assemblage of the moment. This intra-connection and intra-separation underpin a number of non-western Indigenous worldviews. I have discussed in more detail elsewhere knowledges that have long articulated perceptions of intra-action: woven ontologies, that Western sciences such as quantum physics are just starting to articulate.[8]

Western-framed language from Dewey and Barad integrated above convey the co-created nature and multiplicity of lived experience. Acknowledging that Indigenous worldviews articulate intra-action best for me, Western philosophies such as pragmatism also act as an antidote to the misguided Cartesian separation of subject from object. The lesions created by Cartesian separations of subject/object contribute to the infection of European expansionism that many in the twenty-first century hope to heal.

Colonialism attempted to cull Indigenous dances, countless of which offer passageways into awareness of being a part of it all. Elements of human, non-human, and more-than-human coming together in the temporary framework of dance offer a medium for noticing co-creation: for noticing *all my relations*.

My Practice

In the everyday movement of my contemporary practice, dance has a double edge. It can contract my sense of Self in the shackles of the racist environment of my dance technique training, but it also expands me, particularly when I use the physicality of my trained body to express and connect in choreography and improvisation. Because of my experience in traditional dancing, I feel the potential for vastness, pulsating, changing, and shifting in response to the sensations that dance movement produces. In the somatic physical meaning-making of sensation, I and World are conjoined as dance.

The dancing I discuss below is a moving mind-full-body-in-environment practice informed by the spatial politics of the city. In this contemporary urban practice, I sense the edges of mySelf as the skin sack of the body, and yet I also find parts of mySelf in the memory and familiarity of the streets and buildings around me. Here I am, writing and dancing from the historied urban experience of my Black, female-presenting, dancing body in the Western cities of London and Cambridge, Massachusetts. I dance around the rivers (Thames and Mystic) that run through and define these metropolitan areas: dancing in a continuity of meaning-making and fabrication that is comprehended in Us (city and me). We become a fascinating entanglement to explore the perceived edges, absorption, expansion, and integration that are revealed in the us-ness of dancing.

As I become more familiar with the sensation of dance along these rivers that thread through the heart of the cities we inhabit together, they begin to feel a part of how I locate mySelf in London or Cambridge. Because of this, in the past, I would write "she" as I talk about the rivers as extensions of myself—that is the river starts to become a part of the location of me in the city, an extension of how I identify mySelf in the city, and I spoke of mySelf as she. But this last year, I have been using they/them as my pronouns. I do this in solidarity with those who have their chosen pronouns challenged or ignored or for whom he and she are neither enough or appropriate. As I write this chapter, I start to habitually write she/her, then go back and change the text to the they/them that I now use to identify myself. This, the they/them (me) and they/them (the rivers), unexpectedly captures the sum of Us in a new collaborative turn of phrase. I find the they/them of river, and of me, write me into more possibility.

Thames

I am walking along the Southbank esplanade in London. The city is laid out before me across the expanse of the Thames. I am muttering about a meeting. "I am going to …. I will say this to show…." I am practicing my positive self-talk!: "I am going to show them who I am." Muscles tense, I am clenching my right fist, walking fast, eating the city pavement with every step. Regardless of my summer dress, I am burning under layers of bags, unanswered emails, and air conditioning vomiting out of buildings onto the concrete pavement. Then suddenly like ice touching the tip of my tongue on this hot day, I hear them. *Recognition*, the lap of a voice that is a grey/brown upsurge curl, breaking on the sand. That rhythm is the sound they make in the wake of a small boat driving upstream. The sound throws me into noticing Us. I breathe in on the tempo of their voice, the tone of the next wave breaking. My fist unclenches, but more, I feel my feet in my shoes; my diaphragm sinks in a deep breath of recognition that they draw from me. Their voice licks at the nape of my neck, and I sigh into a rhythm. Almost involuntarily, I say out loud "oh, there you are."

Despite my hurried, full day of appointments, I am drawn to peer over the embankment wall to look down at the ripples of water they used to call me. Knowingly, I look up to see the trail of foam that is a small boat vanishing. Haunted by dancing with river, I was shocked. I had not realized how well I knew the us-ness of river. I caught the sound of the waves, and they caught my breath despite the bustle of the pavement, the sound of traffic, and the commotion of my worry about the meeting. Demonic, they are within me, I realize, river-me danced up from moving on the beach with Thames over the last few months. Danced up from a twist of foot, swirl of sand, and turn. Caught by the off-balance in fragments of talus over calcaneus, glass, pebble, an incline of shoreline that is the physical memory in the sound of the waves. Standing looking out on the concrete of the Southbank for a moment in the rush of people walking toward the bridge, I am historied by having once bent forward to catch mySelf: hand reaching out for the damp caress of sandy shore, weight transferred from turning foot to shoulder, elbow, wrist, shoreline; equilibrium and sweep of leg round, a crescendo of breaking wave on the Thames beach; up-right again moving back to turn and face the expanse of the water edge. Dancing spills, vibrates, and forms across the flesh of my muscles, in the laps of waves, soft sand, and the bank of a gull's wing.

We—Helen, Andrew, June, Maga, Cheniece, DancingStrong Movement Lab members, Thames, and me—were there, at the river edge. We danced for a few freezing days in December, and then a few bright days in May, and then a week in June. And now here on the Southbank a few weeks into July, the sound of the river ignites in me a realization: our time together is not just a mental memory; it is a physical call bringing me into the awareness of river-me, which I learned or felt when we were together (*see* Figures 12.1 and 12.2). The sound and connection are like a gentle kiss as I balloon into the continuity of this emergent us-ness.

Mystic

Half a year later, in early February, the Mystic River icily whispers daintiness when I visit them in Massachusetts. I am grasped by a sense of openness within, and I find I am gently looking for the invitations to be with... that I learned to notice at the Thames. I am captured by their sameness—as I listen for the sound of the waves of the Mystic, noticing the non-human footprints and seeing the impact of my footfall on the softness of the snowy tide lines. Mystic has some familiar invitations and yet a distinctly different voice from Thames. But I do not want to anthropomorphize Thames or Mystic. I consider how familiarity and unfamiliarity is in the meeting of us: the perceived edges I prescribe as mySelf and prescribe as river. Today river-me reconfigures to the edge of the icy snowbank. I feel roots into the ground reaching, sucking moisture, elongating, contacting, worming my way through the snow and cold earth to meander. Barefoot, my feet feel the chilled caress of river's skin if earthen flesh banks hold river, as my skin indicates an edge. Skin-banks wounded by the cut of industrial division, scarred by tight concrete holding them to a course. Like those on my body, our scars are the sum of human intention, the result of the ebb and flow of growth that documents a history of interfaces. Mystic is lashed into shape by the city, no longer able to meander. Yet despite the control of their movement, Mystic has a persistent flow, an artery of the city, a pulse on this snowy morning; I feel an ancestral chill in my veins.

Returning to Mystic later in April, my bare feet stand on these banks again, skin to skin as we are of this moment; the us-ness of the dance that emerges is a continuity of the knowledge of river-me: "I love the way you make me feel." How I feel into the world and become aware of self-extension at the threshold of my perception and the onset of what I perceive to be you: the connection. What I love as I move alongside you is the sense-making of mySelf in you. I feel different because I am ignited by the extension of me into you. Us offers me insight into the instances when I tolerate the isolation of the notion of "I." Here I decide I end, and

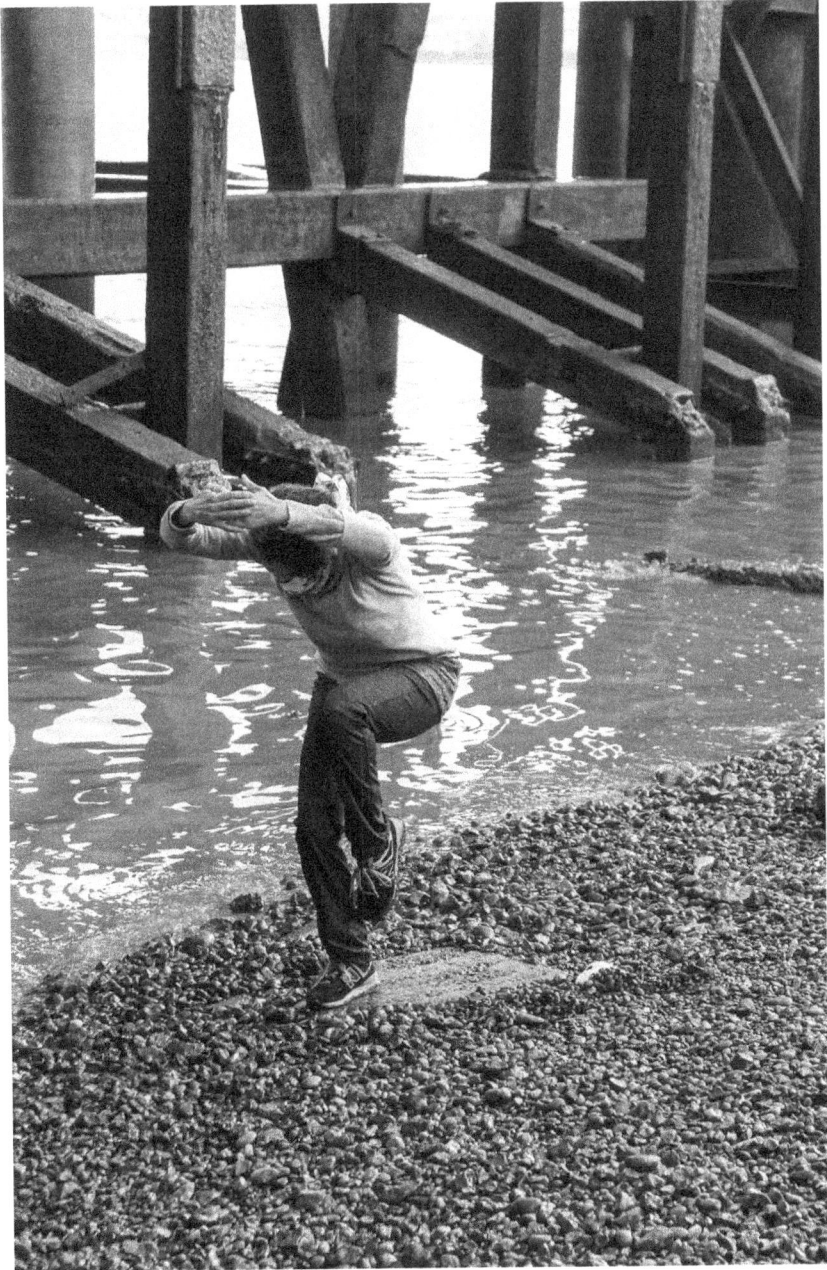

FIGURE 12.1   Raising tide with Thames River and Adesola. DancingStrong Movement Lab. Photo by permission of Cheniece Warner.

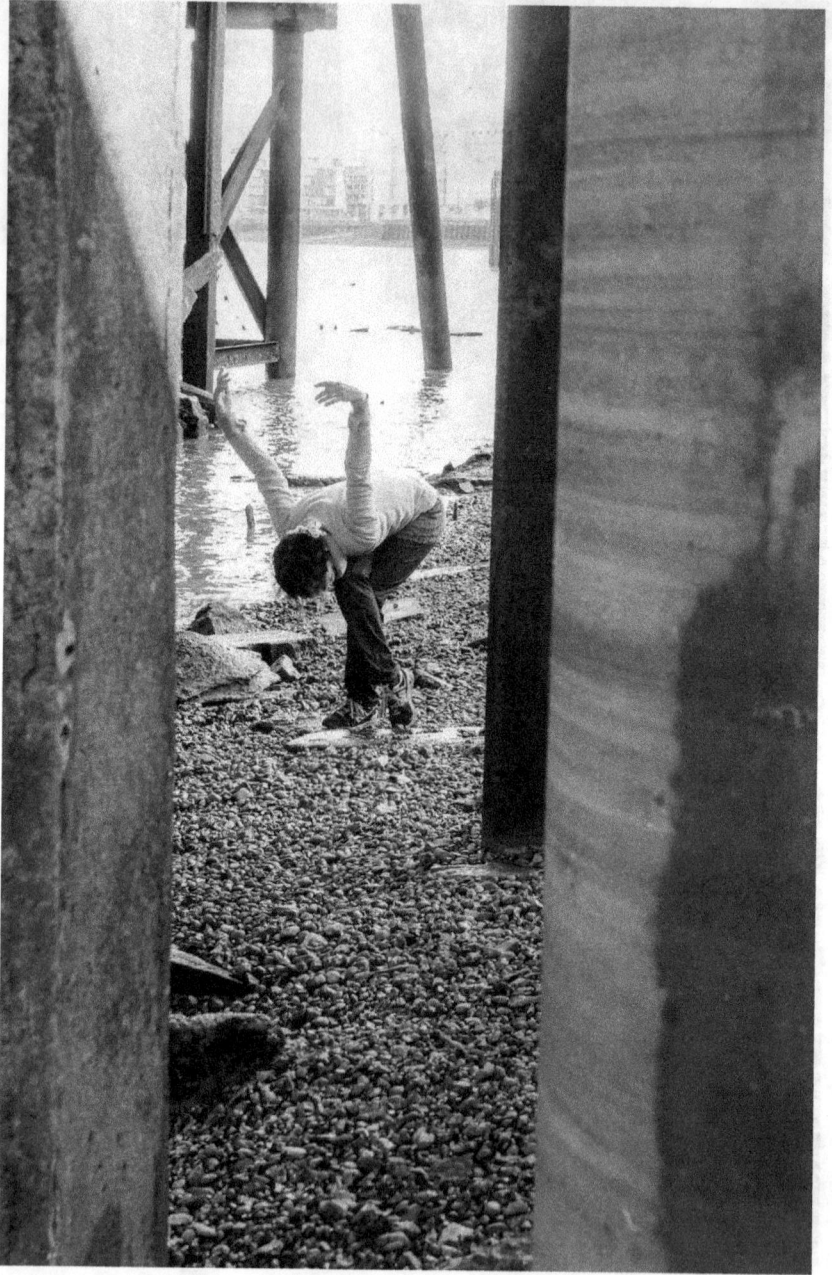

FIGURE 12.2 "River-me" assemblages: the bridge, distant architecture, the Thames river and its bank, a single hand, and Adesola. DancingStrong Movement Lab. Photo by permission of Cheniece Warner.

you begin. And what I love is the complication—the difference you are, the history I do not know, the ideas I had never thought of until I noticed the assemblage of us. I do not want this to unfold into the limits of my habitual prediction for mySelf.

Awareness of Belonging

River-me manifests my desire for connections, adding to, responding to, and moving with, all of which I notice as a sense of belonging. River-me is danced into my intrinsic awareness of being a part of the assemblage. I am aware that I belong to the assemblage, which is the dance of us-ness. I enjoy the sense of want. I describe this above as a sense of being in love: to want to be a part of, shaped by, and sculpted with a perceived other. Part of this belonging, then, is to be willing to connect to the very strangeness that is the other we defined as not being ourselves. This belonging requires the ability to challenge our sense of distinction, to reach out to the other in embrace, and yet not lose ourselves in otherness.[9]

I have been making night dances as an exploration of the other—with the goal of not losing the self in otherness. Back at Texas Woman's University where I teach, there are no local rivers, and so dancers (Barbara, Elaia, Gracie, Holly, Victoria, and I) have been making a dance for the land by a small rain-overflow stream and a cottonwood tree growing from that spot. This is a night dance; otherness is in the dark threshold of the felt. In the night dance, our sight is regulated by how many car headlights pass us as we dance. Over the weeks of developing the scored dance, the otherness of the dark unseen slowly loosens. The night starts to illuminate the textured sensation of the ground, the evening air, each other's breath, and a sense of recognition in our dancing bodies. The otherness in the darkness around us opens as we dance there. First, otherness felt like a threat exacerbated by our inability to see (during our early rehearsals there), followed by otherness as a promise (as we started to open to the space), and then otherness as a possibility, as we allowed ourselves to belong to the space.

The hairs on my arms extend as I practice the night dance score with these five dancers. This score directs us to move forward across the grass on the unique trajectory we face, but also with the commitment of touching the dancers beside us in the line, since each one has their own trajectory. The dance becomes the shape and movement generated in our bodies through the compromise between our individual tracks forward, the connection of weight to the uneven earth below us, the plants and trees we encounter, and our arms reaching out to stay connected to each other at our fingertips. As our fingertips lightly touch and brush against each other, we separate and re-find. As we move forward, the intention of the movement is expanded: my knee deepens when Barbara's foot is off-balance. I bend to offset the swell of Barbara's weight change as we balance again. I *belong* to the line of us moving together in the dark. I feel the swerve of Holly stepping up the incline of a rock on the grass at the other end of the line; stretching my ribs and arm out away from them allows me the freedom to track the ripple of movement that Holly's step produces in us.

I am aware of the distinctiveness of the other dancers in terms of the familiarity of their movements, but my response (knees, ribs) maintains mySelf, not being lost in otherness, which would pull me off balance and stop the movement of the line. The us-ness of belonging requires a threshold state where self/other is both a distinction *for* and promise *of* connection. This danced line is a sharing. In dancing this score, each dancer must accept a sense of incompleteness, that their movement intention is transformed through sharing in the line; that this sharing of the agentic power of their movement opens their dance to potential movement horizons created through the common co-creation of us-ness.

This belonging thus involves the porosity of a boundary that comes from an awareness of the temporary accumulation that is the moment. From the dance score to the city journey, us-ness requires belonging and otherness to be reconfigured, moving the demarcation of the "other" from sovereignty-threat to potential-and-possibility.

Expanding within Belonging

Mystic River is in a May spring swell. At first, I move very slowly by Mystic, stand and watch, smell and feel the ground, taste the air swept up from the ocean, carried in the wake of the water to reach my skin. Today I move slowly so that I do not impose a set of preconceived movements on my asking body. In the past, I have moved fast for the same reason. When I move fast, I do not have time to preconceive, plan, or reproduce; the fast movement becomes almost like an elongated falling. The movement is incidentally carved out of the physics of the flexibility of the sinew and fascia of my body and the scale and incline of the shoreline. But today I feel the need to move slowly as an act of physical listening. Here, I close my eyes to find my way out of thought and to feel the sensation of sunlight on my closed eyelids. Presently, I feel moisture in the air and on my skin, crescendos of sounds, smells, and breezes.

I relish the feeling of possibility the idea of this river-me dancing practice anticipates, and when I am there in the sensations of river smell, a reflection of sun, faint dampness in the air, I extend into river. Then when I remember my there-ness with river I juggle recall with hesitation. Was this ever anything more than my fancy, the misgiving that my awareness could even capture or recognize anything except my own projections?

But us-ness is confirmed by the point that as I danced by the river, I could not have imagined the sensation. I was surprised as I encountered innate movement opportunities. I determined mySelf in ways I had not realized before. I felt an unexplored possibility for the change of weight from one foot to the other, finding a sense of self in shapes I had never considered. I could not have imagined you: river-me is a surprise. Yet, it is only my astonishment that tempts me to doubt us: where could we have come from? Where else was that unexpected movement but a dance held in the congregation of Us?

A sparkle of sunshine flashes across my closed eyelids. Nuzzled in the pitch of the breeze, a sound of lapping waves steadies my leg on the incline of the shoreline. This moment is a physical sensation blazed in muscular remembrances of river-me. But the recall is of the *affect* of the moment, given form and bound by a myriad of senses and meaning-making I identify as my life history. I am astonished at the porousness of my edges. At times I extend so bravely into what I perceive to be the city, I cannot recognize mySelf anymore. But sometimes, disillusioned, I guard my edges like a duty soldier, I guard what I perceive to be mySelf so well that responding almost feels unattainable.

As I recall the dances that become the events of river-me, I do not have images in my memory of the aesthetic of my body. Correspondingly, I do not have an inner-image of the river. Rather river-me is the feeling of the grass and splintered concrete, the sparkle of the sunlight on the water. River-me is made from recallable meaning within the area that contains my felt/danceable construct. My recall presents a kind of shroud that blankets the moment of dancing woven out of hundreds of senses and the meaning-making that binds them in a memory thread. This is where I feel my belonging to intra-action, the transaction of being river-me: the border of being lost in the sensorial that marks our us-ness.

My proposition for this belonging is that I have learned to know mySelf differently rather than understand "others" (rivers) better. I remember that in this transactional ontology, "Self" is the integrated experience of Us (river-me). River-me is the awareness of belonging to us-ness in which I learn to notice a new deeper understanding of the possibility of "us" of which "I" is just the familiar part. I am trying to avoid ecological anthropomorphism. I consider I am not learning more about rivers, I am learning more about the ability of river-me to respond and be responsive with Thames and Mystic.

Expanding understanding of my responding Self includes noticing where river flows into me, as we are part of each other, "I am because we are." This is a phrase associated with the African philosophy of Ubuntu. I understand that my ecological responsibility in the twenty-first century includes the bravery to extend my sense of Self enough to comprehend the histories and consequences of my actions: how bravely I allow my perception of Self to be changed by my responsibility to the assemblage of us. We are inevitably connected. Rather than separate us through "caring for river as I would care for myself," I must learn what it means to "care for us."

How bravely I grasp that *us* is somewhat measured by how far the accumulation of my experience is integrated into meaning-making and responsivity. To be river-me, I am charged to live beyond the limits of my imagination for what I see as mySelf, to be at the edge of knowing and yet incompleteness in a state of activation and responsiveness. I do not see my agency as the privilege to determine lines of otherness; rather, I hold agency as the awareness of the impact of my response and responsibility *in* us-ness.

Boundary and Belonging

The ground is uneven here, not a slope so much as a cartography of deposits washed up by the tides. The unlevel lie of Mystic's banks caresses the pull of the ocean, corseted by the strips of concrete roads that run alongside them. I am standing on the buckle of the shore dancing, and my movement score directs me to stand on one leg. The score reads: "Balancing—push balance until you lose balance and balance again. Noting the process and connections and separation that involves balance and off-balance."[10]

I feel mySelf in the ankle of my balancing foot, the knowledge/awareness and meaning of being here pulsating between the tendons and muscles of my toes, and the softness of the soil. The responses, alterations, and improvisations I am making as I dance this *balance score* become the apparatus that draws a kind of blueprint of the topography of the bank. While dancing, I find I am as aware of the tufts of grass, silt, discarded bottle tops, soap boxes, and plastic cups around my feet as I am of the rise of my shoulders or the curve of my back adjusting to my ankles. It is as if I extend and pour into the bank below my feet.

Then I see them, five ducks swimming enthusiastically toward the riverbank and behind me four geese and some more ducks. They coercively waddle down the uneven bank toward my feet. I protectively sweep energy into the upper part of my body; the geese are at hand height, and all the beings around me seem to expect my hands will be fruitfully holding food! Losing bravery, I shrink into a memory of being bitten by a goose when I was three years old; I lose the extension into the soil I held moments ago and almost stumble up the bank to the pavement: "I don't have anything for you!"

In *The Body Keeps the Score: Brain, Mind, and Body in the Healing of Trauma*, psychiatrist Bessel A. van der Kolk describes how the neurology of the brain is shaped through the feedback of the tactile register of the body.[11] This sensing shapes the reality of consciousness as the brain draws on past experiences of senses to determine the meaning of the tactility of now. Similarly, The Franklin Method, through scientific research, concludes that the physicality of the body changes the physicality of the brain.[12] The physiology of the brain reshapes itself to notice and become attuned to the world around, according to the focus (or meaning-making) of sensation. Reality changes as sensation changes. We all change (river, body, brain, ducks) when awareness of brave belonging shrinks to stumbling up the riverbank—a riverbank made steep by the memory of my three-year-old goose bite.

There is a construction of reality as static and existing at the boundary where the "I" ends and the world around starts; demarcated by the membrane of the skin. But this I/world boundary is not fixed and the skin is osmotic. I/world is a contracting and expanding sense of awareness resulting from the mediation among sensing mind-full-body (corporeally present and making meaning), habitual mind-full-body (historically shaped by activity and retort), anticipatory mind-full-body (sympathetic and arranging), and cultural, social-political mind-full-body (historied and regulated), all of which conjure the reality of the brain. Reality is thus a distinctive boundary expressed only through emergences from the throng of integrated experiences. Realization and reality are thus ever-shifting within a transactional relational ontology. I have recently suggested that boundaries are therefore temporary and have a porosity prompted by expansions and contractions within us-ness.[13] Thus, boundaries do not mark the edge of belonging; they merely suggest "the how" of belonging.

The Politics of Belonging

I wonder what belonging, experienced through my dance practice, says about belonging from the perspective of practices in geography and social science, where the word *belonging* is often used as a synonym for identity, particularly national or ethnic identity.[14]

The identities attributed to my moving body situate where and how my contemporary dance practice is expressed. My dance practice renders different scales of belonging within the social and political site of the city. I experience belonging to an us-ness that informs a shift of weight to my toes with the riverbank. However, I am part of a range of cultural, ethnic, social, and political *us-es* that inform where and how I walk across the expanse of the city. The city offers a keen sense of territorial belonging through boundaries of governance, which include and exclude. At this scale, belonging is interpreted as *access*.

My Black female-presenting body has access to some parts of the city, and other parts of the city are inaccessible or dangerous. I learn to stay where I "belong." Belonging, here, being limited to the areas of governance that offer me a right to respond, move, and be a part of the city. Walking down an unfamiliar street in Cambridge at night, I scan the doorways for people, and I look for political signs, rainbow colors, Black Lives Matter signs, or Confederate flags, and I also read the graffiti. I assess how I belong on the street. I read the indications for how I should respond to the street. In London, where I grew up, I am historied into the socio-spatial matters of my belonging. I know the places young Black men have been stabbed, the shops where security guards will follow me, and the office buildings where financial decisions are made. I know my presence is not welcomed there, and I do not feel safe there.

Wandering the city banks of the Thames and Mystic and the darkness of night dances has overtly highlighted my dancing body, which is read as small, Black, and feminine. In order to dance in the city, I claim that my dancing Black femininity belongs—rather than looking for social permission. But when I am alone on the riverbank, my sense of belonging/not belonging is multimodal. In contrast to the anti-belonging aesthetic of being Black, female-presenting, and dancing on riverbanks of a Western metropolis, I feel the intimate place of me in the assemblage of dancing-riverbank. I manage the threshold of us-ness when my presence in parts of the city must negotiate my body matter-ing. But us-ness exists in relationships of care, response, and responsibility; us-ness is a felt intra-action I call river-me, danced into awareness over the last year.

Fearless Belonging

I dance both being within and witnessing us-ness. River-me emerges from the dance of us-ness as something recognizable through the actions generated from the accumulation of the moment where the dance emerges. Us-ness is revealed in the response of the dance—the ethics of my responses within the moment of river-me and also the realization that my Black female-presenting body is as objectified as the riverbank. The subject/object divide is embodied in the city through the notion of boundaries (and governances). Ironically, I am aware that while I am peacefully dancing to manifest an inner awareness of my belonging to us-ness, to some, the sight of my presence on the riverbank is an object to be rejected.

Drawing upon knowledge arising from my river dances, I notice generosity as a propelling force in the physicality of river-me and dancing in my darkness. Generosity emerges from the bounteous possibility of the assemblage of which my dance is a part. I notice generosity in the threshold of accepting and giving, shaping, and being shaped. The somatic endeavor of the *balance score* on Mystic's bank expanded us from ankle to soil and into the historical lapping of the water that created clumps of earth, fed the tufts of grass, and deposited bottle tops. Otherness was pushed into the distance, and I became us-ness with the bank. Then when the ducks and geese got close to me, I contracted my fingers and shoulders in the memory of being bitten when I was three. Remembering this, I withdrew the edges of mySelf so that I stumbled, unable to respond to the very bank I just spent five minutes feeling so connected to. The relationship was changed by my response and responsibility to the us-ness of the moment.

I have been tending to the life force of the edges of mySelf where others become a focus. But the potential of us-ness is not useful if it becomes an obsession or even a good-natured preoccupation of demolishing the edges or boundaries of Self. There is much more potential in recognizing the shifting nature of the composition of us-ness. It seems vital to consider the permeability in us-ness, and therefore the effect we all have on the moment through our care and responses within us-ness.

The consciousness of felt sensations of Thames or Mystic in my dancing is developed from the multiple dance experiences of responding with rivers. These places of knowledge are also a form of learning to belong with rivers. In belonging, I sense beyond the apparatus of breathing or swallowing; river-me lingers in sensations learned and recognized through dancing river's edge. Belonging creates sensations of fearless dancing in the moment of us-ness. Arabesqued out along the sandy shore of Thames or the muddy bank of Mystic, I question how fearlessly I can belong to us-ness in order to manifest the dance. Belonging becomes dancing responsible-ness within and for the accumulation of now. Can I be vulnerable to the common creation of the moment, allowing reinterpretation and the possibility of being unrecognizable for a moment, not letting that freeze me into fixing mySelf but continuing to respond despite any sense of incompleteness?

Belonging means careful consciousness of being in relationship. It is *us* as the dance, us-ness, the sensation of the moving porosity of being a part of everything and all. Fearless belonging allows the sensing self to generously feel incomplete, responding in the moment of now in fearless concern for the whole assemblage of us. *Mitákuye Oyás'iŋ.*[15]

Notes

1 The term translates variably as "we are all related," "all my relatives," or "all my relations" and considers all life forms to be related. Daniel P. Modaff, "Mitakuye Oyasin (We Are All Related): Connecting Communication and Culture of the Lakota," *Great Plains Quarterly* 39, no. 4 (2019): 346, https://doi.org/10.1353/gpq.2019.0055. For the traditional spelling, *see* Rani-Henrik Andersson and David C. Posthumus, *Lakhota: An Indigenous History* (Norman: University of Oklahoma Press, 2022), 22. My understanding of the term is particularly indebted to Victoria Chipps. *See* Victoria Chipps, *Pray from Your Heart: Teachings of a Lakota Elder*, ed. Susan Chiat (Alma: Morningstar Publishing Company, 2001).

2 Jane Bennett develops Gilles Deleuze and Félix Guattari's concept of "assemblage" in order to describe a new understanding of the "part-whole relation." For Bennett, assemblages are "ad hoc groupings of diverse elements, of vibrant materials of all sorts." Jane Bennett, *Vibrant Matter: A Political Ecology of Things* (Durham: Duke University Press, 2010), 23–24. This is the spirit in which I use the term in this chapter. *See also* Gilles Deleuze, *Expressionism in Philosophy: Spinoza* (New York; Cambridge, MA: Zone Books; Distributed by MIT Press, 1990).

3 Chief Seattle, *Chief Seattle's Testimony* (London: Pax Christi International Catholic Peace Movement; Friends of the Earth, 1976).

4 John Dewey and Jo Ann Boydston, *The Later Works, 1925–1953*, 17 vols. (Carbondale and London: Southern Illinois University Press; Feffer & Simons, 1981), 97.

5 Karen Michelle Barad, *Meeting the Universe Halfway: Quantum Physics and the Entanglement of Matter and Meaning* (Durham: Duke University Press, 2007), 178.

6 John Dewey, *Art as Experience* (New York: Capricorn Books, 1959), 38.

7 Pā'utu-'O-Vava'u-Lahi Adriana Māhanga Lear, et al., "'Atamai-Loto, moe Faka 'Ofo'Ofa-'Aonga: Tongan Tā-Vā Time-Space Philosophy of Mind-Heart and Beauty-Utility," *Pacific Studies* 44, no. 1–2 (October, 2021): 1.

8 Adesola Akinleye, *Dance, Architecture, and Engineering* (London: Bloomsbury Publishing, 2021), 30.

9 Stavros Stavrides, *Common Space: The City as Commons* (London: Zed Books, 2016), 155.

10 I wrote this score for students at ACT, MIT during an artistic residency hosted by Gediminas Urbonas and funded by CAST.

11 Bessel A. van der Kolk, *The Body Keeps the Score: Brain, Mind, and Body in the Healing of Trauma* (New York: Viking, 2014), 46.

12 *See* Eric Franklin, *Grow Younger Daily: The Power of Imagery for Healthy Cells and Timeless Beauty*, 2nd ed. (Minneapolis: Orthopedic Physical Therapy Products, 2019).

13 Adesola Akinleye, *Navigations* (London: Theatrum Mundi, 2022). In this book, I discuss a lexicon of ten "worded-ideas" that seem to resonate across an interdisciplinary interest in how humans create movement and a sense of Place: spatial practices. Boundary is one of the words of the lexicon. In this discussion, I suggest boundaries have porosity because "reality" is responsive

and contingent; thus, boundary is also responsive and contingent. Therefore, boundary is a product of perception rather than a structure that sits outside of the perception to modify it.

14 Marco Antonsich, "Searching for Belonging – An Analytical Framework," *Geography Compass* 4, no. 6 (2010): 645. https://doi.org/10.1111/j.1749-8198.2009.00317.x.

15 Thank you to Garnette Cadogan, Helen Kindred, and Gediminas Urbonas for their presence across the period of dancing up river-me this chapter recalls.

Bibliography

Akinleye, Adesola. *Dance, Architecture, and Engineering*. London: Bloomsbury Publishing, 2021.

———. *Navigations*. London: Theatrum Mundi, 2022.

Andersson, Rani-Henrik, and David C. Posthumus. *Lakhota: An Indigenous History*. Norman: University of Oklahoma Press, 2022.

Antonsich, Marco. "Searching for Belonging – An Analytical Framework." *Geography Compass* 4, no. 6 (2010): 644–659. https://doi.org/10.1111/j.1749-8198.2009.00317.x.

Barad, Karen Michelle. *Meeting the Universe Halfway: Quantum Physics and the Entanglement of Matter and Meaning*. Durham: Duke University Press, 2007.

Bennett, Jane. *Vibrant Matter: A Political Ecology of Things*. Durham: Duke University Press, 2010.

Chipps, Victoria. *Pray from Your Heart: Teachings of a Lakota Elder*. Edited by Susan Chiat. Michigan: Morningstar Publishing Company, 2001.

Deleuze, Gilles. *Expressionism in Philosophy: Spinoza*. New York; Cambridge, MA: Zone Books; Distributed by MIT Press, 1990.

Dewey, John. *Art as Experience*. New York: Capricorn Books, 1959.

Dewey, John, and Jo Ann Boydston. *The Later Works, 1925–1953*. 17 vols. Carbondale; London: Southern Illinois University Press; Feffer & Simons, 1981.

Franklin, Eric. *Grow Younger Daily: The Power of Imagery for Healthy Cells and Timeless Beauty*. 2nd ed. Minneapolis: Orthopedic Physical Therapy Products, 2019.

Modaff, Daniel P. "Mitakuye Oyasin (We Are All Related): Connecting Communication and Culture of the Lakota." *Great Plains Quarterly* 39, no. 4 (2019): 341–362. https://doi.org/10.1353/gpq.2019.0055.

Pā'utu-'O-Vava'u-Lahi Adriana Māhanga Lear, Kolokesa Uafā Māhina-Tuai, Sione Lavenita Vaka, and Tevita O. Ka'ili Maui-TāVā-He-Akó. "'Atamai-Loto' moe Faka 'Ofo'Ofa-'Aonga: Tongan Tā-Vā Time-Space Philosophy of Mind-Heart and Beauty-Utility." *Pacific Studies* 44, no. 1–2 (October, 2021): 1–12.

Seattle, Chief. *Chief Seattle's Testimony*. London: Pax Christi International Catholic Peace Movement; Friends of the Earth, 1976.

Stavrides, Stavros. *Common Space: The City as Commons*. London: Zed Books, 2016.

van der Kolk, Bessel A. *The Body Keeps the Score: Brain, Mind, and Body in the Healing of Trauma*. New York: Viking, 2014.

13

HOW TO APPRENTICE WITH LAND IN ENCHANTED KINSHIP

Christine Bellerose

Rockcliffe Park, Ottawa, Ontario, Canada: 45.4471° N, 75.6847° W

This set of practice pages documents ecosomatic preparatory work for ecoperformances that took place in the summer of 2022 at the junction of the Ottawa and Gatineau Rivers on unceded Indigenous land in Rockcliffe Park, Ottawa. Somadance process is how I access the dynamic meeting of body and land from which a touch, a shared breath, a contemplative gaze serve as my introduction to the surrounding somas. And by ecoperformance, I mean a conversation with the land across species. My ecosomatic approach to apprenticing with the land signals a willingness to bond with living place-memories. This creative *démarche* repatterns my ecological awareness in relation to a particular site.[1] The chapter concludes by offering the reader several short, accessible activities toward developing enchanted kinship with the soma of place.

I have lived in Ottawa-Gatineau on and off since girlhood. I am the daughter of a soldier from the Carignan-Salières Regiment and a *fille du Roy* (King's daughter).[2] Has anyone in my ancestors' line taken up canoe and paddle? I am the daughter of hopeful farmers. I am the daughter of an Irish girl who shipped herself across the ocean. I am the daughter of a servant maiden and seamstresses. I am the daughter of ice sheets, river water, and fertile soil. I am the daughter of history. I am the daughter of Land. My roots plunge deep into dreams of a better world. My feet dance on the shame of colonial crimes.[3] Apprenticing with land sits me at the table with my kin. We don't always get along—how ordinary, the tale of enchanted family gathering is! In her own way, Land teaches me about who I am.

DOI: 10.4324/9781003390985-17

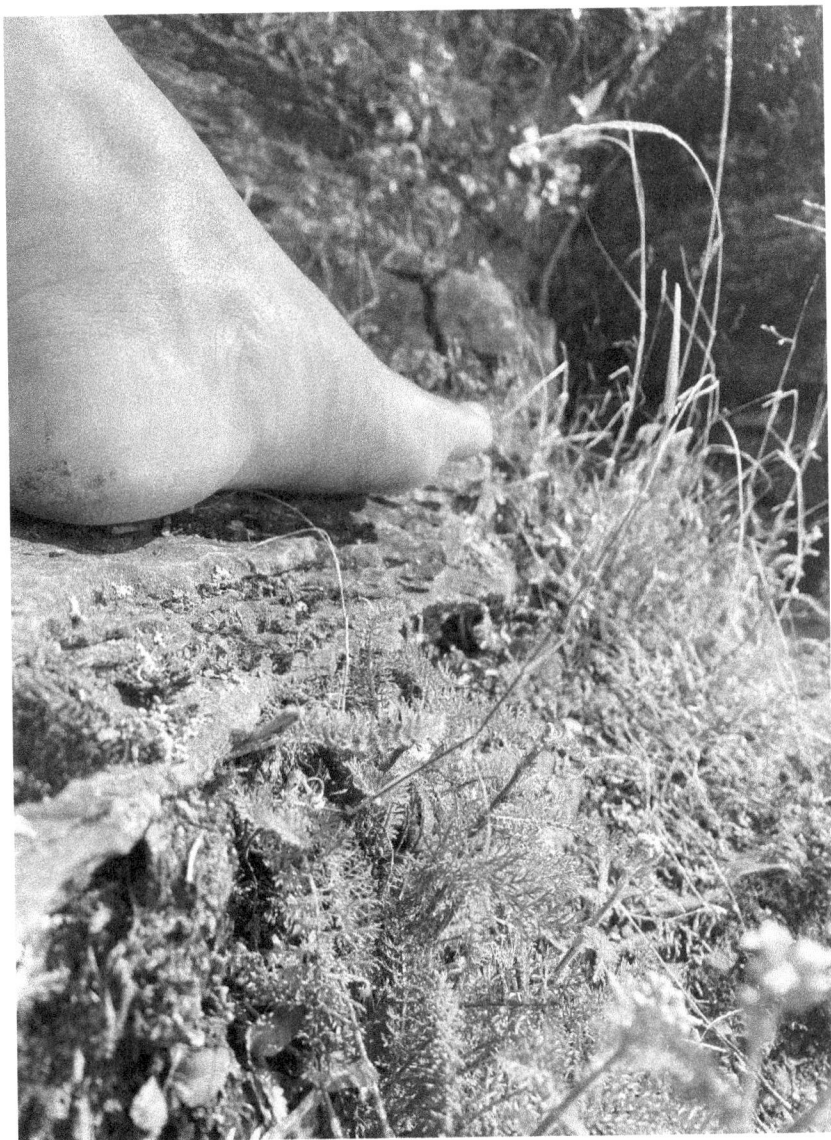

FIGURE 13.1   Bellerose's left foot meets the cliff's rough edges of eroding sandstone, tufts of yarrow, and moss colonies. Rockcliffe Park, Ottawa, Ontario, Canada (2022). Self-portrait, courtesy of the author.

My ecoperformances stem from a desire to repattern enchanted kinship by way of attuning to the stories that are already and always existing in place. For me, enchanted kinship patterns a partial story between the personal and the land, somatics and dance. I interpret the story as somadance through somatic practices and dance performance. In kinship with nature and history, my personal inquiry positions me as a Canadian settler, a woman, privileged, neuroqueer, and more. *Knowing otherwise.* Yes, "knowing otherwise," is an argument Canadian philosopher Alexis Shotwell makes of the knowledge accrued from an aesthetic experience as "know[ing] the world otherwise through this sensuous knowledge."[4] In kinship with somas existing in lands where I dance my body shapes ecosomatic experiences of the lived encounters with other species. This approach to enworlding interlaces stories of a complex dancing pair: soma-me and soma-land.

Always before entering historical land and territories, I do my homework. I may be intuitively called to a place, enjoying a path of least resistance to a site that will sing my soul alive. What follows the initial impulse leads to a thorough introduction to the land, includes honoring local Indigenous protocols of place, being informed of the colonial municipal laws and regulations, keeping my body safe while on site, asking permission from the land before entering the place of performance-calling, not overextending my stay, and not encroaching on places of worship or reverence.

Rockcliffe Park is the site where, in 1613, the geographer to King Henri IV, Samuel de Champlain, traveled the land. Accompanied by Huron-Wendat paddlers, the canoes journeying up the Ottawa River were tasked with locating an alternate waterway to the Indian Sea and its profitable spice resources. The site is first and foremost the ancestral home to an entanglement of Indigenous settlements—Nishnaabeg and Kanien'kehá:ka Peoples—and to many transiting visitors *en route* to other thriving settlements on Turtle Island. Oral history tells of a human presence as far as the Paleo-Indian Period (approximately 8500 B.C.E.). *Akikodjiwan* and *Akikpautik* (Pipe Bowl Falls) are mighty waterfalls just upstream from Rockcliffe Park. "Through colonization *Akikpautik* was eventually dammed."[5] Water rose, thereby defacing the waterscape. Some say, the waterfall remembers her original sound. Thousands of years later until the 1990s, here, is also the story of soft wood timber floating their way to the pulp mill located in Thurso Q.C., on the Ottawa River about 45 km east of Gatineau.

On the day of the preparatory work, I come as I am and yet, a little nervous in committing to attune to this here/now land. What/who will I attune to? Where in my body will I hear the stories of the land? What act of apprenticing will the land task me with? In order to release some of the tension, and in a preparatory grounding posture, I remind myself that through layers of interlaces, our somas have been dancing with each other for a long time already.

As soon as get out of my car at the parking lot, I feel called to a place. The caller comes as a bark-to-skin tactile personae, a ghostly tree-finger that taps my left shoulder blade; *hello you, come over here!* I recognize the interspecies skeletal poke from an earlier ecoperformance with the tree that saw me born in Montreal. *Ah*, I hear myself say, *the apprenticing begins here/now*. I had planned to bring my hiking boots but have forgotten. I take note of the broken bottles and other human-caused litter and make note to bring gloves and garbage bags on my follow-up visits. A few Google snapshots identify the flora for me, indicating that the foliage is of non-poisonous species. All is set for deeper grounding now. My breathing slows. My nervous system calms. I tune in much more clearly to ecosomatic senses of existing, spatio-temporal-kinetic awareness, attuning to my own and the environmental somas, and an empathy to stories already existing. *Ok.* As I will come to the end of the downhill path and stand on the edge of a cliff, I will experience a series of instinctive impulses. In the runners-up for first impulses in performance-related choices, we have: (1) throwing myself off a cliff and into the water (a common impulse in cliff and water callings); (2) flying into the open sky (another common but misleading impulse in cliff-hanger encounters); (3) precariously treading the side of the cliff (of the many unsafe somatic callings eroded rocks engaged in millennial-old storytelling call for); (4) and finally, my honed performance training sense kicks in warning me of the dangers of accepting a no-return ticket to the eternal light. This meeting of somas in storytelling is indeed a deliciously addictive liminal line between depatterning somatic amnesia and repatterning ecosomatic senses. Thus, at the end of the pathway, I will have reached a calibrating point where free-bodied somas acknowledge bodied-soma's limitation. That is what I come here to seek; this enchanted kin's waltz.

But for now, let's begin at the start of the path when/where we begin to dance together—the land and I—to co-create. Camera in hand, I begin documenting my dancing down the path leading to the open riverscape (*see* Figure 13.2).

During the preparatory work, I deviate from the nearby Pinesi canoe portage trails where I had expected to dance. Instead, the place-calling pulls me to the cliff edging off Ottawa River. As usual, the somas of this place are quite windy and even more so as the daylight will draw near its end. Down the path I go, stepping over a fallen tree and pausing to feel its smooth body where many hikers have pressed their hands for balance (*see* Figure 13.2). The raised foot feels cool air on its sole, while the other

FIGURE 13.2   Bellerose's stepping over a tree on the downhill path, a hand pressed down for balance, Rockcliffe Park, Ottawa, Ontario, Canada (2022). Self-portrait, courtesy of the author.

foot, pressed against the gritty dirt, feels the discomfort of supporting my weight (*see* Figure 13.2). I am in no hurry. For half a day, my human body integrates encounters with the shadows and streaks of light through the canopy of a few of the Ottawa area native trees: oak, balsam willow, and fir. I take in the scent of river water, feel the slick slippery dark granite slabs, prickly wild grass, tufts of yarrow, and moss colonies, the smooth-polished sandstone ridges (*see* Figure 13.1), and I marvel at a spine-like driftwood shape (*see* Figure 13.3). As the motion of time sets in, my senses pick up atmospheric nuances: the smell of fish decay mixed with organic debris finding their way into the main watershed and the gentle clip-clopping sound shallow waves make as they meet the sandy shore. At one long and slow moment in time, this shore was a glacial ice sheet. When it retreated, the Ottawa River valley emerged. Sweet and salt water pooled together into the St. Lawrence River, and into the gaping mouth of the Atlantic Ocean. Right here and now, it is like a family portrait. Me and the land, we stand still in-between times for as long as I can to commemorate our union, today. For what seems like a time stretching across the ages, I delight in this earthy realm (*see* Figure 13.3). In between here and a dream state, I can make out a white cloth the size of a ship's sail rising out of the flow and

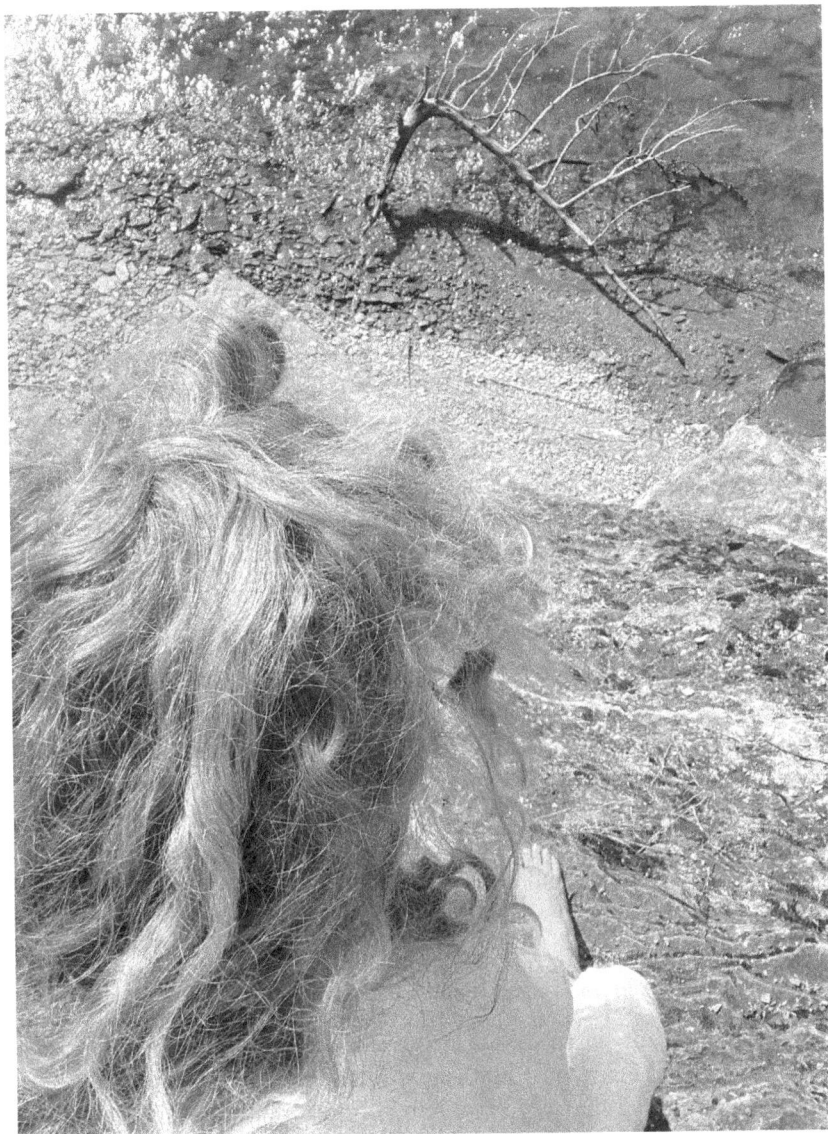

FIGURE 13.3 Bellerose's contemplative communion with the river water and a spine-like driftwood shape, Rockcliffe Park, Ottawa, Ontario, Canada (2022). Self-portrait, courtesy of the author.

heavy with water, sucking at the dark granite slabs lining the river shore. In the process of attending to performative listening to the land and to my somatic drives, I also attend to being playful. I allow observing without judgment. And because of this, ecosomatic synchronicities between the present and dream states emerge.

One afternoon of playing with the land reaped plentiful material for a later ecoperformance session. Not overstaying my welcome, I bid farewell to my kin.

I have selected three images to accompany this short chapter: my hand and my foot (self-portrait) on the path down the edge of Ottawa River's cliff because much of the experience of apprenticing with the land this time is one of grounding. This documented preparatory work by way of a somadance approach to engaging with cross-species kin interlaces my roots with the storytelling of the land as shared stories walk me into my own present time (*see* Figure 13.1). And I have selected images in a *Land to Water* framework; these amazingly evocative images are snippets of visceral encounters with elemental somas of earth, water, and wind (*see* Figures 13.2 and 13.3).[6] The last one is an image of the powerful contemplative state an ecosomatic aliveness can procure.

I share these stories because they are foundational to my repatterning a relationship with enchanted kin. By remaining open and curious, change will take place in bodies from where and how storylines live anew.

Seven Somadance Practices

The seven activities suggested below were developed through successful and less effective somadance attempts. They serve as an approach or template for how to apprentice with land in an enchanted kinship. The task is to unwind and set patterns of expectation in dynamics of relationship. In part, this can be done by suspending judgment while entering the realm of the wondrous. Note that by "land," I mean all elements, all species in the land's family. See what land means for you! Trust in finding threads which lead to unthought questions.

See Yourself Safely in and out of the Place of Calling

You do not owe anyone the promise to bend backward like a blade of grass. Negotiate with the land just like you would if you had to find the proper translation channel during a United Nations multilingual conference. Remember that somadance ecoperformance is an interspecies conversation in movement; everyone gains in listening attentively to the story so as to respond authentically.

Enworld(ing) Ecosomatic Continuity by Honoring the Sovereignty of Different Species

I was introduced to ecosomatic work through the writings of ecofeminist philosopher, Val Plumwood, and her articulation of continuation across differences, which she developed in the 1970s.[7] Honoring the sovereignty of different species is my baseline approach. Honoring the sovereignty of different species is a practice that is also present in Indigenous cultures. Canadian lawyer, John Kegedonce Borrows, speaks of a historical marker woven into the "Wampum Belt of the 1764 Niagara Treaty which was negotiated between the British and the Haudenosaunee Nations as Canada's baseline for a co-habitation between Indigenous and non-Indigenous Peoples."[8] This baseline negotiation has been swept under the colonial rug. Engage in reviving an ecosomatic continuity by honoring the sovereignty of different species.

Practice Performative Listening by Trusting Your Felt and Imagined Senses

Go to a place of interest to you. Dance the encounter, not to prove imagination, but to be imaginative. Accept and learn from the result. It can unwind later in your life in unexpected ways. Stó:lō Coast Salish storyteller Lee Maracle directed me to "look inside" when I told her about my worries about not speaking River (nor the river, English).[9] I was fortunate to be mentored by her and have subsequently been spending years practicing listening inside of me. Somatics assists in finding operative and experiential inner talks, and so does phenomenological descriptive writing in translating sub-verbal (or pre-verbal) phenomena.

Homework Time! Land Is Sovereign to Itself, but Territories Are Not Ahistorical

Be humble, respectful, educated, and genuine in your desire to dance [with] land. It may not be possible today. Nations, states, and *man*made borders have made getting to know each other through movement, a practice of "radical tenderness."[10] Use discernment all the way through to selecting photographs. Do the best you can to be true to your apprenticing with the land. It is a practice with a steep learning curve harboring serious responsibilities. Guilt, martyrdom, self-flagellation, and colonial reconciliation are dominant institutional dogmas that will not serve you at the table of enchanted kin. Find your own liminal tightrope.

Introduce Yourself and Your Art to the Land

Let her know of your intention. I often ask to learn about my roots and lineage.

Observe the Speed (Time-Space) and the Kinetic Lines of the Land

The more observant you become, the more synchronicities you will observe. Kinetic lines will appear even as you do not *see* through your eyes, you will *see* them through your skin. These observations will move not only your mind but also alter your nervous system all the way to shape-shifting your body into archetypal figures of a story being told.

Have Fun!

Get acquainted with the land's various personalities and see how each has its own sense of humor. The wind is a trickster and will strive to continuously startle you; the water pulls you in its embrace; but remember, you are not a fish and you will need to breathe oxygen; winter fir loves to drop snow piles on your head; each and every snow crystal demands you notice their individual song—if you don't, they will consort to trap you, make you fall head over heels, bury you alive where/when they will, then sing and laugh at you out loud. Laugh too! Experience for yourself how animated the land becomes.

Notes

1 Démarche (French language, f. noun) Dé = unwind; marche = walk. Unwinding is part of somatics' healing-attuning method from depatterning to repatterning processes. Démarche is translated to English as an approach, a manner, a way of being, and a methodology. Here, I use the word compound of creative *démarche* to highlight a creative deambulatory somatic on-impulse approach.

2 France's King Louis XIV acceded to the throne in 1661. A first contingent of the *filles du Roy* (King's daughters) was sent in 1663 in order to increase the population of the colony. Two years later, the Carignan-Salières Regiment disembarked in New France to ensure the safety of the colony and, more specifically, to deal with the Iroquois (Haudenosaunee) "threat." *See* Louis Richer, "Carignan-Salières 1665–2015: le Québec se Souvient," *Histoire Québec* 21, no. 2 (2015). www.eru dit.org/fr/revues/hq/2015-v21-n2-hq02293/79979ac/.

3 Canadian philosopher Alexis Shotwell writes about the complexities of shame specifically addressing whiteness and makes a case for "why white people should embrace shame about whiteness." *See* Alexis Shotwell, "Is It White Shame?," *Alexis Shotwell*, alexisshotwell.com, August 31, 2017. https://alexiss hotwell.com/2017/08/13/white-shame/.

4 Alexis Shotwell, *Knowing Otherwise: Race, Gender, and Implicit Understanding* (University Park: Pennsylvania State University Press, 2011), 49.

5 Lynn Gehl, "Akikodjiwan: The Destruction of Canada's Heart of Reconciliation," *Watershed Sentinel*, March 8, 2018. https://watershedsentinel. ca/articles/akikodjiwan/.

6 Sondra Fraleigh, *Land to Water Yoga: Shin Somatics Moving Way* (New York and Bloomington: iUniverse, Inc., 2009).

7 Val Plumwood, *Feminism and the Mastery of Nature* (London and New York: Routledge, 1993).

8 John Kegedonce Borrows, "Wampum at Niagara: The Royal Proclamation, Canadian Legal History, and Self-Government," in *Aboriginal and Treaty Rights in Canada: Essays on Law, Equality and Respect for Difference*, ed. Michael Asch (Vancouver: UBC Press, 1997), 169.

9 Stó:lō is the Halqemeylem word for "river." Lee Maracle, *Memory Serves: Oratories* (Edmonton: NeWest Press, 2015).

10 Dani D'Emilia and Daniel B. Coleman, "Ternura Radical es…. Manifiesto Vivo," *Hysteria* 37 (2015), https://hysteria.mx/ternura-radical-es-manifiesto-vivo-por-dani-demilia-y-daniel-b-chavez/. Available in translation at https://danidemilia.com/radical-tenderness/.

Bibliography

Borrows, John Kegedonce. "Wampum at Niagara: The Royal Proclamation, Canadian Legal History, and Self-Government." In *Aboriginal and Treaty Rights in Canada: Essays on Law, Equality and Respect for Difference*, edited by Michael Asch, 155–172. Vancouver: UBC Press, 1997.

D'Emilia, Dani, and Daniel B. Coleman. "Ternura Radical es…. Manifiesto Vivo." *Hysteria* 37 (2015). https://hysteria.mx/ternura-radical-es-manifiesto-vivo-por-dani-demilia-y-daniel-b-chavez/.

Fraleigh, Sondra. *Land to Water Yoga: Shin Somatics Moving Way*. New York; Bloomington: iUniverse, Inc., 2009.

Gehl, Lynn. "Akikodjiwan: The Destruction of Canada's Heart of Reconciliation." *Watershed Sentinel*, March 8, 2018. https://watershedsentinel.ca/articles/akikodjiwan/.

Maracle, Lee. *Memory Serves: Oratories*. Edmonton: NeWest Press, 2015.

Plumwood, Val. *Feminism and the Mastery of Nature*. London; New York: Routledge, 1993.

Richer, Louis. "Carignan-Salières 1665–2015: le Québec se Souvient." *Histoire Québec* 21, no. 2 (2015). www.erudit.org/fr/revues/hq/2015-v21-n2-hq02293/79979ac/.

Shotwell, Alexis. *Knowing Otherwise: Race, Gender, and Implicit Understanding*. University Park: Pennsylvania State University Press, 2011.

———. "Is It White Shame?" *Alexis Shotwell. alexisshotwell.com*, August 31, 2017. https://alexisshotwell.com/2017/08/13/white-shame/.

14

FEEL THE CARBON UNDER YOUR FOOTPRINT

Indigenous Approaches to Grounding

Nathalie Guillaume

Kūkaniloko Birthstones State Monument, Oʻahu, Hawaii: 21.5048° N, 158.0364° W

Port-au-Prince, Republic of Haiti: 18.5358° N, 72.3331° W

My work in healing and somatics is informed by my personal experience of Haitian cultural knowledge, my work as a Doctor of Acupuncture and Oriental Medicine, and my background in dance and somatic movement education. More recently, it is also informed by Indigenous knowledges particular to Hawaii and my physical experiences at some of its sacred sites.[1] This set of practice pages offers a series of simple movement patterns that can be utilized in order to explore and strengthen the connection between our feet and the ground and to deepen our sense of belonging, no matter where we are. Originally inspired by a visit to the sacred Kūkaniloko Birthstones State Monument in Oʻahu, Hawaii in 2017, this work can be done in *any* outdoor location; you can begin wherever you are.[2] Most simply, we will use a movement pattern based on the spirals of a labyrinth in order to embody a purposeful journey toward our own center and back out into the world. Through developing a stronger connection with place through our feet, we learn to root ourselves from within and thus can find ways of belonging in every place we step and with every step we take.

When we assert every footprint we lay with confidence, we become in tune with our surroundings—we are rooted with/in nature. With each step I take, I find I am deeply rooted to my homeland, Haiti, where practices that connect us and bind us to the land still take place. In the Haitian folk tradition, the Kreyòl proverb, *"twa fèy, twa rasin, jete bliye ranmase sonje,"* roughly translates to "any three leaves and three roots from your garden can heal you." From the moment your mother plants her/your placenta in the soil, all pertaining greenery is thought to have stem cells from your amniotic source—thus any three leaves or three roots from *your*

DOI: 10.4324/9781003390985-18

garden are *your* pharmacy. This notion of rootedness and connection has caused me to consider how I can belong at that deep level of healing and connection when I'm away from my homeland.

The first time I visited the Kūkaniloko Birthstones State Monument I could feel the power of the historical and sacred place, which consists of 180 large stones over 511 acres.[3] Some of the stones are smooth, others are carved with long grooves or bowl shapes. All are spectacular and contain tremendous energy; the Birthstones have a quality that is somatically reminiscent of a mother's body carrying her children. For seven centuries, the site was used as a birthplace of many of the Island's High Chiefs. The stones constitute a central component in the ritual practice of what can be termed the Indigenous religion of the Hawaiian land involving a belief in the bodily incorporation of primordial feminine energy for the creation and the channeling of information from the divine world to the material world, often for ceremonial therapeutic purposes. Some suggest that the site may have served an astronomical function given its place as the geographic *piko*, or navel, of O'ahu. Kūkaniloko also means "to anchor the cry from within"—another metaphor of birthing and connection to place.[4]

One of the traditional songs about Kūkaniloko recounts how "*O Kapawa, o ke alii o Waialua I hanau i Kūkaniloko,*" or "Kapawa, the chief of Waialua, was born at Kūkaniloko." It goes on to say that the placenta from this birth was placed at Lihue ("*O Lihue ke ewe*"), the navel cord at Kaala ("*O Kaala ka piko*").[5] This connects so deeply with the Haitian practice of burying the placenta and the subsequent idea that one's own healing garden will grow from that site. The fact that this practice is found in both cultures is very visceral to me.... Perhaps one reason I make connections across cultures so easily is partially attributed to the fact that I am a Black woman of African heritage with a natural inclination for Asian philosophy. I like to call this hybrid perspective *AfrikAsia*, a blend of two continents which share ancient wisdom through their healing and creative practices. While I write and often live in English, I still think and feel in French and Kreyòl, a multiplicity that shows up in my definition of global indigeneity.

I consider all Indigenous people to be united through painful histories of racial oppression, land occupation, or cultural appropriation in the name of political, religious, or intellectual correctness. It might seem unusual for a Haitian—a person of African descent—to identify as or with Indigenous populations. However, there is a background for this in Haiti. When the father of our nation, Emperor Jean-Jacques Dessalines, renamed the former French slave colony of Saint Domingue, he reclaimed the Indigenous Taino name for the island, "Haiti," which means mountainous land.[6] And the first pro-African cultural movement in Haiti was called *Indigénisme*, or "indigenism."[7] It was the first cultural movement to celebrate Vodou and other African aspects of Haitian culture, which had always been defiled while the French influence was favored.

As a Haitian woman, I feel deeply connected to the Indigenous culture of Hawaii, which weaves a rich thread to my disciplinary roots in Traditional Chinese Medicine and to ancient wisdom keepers on all sides. As a result, I have developed my "Aloha Oshun" program, which is a blend of Haitian and Hawaiian ways of knowing and healing established in my primary work as an acupuncturist. The name of this personalized eco-retreat is a mix of *Aloha*, the Hawaiian word for greeting with the breath, and *Oshun*, the Yoruban goddess of sweet waters, fertility, and love.

In my medical training, I learned that in 1950s China, the acupuncturist became known as the "barefoot doctor"—one who walks through the fields bringing healing to rural communities.[8] For me, the barefoot doctor is a kind of *Kouzen Zaka*, an agricultural *lwa* (deity) in the Haitian Vodou pantheon. He lives on and with the land and is in touch with and grounded in place—a bit of a trickster in a straw working hat, he is a master herbalist and works very closely with the cycles of the moon. That is why we celebrate harvest during his holiday season in the spring.

Metaphorically relating with the moon cycle of a fertile woman and the harvest of her pregnancy which is birth, let me return to my experience at the Birthstones. As my feet compressed the carbon of the red clay soil, I moved in a circular pattern near the stones. As I moved, I allowed my imagination and my breath to be carried back into my mother's womb, which can quickly transform into a portal for all the maternal forces that have carried me into and through this world. When I do this, I tap into my ancestral energy, my vital force (*Qi*) activates, and my congenital intelligence circulates between physical and emotional states as I let go of my thoughts and simply allow my womb to feel through my feet. I experience my sacral chakra getting warm... the movement starts slowly from the internal aspect of my anterior leg to reach my pelvis where I allow a slow swerve to invite me to swing side to side with the soft breeze (*see* Figures 14.1 and 14.2).

From the moment we take our first steps as toddlers, we are constantly exploring and investigating the world around us through our senses, more specifically our tactile signals. As we explore a somatic mindset and approach how we stand and walk, we heighten our awareness of the ground on which we stand. We learn how to firmly plant our feet into the carbon stored in the Earth's rocks and sediments connected with the planet's roots; this is how we evolve into moving with greater ease and strength to improve flexibility and enhance structural integration while negotiating gravity.

FIGURE 14.1   Photograph of the author grounding into the earth as she reaches through the arms and hands, from the series, "Vodou Alchemy," which features original music and dance in multiple sacred sites, including O'ahu (2017). Courtesy of the photographer, Becky Yee.

Grounding Practices: Visualizing and Walking a Labyrinth

I invite you to dive into a somatic experience through Qigong, yoga, meditation, and self-care. In what follows, I offer simple movement patterns to help you firmly plant and ground your feet to the earth. These patterns flow through the acupuncture meridian pathways, each belonging to one set of cardinal directions that assist in magnetizing organ frequency in walking. Geographical instructions for visualizing natural elements, emotional play, and release related to these cardinal pathways are also included. Create your labyrinthian walk any place that offers an expanse of space to walk and trace through—grass, sand, dirt, or any terrain that invites you to become better acquainted with yourself through the ground. Read these instructions through a few times, then just flow into your labyrinth with what you remember of the four directions. If you repeat this process a few times, you will remember more each time and learn to improvise your personal labyrinth more richly.

Turn to the North: Water

In Chinese medicine, North is the location of the *water element*: that which soaks and descends. First, stand up slowly, and feel the soles of your feet on the ground. The sole of the foot is where the Kidney meridian begins along with *the water element of life*.

FIGURE 14.2   Photograph of the author moving in a circular pattern, from the series, "Vodou Alchemy," (2017). Courtesy of the photographer, Becky Yee.

As you walk North, bring your attention to your legs and the soles of your feet, allowing the deliberate act of walking to align your breathing with your steps. Do this by planting your feet into the ground softly, first reaching the heel forward, then rolling your weight toward the toes while visualizing the suffering of your mother giving birth to you. Birth and the yielding of the mother is associated with earth and water.

Begin walking in a circle spiraling inward toward the center in smaller and smaller circles. This creates a spiraling labyrinth of a life journey. Please pay attention to your feet as they come in contact with the ground's surface. What are you walking on? Just notice—is it hard? Is it soft? Are you on dirt, stone, grass, concrete, or something else? Pay attention to the pace of your walk. Notice the shift of weight in each foot as you move.

Explore spiraling in and out of your circular labyrinth for a few moments. By being playful with your pace and your feet, you will discover that one way leads you into the center of the circle, and—reversed—the same path brings you out again. This improvised movement represents coming close to and moving away from a personal, transformative center—a place of intense knowing and change.

Warm to the South: Fire

In Chinese medicine, South is the location of the *fire element*: that which blazes and ascends. Walking a labyrinth can release stress, tension, frustration, pain, and hurt. As you face South in your circular path, stop for a moment, and let your mind be quiet; feel your body begin to relax. Your individual DNA activates ancient memories of your birth through your mother giving you the fire of life after her water breaks.

Known as the Emperor, the Heart is the main organ within the fire element. It is responsible for circulating blood throughout the body and thus nourishing all of the internal organs. As such, the Heart is the organ that allows oxygen to flow to the muscles and organs and will enable you to feel warmth, empathy, and compassion. Ask yourself a few questions: are you trying to break specific patterns in your life? Do you sometimes feel like you are drowning or becoming overwhelmed? Where do you feel such concerns in your body?

Breathe away any troubling thoughts and gather fresh energy from the ground by slowly bending down, bringing your hands toward your feet to connect with your roots; then gently bring them up following along the inside and outside your legs in any way that comes to you. Don't think about it too much. Relating your arms and legs helps your feet feel the warmth of the ground getting sweet and thick.

Pause in your walking to visualize a steamboat with coal burning in your Lower Dantian about 4 inches under your umbilicus. Imagine your motor running. Take four to five deep breaths as your engine traverses the sea of your spinal fluid and bursts into an ocean of emotion at your heart center. Now bring your hands up to your chest center and connect with your heart space. Wait to feel your heartbeat; then start to spread your fingers a bit.

Release any excess emotions with light tapping along the Kidney and Heart points around your sternum, breastbone, and under your clavicle. Here you can apply a variation of the Emotional Freedom Technique (EFT), which stimulates acupressure points by pressuring, tapping, or rubbing these points while focusing on situations that represent personal fear or trauma.[9] Feel the roar of waves brushing on the shore. Allow your exhalations to be long and deep. Through your kidneys, release fear, and receive love through the heart.

Face the East: Wood

Prepare to face the East. In Chinese medicine, East is the location of the *wood element*: that which can be bent and straightened. To prepare for this direction, you may want to stand still for a few breaths to integrate the North and South before continuing on your path.

The wood element is a place of bending and reflection where many people experience depression, anxiety, and anger due to repression of their Liver Qi (energy). Pause here facing eastward and stay as long as you like with any feelings that may come up.

Visualize deeper roots as you continue to walk this labyrinth practice, then reach toward heaven like a tall tree! Slightly bring one foot up to embody a yoga tree pose. Yoga *Vrksasana* (tree pose) teaches you to simultaneously press down and feel rooted as you reach tall like the branches of a mighty tree. In this pose, you find a strong sense of ground through the strength of your standing leg. Bringing the sole of your opposite foot to your shin or thigh challenges balance. Continuously engage your ankles, legs, and core strength, and notice what tiny movements your body might make to help you stay balanced. Your tree can be as big or as small as you would like. It can be a robust, still tree or like soft grass moving with the wind.

Open your arms wide and receive environmental frequency from the tip of your fingers into your heart chambers. Slowly feel the warmth in your chest, following a drum-like rhythm undulating your thoracic spine and emancipating your emotions. What feelings are screaming to come out? Do you want to sing, laugh, or cry?

Most of us carry our frustrations in the upper body, and a good way to release them is by making side-to-side motions to massage the liver. Make wing-like motions with your arms and stretch wide from your shoulder blades. Allow the tension from your rotator cuffs to start dissipating. Loosen the weight of the world pressing onto your neck as you rest your shoulders downward.

Slowly raise both your hands to chest level, palms facing out, and ground yourself for a few breaths at the center. At the last breath, move both hands to the left and allow your head to follow your fingers slowly. Repeat this to the right. Keep your core strong and your stance firm. Release any excess thoughts while visualizing the many acupuncture points of your palm, and return to the center when ready.

Remember that your shoulders and hips are symmetrically connected, and light twisting motions of the hips and shoulders in opposition not only massage the internal organs but also align the spine, adjust your posture, and promote effortless balance. Give attention to your feet as you twist and press your big toe into the ground where the Liver meridian begins. Allowing the Liver Qi to circulate freely stimulates blood flow in your lower extremities. Point and flex your feet as you finalize this section and prepare to step in a new direction.

Walk toward the West: Metal

In Chinese medicine, the West is the location of the *metal element*: that which can be molded and changed. When you circle toward the West at a comfortable pace in your labyrinth, place your attention on what you see. Notice the color, shape, and texture of objects. Notice also your reaction to what you see. For example, a beautiful flower may elicit a deep sense of satisfaction, or the sight of a slug may give you a little shiver. Rather than passing judgment on what you see, try to be impartial. Remind yourself you are just visiting, not judging. When thoughts and words come into your mind, release them, bringing your focus back to seeing the world around you. Once we remove shame and blame, we finally allow ourselves to express our genuine emotions.

The metal element reminds us that nostalgia and grief pouring out of our lungs come from longing for things, places, or people we miss—yet relive instantly once we look for them.

During this section, pull your gaze onto the horizon and use your sense of inner sight and body awareness to experience the dance happening in your feet. Absorb the colors of green, red, and brown that you see around you. Inhale the aromas of the trees or flowers, incense or candles around you, and with warmth, remember where you are.

Embrace yourself, and as you separate your hands, start to form a pattern with your fingers called the "Sword Finger," a hand and finger position based on therapeutic Qigong that is quite powerful in emitting energies. Simply stretch out your middle finger and your index finger while pressing them slightly together. Then curl in your ring finger and your little finger and cover their nails with your thumb. Your thumb and these two fingers form a "ring," while the other two are stretched out as one finger. The index finger and middle finger build the "blade" of the sword, and through the "tip of the blade," emit the energy. This position can clear unwanted energies by shooting them out through the tips of the Sword Fingers. The emitted flow of energy is like a laser beam coming out of the Sword Fingers, which is used to "cut" blockages and energy jams that could be blocking your respiratory tract.

The Lung Meridian runs throughout the thumb, and when activated, your inner weapon can help soothe grief and cut out any unwanted toxic thoughts. Make sharp movements with your hands coming from your body's center toward the outside, releasing excess emotions. Now that you have completed this energy-clearing circuit, you can return to your center with a fresh perspective and clear mind.

Finding Center and Ground

Find a way to rest your stance and breathe into the center of your labyrinth wherever you are. Place the palms of your hands over your belly and pause for a while. In Chinese medicine, the center is the location of the *earth element*, which allows for sowing and harvesting. The center of the body is where the spleen and stomach are the most active. You can digest difficulties more easily and find the metabolic energy for breaking down nutrition into action. Stimulating the earth center makes you hungry for life and evokes a sense of taste for a more refined palette.

When we ground ourselves, we are calming or slowing down our emotions and getting more in touch with internal and external worlds. Allow this transformation and transportation of the final step of the practice to ground your energy fields, making you more confident in your decisions. Grounding as centering can be helpful when you feel either unbalanced or nervous. Being grounded also means being more mindful of nature and your environment—considering your own "carbon footprint."

Closing to Begin Again

Each labyrinth experience is different. You may not feel much, or unexpected emotions and memories may surface. Regardless, listen to your whole-body consciousness, and take time to breathe. If you are excessively troubled by thoughts or emotions that have arisen, consider journaling, listening to your favorite song, shaking your body, tapping your limbs with the palm of your hands, making sounds of release, or simply lifting your hands to the sky and looking up. If it is convenient for you to sit and rest for a while, remember that the sitting bones are another set of feet, and you can also practice center, ground, and breath in sitting. When you are ready, reverse your path in the labyrinth to find your way to the point where you began.

Notes

1 I acquired my dual acupuncture license in Hawaii and have been working with Hawaiian *Kumus* (teachers) on several of the islands, including Maui, Kauai, Big Island, and O'ahu, since 2016 when I accepted a faculty position at the World Medicine Institute. Without the blessing of a *Kumu*, one cannot enter certain dimensions of healing work on the islands. I highly value the blessing of the land and would not misuse it for my own benefit.

2 My visit to the site was in 2017, when public access was more available. The Park has been closed to public access, including a prohibition on the use of drones, since approximately 2019. This chapter neither suggests nor condones making a pilgrimage to the Kūkaniloko Birthstones State Monument.

3 Casey says that emotions surround us, which is why we can feel things in places. Edward S. Casey, *Turning Emotion Inside Out: Affective Life beyond the Subject* (Evanston: Northwestern University Press, 2022), 7.

4 Office of Hawaiian Affairs Research Division, *Kūkaniloko (Information Sheet, September 2013)* (Office of Hawaiian Affairs Research Division and Land, Culture, and History Section, 2013). https://19of32x2yl33s8o4xza0gf14-wpengine.netdna-ssl.com/wp-content/uploads/Attachment-6-OHA-Kukaniloko-Information-Sheet.pdf.

5 Ibid.

6 Laurent Dubois, *Avengers of the New World: The Story of the Haitian Revolution* (Cambridge: Harvard University Press, 2004), 299.

7 Jean Price-Mars, *Ainsi Parla L'oncle/So Spoke the Uncle*, trans. Magdaline W. Shannon (Washington: Three Continents Press, 1928/1983).

8 *See* the government training guide developed under Mao Zedong: The Revolutionary Health Committee of Hunan Province, *A Barefoot Doctor's Manual*, revised and enlarged ed. (London: Routledge and Kegan Paul, 1978). Barefoot doctors reached almost every village and were the backbone of the rural cooperative medical service, which was operated by the peasants on a collective and mutual aid basis. *See* Chi Wen, "Barefoot Doctors in China," *Nursing Digest* 3, no. 1 (1975).

9 Morgan Clond, "Emotional Freedom Techniques for Anxiety: A Systematic Review with Meta-analysis," *The Journal of Nervous and Mental Disease* 204, no. 5 (2016). https://doi.org/10.1097/NMD.0000000000000483.

Bibliography

Casey, Edward S. *Turning Emotion Inside Out: Affective Life beyond the Subject.* Evanston: Northwestern University Press, 2022.

Clond, Morgan. "Emotional Freedom Techniques for Anxiety: A Systematic Review with Meta-analysis." *The Journal of Nervous and Mental Disease* 204, no. 5 (2016): 388–395. https://doi.org/10.1097/NMD.0000000000000483.

Dubois, Laurent. *Avengers of the New World: The Story of the Haitian Revolution.* Cambridge: Harvard University Press, 2004.

Office of Hawaiian Affairs Research Division. *Kūkaniloko (Information Sheet, September 2013)*. Office of Hawaiian Affairs Research Division and Land, Culture, and History Section, 2013.

Price-Mars, Jean. *Ainsi Parla L'oncle/So Spoke the Uncle.* Translated by Magdaline W. Shannon. Washington: Three Continents Press, 1928/1983.

The Revolutionary Health Committee of Hunan Province. *A Barefoot Doctor's Manual.* Revised and enlarged ed. London: Routledge and Kegan Paul, 1978.

Wen, Chi. "Barefoot Doctors in China." *Nursing Digest* 3, no. 1 (1975): 26–28.

PART IV

Place, Plasma, Pluriverse, Potato

15

MY PLACE IS A CHIASMATIC DANCE

Glen A. Mazis

Marietta, Pennsylvania, U.S.A.: 40.0559° N, 76.5517° W

My Place Is Our Place

My place by the wide ever-flowing Susquehanna is not *my place* in actuality but is rather "our" place—a moving and transforming ensemble performance of myriad beings. Like all good performances, this place will become something newly experienced at each moment in the sense that Bergson enunciated when he noted "the continuous creation of unforeseeable novelty which seems to be going on in the universe."[1] By this, Bergson meant that what seems like it is the same as it was in the past can never really be the same if it is fully perceived as having just appeared with a freshness, an openness, and within the spontaneity of the next dance step of the performance. The newness and togetherness of each moment of appearance will only be sensed as such for those with artful sensibilities, like my two dogs, or me after a period of quiet meditation, or in other words, for those who take in the voluminosity of presence through a sensitive resonance. This coming together as a place is a *felt* presence that can wither by being distanced or it can be joined in a playful dance. It can exist and be taken in as a performative becoming or else it can be circumscribed on a Cartesian grid of isolated and permanent points in an indifferent space that we call a map. The first way of access yields a kind of symphony, dance, and/or vision, as a found installation created by a spontaneous coming together of beings, histories, interconnections, and depths that are so much beyond counting or even imagining. It is a chiasmatic dance in the sense that a chiasm is the occurrence of two beings that remain two separate beings and yet also are one simultaneously (in

DOI: 10.4324/9781003390985-20

defiance of binary logic and dualistic thought). For example, a symphony has myriad voices or tones of different instruments coming together in one melody. The "our" of this phrase "our place" in the case of where I dwell means it is an ensemble of flowing water, landing herons, stubbly grasses, railroad tracks, tall black walnut trees, climbing English ivy vines, the large rock behind the house called "Serenity Rock," the cardinals streaking red in the dull late winter landscape, the frogs asleep in the deep mud of the pond out back, the woodpile lining the fence, the cars and bikes that pass by on the street out front, the lazy smoke curling into the sky from the blazing woodstove and emanating a fragrance of sweetly burning hardwoods, and so on indefinitely for an innumerable cast of participants. These participants move within differing figures of "enlacement," to use one of Merleau-Ponty's terms for this phenomenon, "*entrelacs.*"[2]

This is the place of my embodiment, the surround of ongoing sense of innumerable and thickly enmeshed qualities that speak in varied voices, most of them in the voices of silence, but also in rhythmic melodies of sound. This embodied place is described from the phenomenological point of view of Merleau-Ponty. In this chapter, this phenomenological understanding of place will be used as helpful in order to free place from its popularly understood Cartesian fixity. The Cartesian access to this ensemble that fixes landscapes and communities of interacting beings within borders and as set locations on the map is a way of intervening with the teeming life of particular beings to manipulate what is here in such a way that the inexhaustible array of qualities that have become entangled in this coming together can be overlooked for the purpose of achieving a project at hand, such as getting my automobile moving through space in perfect indifference to the specific qualities of the locations I am moving through by using my GPS to arrive at the designated point on the grid. This setting of borders and fixed locations is one that flies free of embodiment and abstracts our experience from a fictional absolute surveying perspective, what Merleau-Ponty calls the "high altitude" [*pensée de survol*] apprehension.[3]

I don't have a body whose sense of the world can be put aside without putting aside my primary way of access to the real. I am a body. The world is an outflow and inflow inseparable from that body, but even that is incorrect. The world and my body are an ongoing emergence of which both are vital and necessary aspects where the supposed outside is the inside and vice versa. There is a circulation of sense that is suggested by Merleau-Ponty's image that "our own body is in the world as the heart is in the organism."[4] This body is a body of so many, within and without, but not really since there is no within and without of its boundaries, just open porous lines that are currents. This body is a flow of senses and energies that spiral in such a way that a space has opened up in a constant movement and transformation of which I am a part and that is what we call a place. I inhabit my body in a privileged sense that creates an asymmetry with the world in the sense that I am anchored in my body, but only in such a way that a constant tide pulls me beyond this anchorage and into the depths of the world only to be continuously returning to myself. It is a set of relations and relationships run through like streaks of certain colors or tones of certain melodies with the particular sense and style of all those beings woven into its choreography and location. It stands in each moment effulgently present like a face of one long ago recognized, as if I had been invited here before I came to be here, but I know this is because this sense of that past time has only come to be after living here for three decades.

This is what Merleau-Ponty calls the time of *institution* (*Urstiftung*), as a wild time, a time that doesn't fit on any grids or chronometers but is the way the present constantly comes forth in the novelty that Bergson recognized and felt our culture does not, in such a way that *the past that never was, now is*, in the present moment. It was as a latency, a coming to be, that truly was not there in the past as if some sort of shadowy half-present as the popular and misleading idea of possibility implies. Time truly newly emerges at each moment of ongoing transformation through the coming together of the forces of myriad beings. This is part of the constant movement as transformation that runs through the Earth and is the marrow of place and of the body's belonging to a place. Even if I have seen the tree outside the window or the river flowing past the house, I have never seen it in this way as I am seeing it right now. This Essay will explore this different sense of the body and the body as place in terms of Merleau-Ponty's very different rendering of embodiment and how all beings are internally related to others within which they form a sense with a deep background of latency. Internal relations are such that each being so related has its sense and identity inseparably from the other beings to which it is internally related versus the external relations of cause and effect that connects discrete and independent beings.

My Place as I Write This Essay in the Natural World

If what I have said so far has any truth to it, I can only describe my place from within the enmeshment of all that surrounds me in this place as perceived, felt, and imagined through my embodiment which is part of this emergence. As Merleau-Ponty says in one of his lecture courses:

> If flavor, odor, touches, sounds of water running, movements of trees, handling of boots, madeleines, cobblestones and ice cubes mediate authentic memory, it is because marginal, pre-objective, non-distant

perceptions, which cover consciousness rather than being objects for it, express a fleshy, body-world nexus.⁵

The perceived world and my body are a nexus. In addition, I can only describe it as how it is coming to be manifest at this present moment as I write this sentence or as Merleau-Ponty says,

> Through my perceptual field, with its spatial horizons, I am present to my surrounding, I co-exist with all the other landscapes which stretch out beyond it, and all these perspectives form a single temporal wave, one of the world's instants. Through my perceptual field with its temporal horizon I am present to my present [...]⁶

Given the constant transformation which is the primordial movement of the world and myself, my place is what it is now. However, Merleau-Ponty finishes this sentence by expanding the extent of the now by emphasizing what it means that time is a single temporal wave: "Through my perceptual field with its temporal horizon I am present to my present, to all the preceding past and to a future." It is not as though literally I exist as an absolute deity would in all times, but that things as they come forward hearken back in their sense and being to the vast horizon of existence that stretches from the deepest past to the furthest future. It is with these dimensions in mind that I can begin to describe my place.

An hour ago, there was a thundering swell of chirping birds flying above the trees, thousands and thousands of them, in some March migration, landing in the trees between this house and the banks of the Susquehanna a few hundred yards away. Their overwhelming cries, their crisscrossing of the gray sky contains their traced lines of flight, their weighing down the tree branches still fills them with presence, and the whole play of their coming still fills this place an hour later with echoes and reverberations. The trail of their flight extends over the years of the many migrations that pass over this river and this spot and contains a variety of voices, like the honking of the Canadian geese yesterday and opens a direction that heads both north and south and toward a future of such v-formations or hordes of swarming blackening of the sky of other types of birds. These movements and vocalizations travel within the periphery of my body still, setting up a sharpness, a tension, an on-edge awareness of imminent leaping into the sky, and their absence an hour later is not the normal openness above the trees but instead is an invisible but yet felt and seen in some sense trailing of their massive movement together that exerts a pull of this place to the southern shore of the river. Yet, the river flows with an urgency that has a different pace after the heavy rains west of here but also has a sense of being as old as thousands and thousands of years of flow as one of the oldest rivers on the planet. This river flow runs beyond its literal banks making its way through the house, of course not literally, but with a pulsation of driven force (as well as a meandering on other days) that creates vortices snaking through each room in this house, along the banks and neighboring forests, and even under the ground upon which the house rests. This is a river house in a river town.

The hill that stands high across the river and looks down on this place makes it seem as if the sky is concentrating on this stretch of land from one side of the hill, but the other side is also present as giving a strong shoulder to oncoming winds and storms, rebuffing them with a shrug and inherent strength. We feel its blocking out of threatening forces as if our house, our place, is embraced by a gentle force looking out for us peeking over the river, massive and strong. The railroad embankment looms right across the street as both buffer to the winds and torrents, and possibly flooding waters from the river, but also has embedded within its face toward us the sounds of the blaring train whistles and loud rumblings on the tracks that will become an overload to our ears and shake our bodies when a train passes but are always also echoing within the boundaries of this place. It, too, has such a presence with a strong shoulder that we know it will press against the tides of the river when they do seek to overrun its walls. It will push the waters back into the flow downriver and away from us. The ditch that runs along the river still has ghostly mules pulling barges by the river downstream from the nineteenth-century busy traffic on the Susquehanna. The cargo made of cousins of trees of this heavily forested little river town came from up north to pass this place by to the larger world. The crumbling stone foundations lining the river are half-presences of the many iron furnaces that surrounded this town on the edge of the river and there almost remains the wisps of their vanished smoke. However, now there is a sun filtered greenness that spills over trees crowding the bank covered with ivy looking down toward the water where lazy herons flap up and take short flights to spots where they wait patiently for a fish to come downstream. The tenuously loping deer, trundling turtles, slithering black snakes, and scurrying skunks give the forest a certain bustle of neighbors who will spill over into yards or ponds or just stay on the edge of town forming a frame that colors what is inside of it.

When previously considering Merleau-Ponty's statement that

> Through my perceptual field, with its spatial horizons, I am present to my surrounding, I co-exist with all the other landscapes which stretch out beyond it, and all these perspectives form a single temporal wave, one of the world's instants,[7]

the focus was on the single temporal wave. However, this statement also points to what we might call the "spatial wave" of this, my place, or any place. The Susquehanna River that flows by my house shows itself as a flow that originated a hundred miles away in upstate New York and also in Western Pennsylvania, journeys through the dams at Holtwood and Conowingo, as well as its emptying into the Chesapeake Bay where my 50-year-old sailboat at Havre de Grace is very much present as I glance at the river. In some less prominent way, so are even the oceans far away and their circulating flows that connect in the planet's great fluid presence. The water echoes with the sky's touch and the rain that continually falls as the evaporating vapors rise from the earth. The locust, black walnut, apple, and maple trees that wave outside our back window and hit our eyes as we look out the window from the bed in the morning are part of the plane of rest upon which we lie down at night and also as a cuddling, caressing presence that are part of the bowerlike dimension of the bed as interwoven with these trees outside. The long extent of the land behind the house gives itself as a continuity with the rich black Lancaster County soil and fertile fields that surround this town as well as with the stark escarpment of Chickies Rock that was pushed up from the river bottom when the tectonic plates colliding near Africa pushed the floor of the river into a verticality that is this wizened cliff face overlooking the town. Of course, the now lived sense of the reverberations of tectonic plates colliding millions of years ago wasn't originally a pre-reflective immediate sense of that cliff but entered my knowledge of it through a lecture walk with an archeologist, but reflection intertwines with perception, becomes sedimented with it, and now is part of how the rock whispers its identity to me on a felt, lived level.

However, even without that knowledge which became now the sedimented and immediate sense of Chickies Rock, its vast face, looming above me for hundreds of feet as I walk along the river down from my house itself like an elder's face has the look of an ancient one who has endured for millennia and plays across its surface as its deeper expression it wears on its face. Indeed, the name of Chickies Rock may come from the Indigenous Delaware name for Chiquesalunga Creek; it is also theorized that a cave in Chickies Rock likely served as a shelter for Indigenous people.[8] The tall trees, many much thicker across that I can reach with my arms, standing more than a hundred feet in the air are also present as long abiding witnesses whose age is part of their meeting me at this moment, as does the flow of the river that has passed by for millennia, as do the clouds overhead, perhaps newly formed this day, but part of such cloud formations reaching way back into the primordial time of the planet. Serenity Rock that stolidly sits a hundred yards behind my house rests on the ground surrounded by grass and forsythia, weighing several tons, yet one can feel that it took a long journey to somehow land in this spot by some sort of upheaval of the Earth or a great push from ice sheets or by something ancient of great power that still emanates as a lost trajectory in its squat sitting there. So, it is not only the continuity of time as a single wave that overreaches any bordering of time or space, but it is the depth of the past, a deep, deep past that plays across the present as its weightiness in meeting me and supporting itself and me. This is not a thought, but a felt presence as one looks at the rock or as Merleau-Ponty says in his Nature lectures: "The process of time is inscribed in our body as sensorality. By speaking this way, we cannot speak of a time of Nature in itself, but of a time of Nature insofar as we participate in it." This depth of time is lodged in our embodiment as we are thrown by perception into the world.[9]

Merleau-Ponty describes how our history, the history of human doing, is present always all around us but is not of the same deeper reach as the history of the natural world and he proposes to "oppose to a philosophy of history [...] a philosophy of geography."[10] In this opposition, Merleau-Ponty seeks to access a more primordial sense of the past, one expressed by the manifest history of the Earth. He asks us to turn to our sensual experience of the natural world and we will open ourselves to this depth dimension:

> Whereas geography—or rather: the Earth as *Ur-Arche* brings to light the carnal *Urhistorie* [....] In fact it is a question of grasping the *nexus*—neither "historical" nor "geographic" of history and transcendental geology, this very time that is space, this very space that is time, which I will have rediscovered by my analysis of the visible and the flesh, the simultaneous *Urstiftung* of time and space which makes there be a historical landscape and a quasi-geographical inscription of history.[11]

As *Ur-Arche*, this felt and perceived sense of a deeper history that surrounds us in the natural world, the history of the planet, is an *Urhistorie*, a more primal depth of time than mere human history that is still felt within the background presence of all we encounter. It is carnal, that is to say, taken in by our bodies, even if it has been augmented through our reflective researches. This depth of time permeates all the beings within our inhabited space, and nothing can appear in our worldly space without having within it this depth of time. Time and space of this great depth as simultaneous *Urstiftung* are instituted or opened up, or the translation that I prefer, are *an initial endowment*, of the planet to give us a place and to allow all beings a place. All places open beyond any borders to the expanse of space/time, but every perception returns to be grounded in what is our current situation of space/time in this emerging place.

My Place as I Write in the Cultural World

We have not yet come to the many human inhabitants who form their own nexus encircling and permeating the space that is this place of mine. This is not my town but is rather our town, and my place is only within this town and within this world, which is our world. Merleau-Ponty says that it should be most aptly called an "interworld," *l'intermonde*, "where our gazes cross and our perceptions overlap."[12] He continues that we "are motivated by the intertwining of my life with the lives of others, by the intersection of my perceptual field with that of the others, by the blending in of my duration with the other durations."[13] The traditional notion of subjectivity, of isolated minds and individuals, is a possible intellectual construction abstracted away from the more primary embodied, perceptual access to the world, which is a shared access. We can only find intersubjectivities when we listen to our embodiment and take seriously that our access to being and the world is through perception and its layers of sense in emotion, memory, intuition, and other felt experiences. It is not thought that gives us the immediate intertwining with others, but perception and the body, and the perception that throws us into the world and from which we return or as Merleau-Ponty states,

> But at the very moment that I think in I share the life of another, I am rejoining it only in its ends, its exterior poles. It is in the world that we communicate, through what, in our life, is articulate. It is from this lawn before me that I think that I catch sight of the impact of the green on the vision of another, it is through the music that I enter into his musical emotion, it is the thing itself that opens unto me the access to the private world of another. But the thing itself, we have seen, is always for me the thing that *I* see.[14]

As we perceive the world around us, as I have a place, it is the place and the things of the world that are its larger context that are permeated with the presence of others with whom I am bound up by these perceptions. As Merleau-Ponty puts it, there is a "synergy" among our perceptions and bodies as leaving ourselves in the only way we can by perceiving the world.

This 200-year-old house forming the axis of this circulating sense of place for me is a house that is pervaded with presence of others. From the furniture that has been used by others, so many friends, and family members, such as the chair where my mother reclined or the one by the woodstove where my wife often sits, or the sofa upon which the two dogs are often sprawled or even hidden underneath, or the bowl on the table in front of me given to me by my childhood friend who cast it on a potter's wheel in a college class on a campus we both attended, to the treadmill carried in by those two young guys hired from "College Hunks," to the polished bronze deep cannister holding the kindling that we bought from our two good friends at their stand at the antique cooperative, to the dark wooden end table and bookcase built by hand by the son of a friend from whom I bought this house, to the wooden plank floors stripped of their layers of paint by the man I hired when the house was purchased, to the other floors that were stripped and refinished not as long ago by the gregarious local painter in town whom everyone knows and loves, to the bright yellow in the dining rooms that glows warmly in the morning sun that was painted by the homeless woman who stayed in the bed and breakfast a few doors away during the period she was going through her difficulties, to the wooden upper side of the house just replaced by the newly arrived towering contractor whose wife owns and runs the local café and who holds his newborn baby in one hand as he parades her around town with a huge smile, to the neighbors visible across the yard on both sides, to the parade of friends and town characters who walk by each day with dogs or on errands, and to the so many more whose presences inhabit the house in its atmosphere in a shadowy but pervasive way where one or another might spark forth for a minute from the overall mist of presence.

These presences of others are there each time I look at or sense one of these objects or beings, but they are in the latent background of its sense for me, unless something makes them flash up into more focused presence, but they are always there with me in my place in this house. All the many visitors over the decades also linger in the mists of felt embodied memory inherent in perception. In perhaps one of his most beautiful and touching passages, Merleau-Ponty writes of waking up each day and

> far from opening upon the blinding light of pure Being or of the Object, our life has, in the astronomical sense of the word, an atmosphere: it is constantly enshrouded by those mists we call the sensible world or history, the *one* of the corporeal life and the *one* of the human life, the present and the past, as a pell-mell ensemble of bodies and minds, promiscuity of visages, words, actions, with, between them all, that cohesion which cannot be denied them since they are all differences, extreme divergences of one same something.[15]

The atmosphere that is home in this place is a pell-mell ensemble of many who have meant a lot to us or have touched our lives for an instant or whom we never met but their work or concerns were wound into the unfolding sense of here. The present of the presence of the house in this place is equally at each moment the presence of the past and not through an act of recollection but given in the nonlinear, non-successive thickness of time or as Merleau-Ponty expressed it:

> the visible landscape under my eyes is not exterior to, and bound synthetically to ... other moments of time and the past, but has them really *behind itself* in simultaneity, inside itself and not it and they side by side "in" time.[16]

As embodied beings, we are taken into a time that is found in all that surrounds us in the nexus that is place and within this materiality is held a plurality of temporal instants, eras, and horizons that continually emerge in its upsurge of ongoing manifestation with which we participate through perception's layers of sense.

This simultaneity in time of past and present is inescapable in the makeup of my place, but its horizon goes far beyond my personal past. Merleau-Ponty in *The Visible and the Invisible* discusses at length that there are no isolable persons or places but rather that individuals "are no longer in themselves, in their own place, in their own time; they exist only at the end of those rays of spatiality and of temporality [...] from the first mounted on the axes, the pivots, the dimensions" and "have about themselves a time and a space that exist by piling up, by proliferation, by encroachment, by promiscuity [...] which are the nodes and antinodes of the same ontological vibration."[17] The "rays" of sense are the ways that differing identities resonate, cohere, and are entangled, appearing together in their sense, and the differing senses "pile up" or "encroach" upon each other (temporally) in their moments of manifestation. This coming forth as an ontological vibration runs through persons, generations, through disparate places and times, within the materiality in which we participate and is identified by Merleau-Ponty in this passage as the way that a culture comes forth. Like the subject, Merleau-Ponty says "the space or time of culture is not surveyable from above,"[18] that culture does not hover above or outside experience but flows from within it as the rich background of perception and embodied interconnection with other persons and things. Walking around the outside of my house, as well as walking within its interior, or walking down the street, the sense of the presence of the past is almost strikingly overwhelming in a distinctive way. The houses are made of wood in styles of house-making that belong to the nineteenth century, whether of a Salt Box, like this house, or Gothic Revival, Federalist, Georgian, Victorian, log houses, Italianate, or Colonial Revival. One feels their steadfast presence announcing they have been shielding the occupants of this place for a long time. Actually, a local college has started a project detailing the histories of these old homes and put in their windows an adhesive sign that reads "This Place Matters." This small river town is unique in that it has 430 homes on the national historic register and one immediately feels the depth of time within its streets. The foot wide old plank floors and their pumpkin oak wood used also on the sides of the house breathe their couple century old age into the air of the town as walkers pass by, and it is a town where people still often walk from place to place, which itself hearkens to the past of a time when not every place was connected to others by automobile traffic. However, even in another place, perhaps one with newly constructed homes, there would be a sense of another history that preceded it that had been bulldozed away, whether of patiently waiting forests or older dwellings that had been taken down because of their age and perhaps disrepair or because of dissonance with the tastes of postmodern culture. This deeper background would linger, however, as a felt presence, but one of absent-presence.

When I mention where my place is located, in what town, there is even an echo in saying its name of a distant past. This Borough, as many small towns in this state are designated, was once two adjoining towns, but in 1812 were incorporated into one. The town of Marietta had been laid out by the two adjoining principal landowners, and the name of this town they concocted is a combination of the names of their two wives— Mary and Henrietta—so it resounds with this note of centuries ago. There are many new people moving to this town and a new construction of luxury apartments and townhouses that just replaced the dilapidated old industrial area, but the past lingers and its many interconnections with the wider world and its history. The modern construction was actually designed to echo the old forms of houses in town, since this is a pervasive character to each block, and is named "Riverside Foundry Apartments" since these were the structures that stood there in the nineteenth century. The town's elevation moves from the higher hills along its edge downhill until reaching the Susquehanna River. There is a feel when in the Borough that it is in some way still leaning toward the river as its former lifeblood, at least in the infiltrating sense of life on the banks of the river that runs through all perceptions within this place. This is true in the sense that the town had a cultural and economic importance in the nineteenth century that shone brightly and its refracted light is still present here. The canal boats in the early nineteenth century were loaded with coal, lumber, and other commodities and moved not only on the river but through town on the canal and then later in the century through the railroad cars that ran through town and not only on its edge. The fact that there was a plentiful supply of iron ore and limestone that were the raw materials needed for the smelting of iron by anthracite coal-fired blast furnaces led to eight furnaces being built along the river's edge. Later, when the railroads took over from the canal trade, pig iron was produced and in demand. My neighbor left one of the old railway tracks that remain running through his lawn, as one of the many markers of the distant past, like the horse tie bracket on the wooden step by my front door.

All these commercial successes led to mansions being built in the Borough, the opening of one of the first banks in the area and lavish parties being given and attended by Presidents Grant and Cleveland, Supreme Court justices, and railroad magnates. These proud old buildings give this place a special harkening to the past and provides under one's step a certain ground of felt past accomplishment whose trajectory asks for community members to take up this step and add to its rich history as well as tell its story. Certainly, the period of the greatest energy faded when in the early 1900s, Pittsburgh took over all the steel-making industry in the larger area. Even before this crash in the local economy and fall from its most prominent days, there were days of hardships for many in this town whose presence also lingers. The men who worked in those blast furnace foundries, the mule drivers, the men who worked in the lumber yards were living a hard life and many others in this small town were living existences of exhausting labor while those magnates were having balls in the mansions. Near the mansions are also tiny rundown apartments above storefronts and at the end of town, the now renovated cute little row houses by the river were once factory barracks for workers. The long-ago labors of the workers at the blast furnaces and hauling the wood from the barges also mixes into the atmosphere of the town.

Next door to me is the old station master's house whose second floor is a large ballroom for the entertainment of those who profited from the work of the other citizens of the town who worked long and hard. The street was littered with bars and taverns where the exhausted men drank at night and came home to wives trying to make ends meet and often being the targets for the frustrations of their men. Two doors down is a bed and breakfast and restaurant, which had been a grand hotel and is said to be haunted by a ghost of a woman in her bridal gown who was waiting for her canal man to come marry her. According to the legend, he left her at the hotel, waiting for the groom to arrive, and sailed by without stopping. The women were often abandoned or left in anxiety about their fates, as the large "widow walk" a few blocks down the street on the 200-year-old tavern building attests. When the eight furnaces along the river were shut down when anthracite became scarce, when the canals shut down, and when the lumber had become less plentiful, the suffering of these inhabitants of the town created a sorrowful sediment that the river and the town also have beneath their surface that can be felt as one walks about.

Perhaps, the most moving tale of the town was the fate of the small African-American population that comprised 2.5% of the town. There is a feeling of pride and hope that this was a place of liberation for enslaved people escaping from the South. It was a stop on the Underground Railroad and there are still a few tunnels in town that led from the river to safe houses. However, the intolerance of some of the population to these arrivals, as well as the tenacity and courage of those who stayed to make this their home, is very palpable as one walks down the street on the hill where the tiny Bethel African Methodist Episcopal Church sits surrounded by a small graveyard. In the cemetery are the graves of 34 veterans, including 20 from the Civil War. The first services are said to have been held in 1819, in the nearby woods, under an arbor of trees. The present site was purchased by the church in 1822, and the construction date of the first church is unknown. Probably the ugliest moment in the town's history happened in 1861, when the church was burnt down.[19] It is assumed that it was no accidental fire, but rather an act of aggression toward those seeking a haven. The parishioners did not give up and rebuilt the church, which still stands, although its members have dwindled in recent decades to very few. These times of intolerance, as well as the declining economic base, the struggling factories on the edge of town, contributed to a downward sense to the town's trajectory in the middle decades of the last century.

Yet, the town still had a momentum from its former glory days. As noted before, a place is always transforming and change continued to flow like the river beside the town. Thirty years ago, there was a felt sense in the Borough that the past laid a heavy weight on the town that its vitality had faded and perhaps even evaporated never to be recovered. Yet, like a vein of precious ore, that past history of industry and vitality remained in the ground under our feet and still moved within the shadowy presences of past barges and boats on the river, and many residents picked up on this flow of energy. Happily, the Borough has been reawakening to its flowering heritage, drawing in residents from all over, and restoring its energy. The struggling factories left town. New businesses, restored homes, restaurants, and numerous adventurers on the newly finished 22-mile bike path along the river became part of an infusion of energy, but it is a differing energy that defies the postmodern replacement of historic structures with modular units of commerce and dwelling. Instead, people have been drawn to the town from the big cities a few hours away and even from the Western states, wanting to find a place that still cradles and seeks to uncover its cultural and natural histories.

Place is a happening. Through the intensification of perception and feeling, and through the arts or other meditative and augmenting of felt bodily practices, we can both deepen our sense of place and also infuse it with greater powers of transformation in ways that are expressive of our creativity.

Notes

1 Henri Bergson, *The Creative Mind: An Introduction to Metaphysics* (Mineola: Dover Books, 2010), 28.
2 On the chiasm, *see* Maurice Merleau-Ponty, *The Visible and the Invisible*, trans. Alphonso Lingis (Evanston: Northwestern University Press, 1968), 130–155.
3 Maurice Merleau-Ponty, *Phenomenology of Perception*, trans. Colin Smith (London and New York: Routledge, 1962), 37.
4 Ibid., 203.
5 Maurice Merleau-Ponty, *Institution and Passivity: Course Notes from the Collège de France (1954–1955)*, trans. Leonard Lawlor and Heath Massey (Evanston: Northwestern University Press, 2010), 195.
6 Merleau-Ponty, *Phenomenology of Perception*, 331.
7 Ibid.
8 On naming, *see* Susquehanna River Basin Commission, "Native American Waterbody and Place Names within the Susquehanna River Basin and Surrounding Subbasins," (2006): 25. www.srbc.net/our-work/reports-library/technical-reports/229-native-american-names/. On the possible use of the cave, *see* Charles Scharnberger, Edward Simpson, Jeri Jones, and Joseph Smoot, "Stratigraphy and Structure of the Chilhowee Group in Lancaster and York Counties," *Field Trip Guide, Annual Meeting of the Northeastern Section of the Geological Society of America*. March, Lancaster (2014): 59.
9 Maurice Merleau-Ponty, *Nature: Course Notes from the Collège de France*, trans. Robert Vallier and ed. Dominique Seglard (Evanston: Northwestern University Press, 2003), 119.
10 Merleau-Ponty, *The Visible and the Invisible*, 258.
11 Ibid., 259
12 Ibid., 48.
13 Ibid., 49.
14 Ibid., 11. Italics in original.
15 Ibid., 84. Italics in original.
16 Ibid., 267. Italics in original.
17 Ibid., 114–115.
18 Ibid., 115.
19 Earle Cornelius, "Marietta's Bethel AME Church Readies for Its Bicentennial Celebration," *LNP: Lancaster Online* (Lancaster), October 5, 2019. https://lancasteronline.com/features/marietta-s-bethel-ame-church-readies-for-its-bicentennial-celebration/article_3452444a-e6cd-11e9-9b51-3bd3d90c51ff.html.

Bibliography

Bergson, Henri. *The Creative Mind: An Introduction to Metaphysics*. Mineola: Dover Books, 2010.
Earle Cornelius. "Marietta's Bethel AME Church Readies for Its Bicentennial Celebration." *LNP: Lancaster Online* (Lancaster), October 5, 2019. https://lancasteronline.com/features/marietta-s-bethel-ame-church-readies-for-its-bicentennial-celebration/article_3452444a-e6cd-11e9-9b51-3bd3d90c51ff.html.
Merleau-Ponty, Maurice. *Phenomenology of Perception*. Translated by Colin Smith. London; New York: Routledge, 1962.

————. *The Visible and the Invisible.* Translated by Alphonso Lingis. Evanston: Northwestern University Press, 1968.

————. *Nature: Course Notes from the Collège de France.* Translated by Robert Vallier and edited by Dominique Seglard. Evanston: Northwestern University Press, 2003.

————. *Institution and Passivity: Course Notes from the Collège de France (1954–1955).* Translated by Leonard Lawlor and Heath Massey. Evanston: Northwestern University Press, 2010.

Scharnberger, Charles, Edward Simpson, Jeri Jones, and Joseph Smoot. "Stratigraphy and Structure of the Chilhowee Group in Lancaster and York Counties." *Field Trip Guide, Annual Meeting of the Northeastern Section of the Geological Society of America* March, Lancaster (2014): 53–72.

Susquehanna River Basin Commission. "Native American Waterbody and Place Names within the Susquehanna River Basin and Surrounding Subbasins." (2006). www.srbc.net/our-work/reports-library/technical-reports/229-native-american-names/.

16

COSMIC PLASMA ECHOING IN (OUR) PLACE

Debra Lacey

Conneaut Lake, Pennsylvania, U.S.A.: 41.6033° N, 80.3058° W

The universe at present is estimated to be roughly 13.7 billion years old.[1] Humans, in comparison, have been in existence for a brief moment.[2] The awe-inspiring immensity of our living cosmos leads us to ponder our relationship and place within such boundless time and scale. We find a pathway when considering one of the fundamental building blocks of the universe that has remained constant. The atom, defined as a particle of matter, exists now as it did 13 billion years ago. All that was, and is, in our known world has been created from the same building block that forms each of us now. Humanity and all of more-than-human nature are indelibly linked to, and a part of, the cosmos.[3]

The most prevalent form of matter that permeates our universe is plasma. It constitutes 99% of our visible universe and is of a different order than solids, liquids, and gas. It is regarded as the fourth state of matter and it is found predominantly in cosmic space, in our atmosphere and that of the other planets, in the tails of comets, and solar winds. It exists at very high temperatures and is not naturally present on Earth but can be produced artificially. All matter becomes plasma if it is heated sufficiently. The Plasma Science and Fusion Center at MIT notes that the properties of plasma have been used to confine it within magnetic fields where it can attain temperatures hotter than the core of the sun. Contemporary plasma research is related to everything from the computer chip and rocket propulsion to cleaning the environment, destroying biological hazards, and healing wounds.[4] Plasma is a part of our ecological system. It comprises our sun, which provides for our sustenance on Earth. Our fascination with

DOI: 10.4324/9781003390985-21

the sky resonates within us because it *is* us. The atoms with which we are composed share the same origins as the atoms of plasma in cosmic space.

I suggest that ecosomatic movement offers ways to explore these connections; it can be a doorway through which the embodiment of commonalities can foster new sensibilities. The cellular reawakening can excite and rekindle our personal relationship with our ecological self. This is our entangled ancestry.[5]

When I explore ecosomatic movement practices, I resonate with my immediate environment and atmosphere, but I also exist in a freer, less dense state—I feel as if I am embodying my plasma origins. I sense greater space within my body. My vertebrae feel aligned differently, my lungs feel lighter and more expansive, and I tingle with an alertness as though my cellular body were receiving impressions from the air and cellular world around me. As I engage my soma body, practicing actions such as mindfulness, witnessing, curiosity, mirroring, self-observation, or other somatic tools, my physiological world changes instantaneously. These transformations have uncovered a range of physical, emotional, and psychological states within me, and I draw parallels between them and the qualities of the different states of matter (*see* Figures 16.1 and 16.2). I resonate with these qualities inwardly and outwardly, creating new body/mind impressions.

What also exists within me is an opportunity to realize my place within more-than-human nature. That means, as Andy Fisher, a leading figure in

FIGURE 16.1 My body explores different states of matter: the gravitational world of solids, the uncongealed fluidity of liquid, the molecular lightness of air. Photograph by Debra Lacey, 2023.

FIGURE 16.2 The expansiveness of plasma, the fourth state of matter, resonates within. Photograph by Debra Lacey, 2023.

the field of eco-therapy says, "granting the natural world psychological status; regarding other-than-human beings as true interactants in life, as ensouled 'others' in their own right, as fellow beings or kin."[6] Inviting and developing an embracing view of nature is a step toward realizing that we are participants in a unified whole. In their book, *Active Hope*, Joanna Macy and Chris Johnstone quote a fellow activist speaking about the despair and destruction of the rainforest, "I try to remember that it's not me [...] trying to protect the rainforest. Rather, I am part of the rainforest protecting itself. I am that part of the rainforest recently emerged in human thinking."[7] The insight of this quote highlights our interconnectedness with nature and reorders perspective in regard to our place. It brings us back to our shared origins, reminding us of our ancestry.

The building blocks of life are all around us, are within us, and are us, but our sentient abilities remain underused and unrecognized. The perception of our natural place in the universe becomes lost. Acknowledging more-than-human nature through ecosomatic practice enables connection with the shared and sustaining world around us. Ecosomatic movement can be as simple as standing still, placing attention on one's breathing, and witnessing the surroundings. Curiosities inspired by all the senses shift one's consciousness, enabling new pathways of embodiment of nature and self, and new relationships to our present world and the universe sustaining our world. This is the place of plasma, echoed in our place.

Speculative Plasma Practice

Slowly read these instructions out loud to others or record and replay it for yourself. Make sure to give plenty of time for the mover to respond. This speculative movement practice is designed for an indoor space with windows but can be converted easily to an outdoor setting. It can also be modified and adapted to accommodate your physical needs. Simply follow along in a way that makes sense for you.

Engage in a simple walk around the room
Sense the weight of your feet on the floor
Observe your breathing
Bring awareness to the movement of your eyes and head
Notice what you see inside the room

When you pass a window, pause
Gaze outside the window and register what you see

Invite something in nature to catch your attention
Curiously investigate what you are drawn to
What is its shape and color?
How is it placed within nature?
If it moves, how does it move?
How does it make you feel?
Spend a few moments experiencing your connection
Remember the image and take it with you internally
Continue your walk

When you find an open space, pause

Arrive at a comfortable standing position
Feel the solid bones of your feet touching the floor
Slightly bend and straighten your knees
Sense how your bones stack and connect to your pelvis
Gently move your pelvis
Initiate an undulating movement up and down the bones of
    your spine
Guide the bones of your arms to dance upward
Circle your bony wrists
Paint the air with your articulate fingers
Feel the weight of your skull rotate on the top of your spine
Bring your body back to a standing, resting position
Come to a pause

Experience stillness

Sense the interstitial fluids running through your body
Slowly begin to move as liquid
Taste the watery saliva in your mouth
Invite your mouth and tongue to move
Place your focus on your digestive system
Let the juices of your digestive system initiate movement
Experience the pulsating fluids of your liquid legs
Continue your liquid movement exploration

Allow your expression to slowly take you to the floor
Find a pathway to lying on your back
Enjoy your arms and legs in an "X" shape
Slowly come to a pause

As you lay quietly, observe your breath
You are breathing the gas known as air
Sense the air filling all parts of your lungs
Imagine the air moving throughout your entire body
Begin to float
Explore how floating makes your body move
Take your time with this
When you are ready, find a resting position

Remember the image from nature that you embodied earlier
Invite your body to become that image
Mold yourself into its essence
Become still and experience your relationship
Deepen your connection

Imagine a beam of sun shining down upon you
The beam is hot—like plasma
Your image begins to transform
The intense heat enlivens the molecules of your body
Your molecular world becomes rapidly excited

Explore transforming into an etheric cloud of plasma
Sense a new freedom
Morph, expand, and condense repeatedly

Imagine your cloud moving within the solar winds
Outer space is awesomely vast
There are rarely collisions in space
Enjoy your plasma cloud existence
When you are ready, invite your cloud to stand in a resting position

If your eyes are open, close them
Sense the air around your body
Observe yourself breathing
Allow your breath to deepen
Experience slowly returning into your cellular body

Notice the liquid saliva in your mouth
Quietly move from the interstitial fluids coursing through your body
Arouse the bones of your feet on the floor or ground
Begin to awaken dense bones holding your body upright

Gingerly shift your pelvis, it holds your digestive and reproductive systems
Engage your ribcage, it protects your heart and lungs
Slowly move your cranium, it safeguards your brain

When you are ready, come to a relaxed and quiet position
Experience your stillness

Slowly open your eyes
Observe your re-emergence into place....

Notes

1   National Aeronautics and Space Administration (NASA), "Then vs. Now: The Age of the Universe," *Imagine the Universe!* (2022). https://imagine.gsfc.nasa.gov/science/featured_science/tenyear/age.html.
2   Our earliest known ancestors existed around 6 million years ago; *Homo sapiens* is a more recent iteration beginning approximately 550,000–750,000 years ago. *See* Brian Handwerk, "An Evolutionary Timeline of Homo Sapiens," *Smithsonian Magazine* (February 2, 2021). www.smithsonianmag.com/science-nature/essential-timeline-understanding-evolution-homo-sapiens-180976807/.
3   The term, "more-than-human," was coined by David Abram. *See* David Abram, *The Spell of the Sensuous: Perception and Language in a More-than-Human World*, 2017 ed. (New York: Vintage Books, 1996).
4   All matter becomes plasma at the temperature required for "practical fusion energy." Plasma Science and Fusion Center and Massachusetts Institute of Technology, "What is Plasma?," (2022). www.psfc.mit.edu/vision/what_is_plasma. *See also* Yaffa Eliezer and Shalom Eliezer, *The Fourth State of Matter: An Introduction to Plasma Science*, 2nd ed. (Bristol; Philadelphia: IOP, 2001).
5   My use of "entanglement" is indebted to the work of feminist quantum theorist, Karen Barad. *See* Karen Michelle Barad, *Meeting the Universe Halfway: Quantum Physics and the Entanglement of Matter and Meaning* (Durham: Duke University Press, 2007).
6   Andy Fisher, *Radical Ecopsychology: Psychology in the Service of Life*, 2nd ed. (Albany: State University of New York Press, 2013), 8.
7   John Seed qtd. in Joanna Macy and Chris Johnstone, *Active Hope: How to Face the Mess We're in without Going Crazy* (Novato: New World Library, 2012), 94.

Bibliography

Abram, David. *The Spell of the Sensuous: Perception and Language in a More-than-Human World*. 2017 ed. New York: Vintage Books, 1996.
Barad, Karen Michelle. *Meeting the Universe Halfway: Quantum Physics and the Entanglement of Matter and Meaning*. Durham: Duke University Press, 2007.

Eliezer, Yaffa, and Shalom Eliezer. *The Fourth State of Matter: An Introduction to Plasma Science*. 2nd ed. Bristol; Philadelphia: IOP, 2001.

Fisher, Andy. *Radical Ecopsychology: Psychology in the Service of Life*. 2nd ed. Albany: State University of New York Press, 2013.

Handwerk, Brian. "An Evolutionary Timeline of Homo Sapiens." *Smithsonian Magazine* (February 2, 2021). Accessed 10/31/2022. www.smithsonianmag.com/science-nature/essential-timeline-understanding-evolution-homo-sapiens-180976807/.

Macy, Joanna, and Chris Johnstone. *Active Hope: How to Face the Mess We're in without Going Crazy*. Novato: New World Library, 2012.

National Aeronautics and Space Administration (NASA). "Then vs. Now: The Age of the Universe." *Imagine the Universe!* (2022). https://imagine.gsfc.nasa.gov/science/featured_science/tenyear/age.html.

Plasma Science and Fusion Center, and Massachusetts Institute of Technology. "What is Plasma?". (2022). www.psfc.mit.edu/vision/what_is_plasma.

17

LANGUAGING BODY BY FIELD

Ecoproprioception

George Quasha

Barrytown, New York, U.S.A.: 41.9998° N, 73.9248° W

Orientation

"We could say that practically all the problems of the human race," writes David Bohm, "are due to the fact that thought is not proprioceptive."[1] There is hardly a moment throughout the day in which we are not in some way orienting ourselves in space and time, yet we're rarely aware it's happening. That's because the ability to physically orient ourselves is built into body and mind so seamlessly that it's indistinguishable from just being ourselves. Self-orientation at the level of proprioception, the complex neurophysiological system by which we position ourselves, is self-evident to us and ordinarily requires no special attention on our part. Bohm's application of such a system to account for what is missing from how "homo sapiens" does thinking, and therefore to many central problems facing our kind, may suggest a way into what is still unthought in how we orient ourselves in larger than personal senses and what that means for how we understand—or rather, perform—our relation to life on planet Earth. And it may also offer a way into how we embody, and can performatively think, ecosomatics.

Bohm goes on to explain his metaphoric extension of neurophysiology to thinking in terms that in other contexts we might call *feedback*:

"Proprioception" is a technical term—you could also say "self-perception of thought," "self-awareness of thought," or "thought is aware of itself in action." Whatever terms we use, I am saying: thought should be able to perceive its own movement, be aware of its own

DOI: 10.4324/9781003390985-22

movement. In the process of thought there should be the awareness of that *movement,* of the *intention* to think, and of the *result* which that thinking produces.[2]

He seems to be imagining here a possibility of thinking that is aware of itself in its very act of coming into being, something like contemplative thinking aware of its own activity. In current emphasis it might be likened to "mindfulness." But this is not quite accurate as to how proprioception works: while using feedback in its mode of operation, bodymind performs its spatiotemporal self-knowing *without* an overlay of conscious awareness. We might consider Bohm's critique of human limitation and its would-be meta-perspective as better represented by our neologistic notion of *ecoproprioception*: an extended self-awareness that actively senses and knows itself in dynamic interrelation with whatever makes up the surround. And we might understand this as an *orientational thinking by field.* It is a field thinking, moreover, that includes both our intimate bodymind knowing in spacetime—proprioception—and its equivalent dynamic field which we call "environment," viewed first as what we move through in the current of life itself, but also its many thinkable extensions. Here the implication is that environment is never fully separate from bodymind and neither is it without its own self-organizational and self-regulatory dynamics and criticalities. Thus, two dynamic systems, self and world, activate one encompassing dynamic system, plus *its* self-knowing, or what we're viewing as a dynamic meta-consciousness. Are we therefore saying that such a dynamic system of interactive awareness—an "organismic event"—might actually exist?

The first step in responding to this question—in fact, the first *ecoproprioceptual* stepping out—is to imagine dynamic systems occurring at just such a meta-level and generating a special kind of self-awareness. Yet such an awareness wouldn't belong to oneself more than one's "self" belongs to it. The awareness, rather, is an event that constitutes our interactive existence coming to consciousness of itself. This may seem abstract to the point of being fanciful, and yet this imagined dynamic system "exists" as much as proprioception itself does. The point is that proprioception, while a term designating a given neurophysiological system, describes something that only exists when it's happening, a specific act of orientation; it's in process at a given moment in time and space and does not exist otherwise. And we can indeed be aware of it when it's happening yet only by paying special attention. We would suggest that such an act of attention may be, or may develop into, a bodymind thinking orientation by field. This should indicate an ecosomatic thinking that is performative within a process of engagement between self and environment as well as leading to further theorizing of its possible enhancement and practice. Here we aim to sketch out some preliminaries toward a possible ecoproprioceptual practice.

It's important to keep in mind that a concept like proprioception is basically a description—a neurophysiological mapping—of "natural" actions performed in being alive; and it's a concept where a core principle, involving self-regulating reflective awareness, attracts us to thoughts of taking it further, as it did for Bohm and the poet Charles Olson among others before us. This has meant extending its principle of participatory self-reflective activity to higher levels of complexity. What "exists" here, as we wish to make implicitly understood throughout, is an *event*, not a thing, and often occurs in the form of an *edge*.[3] Over the years the scientific mapping has gotten finer as well as broader. The proprioceptive system works together, for instance, with the vestibular system of the inner ear, the balancing aspect of orientation, and there is speculation about how these work with "mirror neurons," the recently theorized neuroanatomy behind, say, our yawning when we see someone else yawn. The latter is a very basic kind of ecoproprioceptive event: an action we perform involuntarily as an extension of a happening in the field. The event is mental and psychological as well as physical, much like breathing freshly scented air in a rose garden which makes us feel elated. And if any such event is complex in the sense of being both cognitive and physical, we might consider its act of body-extending interactive awareness, like that of proprioception itself, as always potential and in some way already psychophysically built into what we are—and in some equivalent way built into what the environment, in all its networked complexity, is as well.

Preamble to More

Nel mezzo del cammin di nostra vita..., begins Dante's *Inferno*, *Midway in the journey of our life...*, telling us that all orientation occurs in the middle of things, which is where we find ourselves all the time. Beginnings and endings are either mental projections of the present point in time or inflections of the sense of presence. The event of finding our way involves continuous registration in accordance with many levels of feedback from what we encounter on the road. All the senses are active, from the bottoms of our feet to the corners of our eyes, and somehow we manage, indeed edit to a focus, the chaos of scents, sounds, and signals. If the stimulus is from signals, language tends to be activated in response. If from the trees, what? Immediately we're in the question of what constitutes language. No doubt there are those—some Indigenous persons or farmers come quickly to mind—who *read* the weather with uncommon accuracy; read birds, and not just carrier pigeons; read earth for its precious contents. Here we might ask: are dowsers masters of ecoproprioception? They reportedly work by formulating a sharp expectation in the mind (*water*, *gold*), and then, via a two-handed stick, they align rhythms of the body with "electrical" impulses picked up from hidden substances in a vibratory between. (Asked why another person hadn't located the water he'd just found, the dowser said, "He ain't got the '*lectric*!'")

If language comes in response to anyone or anything, there may be impact—somatic, psychological, social, aesthetic. At the most fundamental physical and mental levels, acts of language are consequential and reciprocal with resonance far beyond any immediate registration. A striking language act flaps its butterfly wings and generates a complex response, much of which is involuntary and quite physical. Ecoproprioception, whether at the physical or linguistic level, should be viewed, at a minimum, as bi-directional interaction between agent and receiver—the sensing and saying goes both ways at once; there's oscillation of effect and interdependence of agency. What we look for in assigning agency, as we'll see further, may be distributed beyond all accounting. No action is without that famous butterfly effect of remote consequences, and indeed the outflow may be omnidirectional and next to impossible to track. This is where the notion of ecoproprioception could help discover something inherent in us—a system including and extending the physiological level of proprioception—that orients itself amidst increasing complexity. Here we owe acknowledgment to Gregory Bateson's "ecology of mind" as the beginning of a new order of thinking that inspires inquiry of the present kind.[4]

Language Selving: Somapoeia[5]

When we think differently out loud or in writing so as to be heard differently, we create another kind of butterfly effect: we induce a further listening and thinking in others. This is also the intention of poiesis, to make language events that turn the mind, even turn it *around* (*metanoia*). Poet and somatics practitioner Cheryl Pallant offers this somapoiesis:

> Her body wishes to describe the indescribable, to light a diamond cut with verve, to phrase salivating tongues and quake flesh to spawn [...].[6]

Any inquiry into extending our sense of self-presence, its qualities and timeliness in the world, may intermingle disciplinary lines in an ecology-of-mind *field* perspective, and it will ask more of language than it's prepared to give, but not necessarily more than it's able to discover. Exploratory thinking—some name it "psychonautics"—calls for singular language, meaning that it has to find its poetics, its unique way of carrying out its unprecedented inflections in thinking. Part of poiesis, as Billie Chernicoff shows, is inquiry into its own occasion:

The truth is, when the sentence begins who knows?[7]

Poets of course know about this fundamental principle, but philosophers do too, at least implicitly, and also explicitly starting with "the first philosopher" Parmenides who did it all in one extended poem! And Nietzsche, Heidegger, Wittgenstein, Derrida, Deleuze—whom to read is to learn a language—an idiolect of thinking. Bateson, who thought innovatively and influentially in the ways of anthropology, psychiatry, social patterns, cybernetics, epistemology, and ecology, invented the "metalogue":

> [...] a conversation about some problematic subject. This conversation should be such that not only do the participants discuss the problem but the structure of the conversation as a whole is also relevant to the same subject.[8]

The key "poietic" point here is that language isn't just "expressing" thought and affect but these are generated and evolved through the very "matter" of language in action, itself worked through the physical body; indeed, as Charles Olson specifies, this happens through the *breath* of the thinker/maker/poet, as should become more vivid below.[9] And in the case of dialogue the generative body is multiple and distributed in space.

"Human bodies are words, myriads of words," wrote Whitman in "A Song of the Rolling Earth."[10] What happens when language enters into a space? Of course it depends a lot on the kind of language, space, and situation. In times of yore the poet goes to a place and sings a song, which adapts to the environment as the story emerges. A narrated mythic hero may take on the qualities, say, of a person famous in the town. This would be the proto-Homeric stage where oral poets presented the "same" poem very differently in different contexts, with local political/social accommodations. So the actual poem is always in dialogue with environment, addressed to a specific *listening*.[11] In fact the most ancient poems were *place*-specific, flexible acts of oral language pervasively conditioned by particular local needs and limits. With the advent of writing the limits become additionally codified and generalized, the conventions spelled out, whereby the "medium is the message," and so on. It's debatable as to how much any language act is ever truly solitary, as the external world lives within the language created in it, and the writer internalizes both the conditions and the inherited *speaking force* of the language in its native condition. In a basic sense, poetry, indeed language itself, is already an expression of, as well as a fundamental instrument for, ecoproprioception. There is feedback from the world even when the world is silent.

Ecoproprioception is always already going on in the unmoving, non-directly engaged maker's body with the arising of the very impulse toward language. This may be, for instance, what the poet Robert Duncan called a body tone, signaling a poem on the way, like a birth. Language is called forth bodily or calls itself into play physically—*somapoeia*, "body making"—and the emerging event is ecoproprioceptive. Language is reaching out to meet the world which, in keeping as well with its own rhythms, indeed its intrinsic music, simultaneously seems drawn into languaging: so much so that the world often comes to us *as*, as well as *in*, language. Among the many related phenomena is Atsushi Takenouchi's butoh aligning with "a cosmic wave or rhythm that is *Jinen*" or the related work of post-butoh and post-traditional dancers Eiko and Koma, where like in butoh the poiesis is dancing in accordance with a place and can merge human with non-human trans-identity.[12] There is impact, action in space, further rousing our faculties to act—to adapt William Blake's implied butterfly effect—among the sentient and sapient.[13]

Let's declare a basic principle: ecoproprioception when languaging, just as when extending sensing, is a *process of self-knowing interactive with space-knowing*. Here space-knowing is what I call "axial" as an "open concept," meaning it's left ambiguous as to whether the knowing agent is only oneself or also a knowing *in*, or distributed in, space independently of oneself. Axial concepts, a kind of bracketing, leave undecidable issues open as a spur to further thinking. And the interdependent/intergenerative principle of living language making—itself inherently axial—shows up variously on a spectrum that runs from typicality to diversity. The language we're thinking in here—in this very essay—is fairly atypical without being disruptively divergent. Even though it's not standard expository English, it nevertheless practices an intentional degree of restraint (especially on the part of its thinking poet) aiming to meet the *con*straints of its *con*text: a chosen field of interaction. However unruly (nonlinear) it may appear at times, it remains clearly within the bounds of what cognitive scientists now call *neurotypicality* (contrasting *neurodivergence*). This makes it a quite consciously dialogical and nuanced extension of the ecoproprioceptive principle operative in language. Its ability to observe itself, project a likely accepting readership in context, and monitor its own progress accordingly, all the while doing the (largely performative) thinking which is only minimally preformulated, speaks to its degree of neuro-adaptability and -resonance with environment.

Now let's consider a contrary perspective. But not without mentioning your readerly complicity, if you're still here, making you co-performative of its ecoproprioceptive principle. You may understand these renderings of principle quite differently from the author's intended meaning, which assures their "axiality," including intrinsic ambiguity. Text-reading is a non-universal kind of togetherness that implicitly raises challenging questions like "how neurotypical are you?"

Divergence

Adam Wolfond, age 20, a non-speaking poet with autism, writes about how he does thinking:

> We "think with sticks" (Wolfond), rubber bath toys, tics, flaps, taps, space, pace and proprioception.[14]
> My body pacing is the task of feeling my body and
> the feeling of the pace of the environment I am feeling.[15]

In the film, *S/Pace* (2019), the young man performs actions that, although superficially dowser-like, are basically uninterpretable by neurotypical standards: he moves with high concentration through a room with a substantial, carefully groomed twig in one hand making it twitch quickly as if doing something invisible to the air, and with no goal or apparent endpoint in sight: a pure activity for its own sake. He calls it *ticcing* and alternatively *scattering*. Wolfond tells us that those intense actions with a "talking stick" comprise one of his two principal kinds of poetry; the other he types slowly, laboriously with one finger on a computer that voices the words for him. His speech-generating device says that poetry is part of his body: "It is nature to me, and I think that non-speakers like me dance with language." It's all "languaging"—"an event in which the body and the atmosphere are related."[16]

This identification of signifying movement with language is intrinsic of course to the ontologies of dance and signing, and indeed it's part of ordinary speech gestures. But ticcing is divergent in having no meaning-corroborating context; it's *alien* language.[17] Yet its maker views it as a coherent action of bodymind in communion with space, and he identifies it as "thinking." Clearly it realizes an inherent linguistic potential, manifesting as a personal singularity. It's one of the clearest enactments of proprioception, a word he sometimes uses; indeed, he says,

> The art is the way of the stick
> and the stick is the way
> of my feeling body

Each wave art moves
is like seconds
that rally movement.[18]

His selving actions—his workings—become ecoproprioceptual in their
degree of performativity and in the energetic feedback relation with the
medium of space as it is realized in a chosen place. They become *somapoeic*
in the body's identification with language action, the somapoiesis, the
realized intensity of unprecedented verbal configurations: "ticcing through
the world is like touching it"; "the way of the hand that likes to see."
Wolfond's space itself is aroused to action: "inside the world is/the question
of easy touch"; "The work is to feel the world/that is touching me."[19]

The self-aware thinking commitment at stake here is to *movement* as
fundamental, enabling a transformative and virtually trancelike state of
mindbody: "I sometimes gain the space/like a backwards spiral/that tries
to find its way."[20] The torsional, spiraling actions with the sticks follows
a principle of movement I call *axial*, where a universal turning on an
axis, along with the generation of wave resonance in matter (cymatics),
is converted into languaging coherence and aesthetic appreciation: "the
inward rotation of the spiral is like amazing tall idea always thinking
around and out," writes Wolfond; "the man of autism is rallying the colors
of life to move."[21]

Much of what Wolfond says applies to intimate human experience
more generally, the mostly undisclosed part, witnessed as if in slow
motion through cracks in the ordinary, or an infrascopic perspective.
A close-up viewing of neurodivergence in its creative manifestation is an
invitation to discover unexposed dimensions that are mostly submerged
within neurotypicality. It can reveal how any high degree of singular and
embodied language realization exposes the intrinsic diversity of experience
awake to its qualities—what neurotypicality typically fails to appreciate.
Wolfond's written poetry invites serious study, and could be related, for
instance, to the work of a major poet of recent time, Larry Eigner (1927–
1996), a contemporary of Olson influenced by his poetics of "composition
by field,"[22]

bird sounds
        make the sunned room pass
            the opened window[23]

Eigner, critically palsied at birth as a result of a bungled forceps delivery,
wrote on a 1940 manual typewriter using only his right index finger and

thumb. The act of writing took tremendous physical effort and commitment to inscribe thought in words on paper, yet he wrote over 40 books in his 69 years. Unlike Wolfond he could speak but relatively few understood his speech well. When I published his poetry, I only got to speak with him on the phone and found it challenging, but friends who knew him better could adapt. He created a singular world experience page by hard-won page, transferring a half-reached physical place onto a parallel materially written space. He composed his reality "by field," and thanks to Charles Olson, he found a poetics to accommodate his vision of lingual space. Eigner and Wolfond are true poets of ecoproprioception.

Middle Voice

"Our taboo," said the poet Robert Duncan, "is at root against unintelligible passions."[24] We fear what we can't understand, and our education prioritizes right interpretation as endpoint over open processual inquiry. We like clear boundaries in the thinkable. Such a bias in thinking has delayed our embrace of possible engagement with "alien" language, including the very idea of intelligence and communication in nature between animals, plants, and fungi, recently foregrounded, for instance, in plant and fungal biology and ethology.[25] Scientists now raise questions previously more common to, say, a dancer and teacher like Rebecca Burrill: "What might be a mode of communication for our human kinship with Nature?"[26] Or a poet like Laynie Browne: "Who speaks the language of trees?"[27]

We can now begin to think in accord with a neurodivergent poet's movement languaging in the same frame with innovative poiesis of the last many decades as well as diverse practices both artistic and therapeutic. These practices range widely as different kinds of bodywork and intentional movement. The latter includes a vast world of dance extending from the sacred, ancient, and Indigenous, some with ancient origins, to innovative contemporary work, while body-centered practices and therapeutic systems have names like Feldenkrais Method, CranioSacral Therapy, Alexander Technique, Body-Mind Centering, Trager Mentastics, Hakomi, Body-Centered (or Somatic) Psychotherapy, t'ai chi, ch'i kung, *ad infinitum*. A strong practice of field thinking is highly developed in the work of artist-philosophers like Erin Manning, Brian Massumi, Sondra Fraleigh, and others, representing the integration of multiple ways of thinking, languaging, and practicing, substantially embodying ecoproprioceptual thinking.[28] This ever-expanding frame invites every kind of interdisciplinary attention but remains trans-categorical like Bateson's ecology-of-mind thinking. The deeper message that *we need to keep relearning how to think* gives the right impetus in appreciating how we ecoproprioceive—an intransitive verb with promiscuous transitive tendencies, articulating verbal action in the middle voice.

Since the 1950s poets like Charles Olson, Robert Duncan, Robert Kelly, and Charles Stein have invoked the grammatical function of middle voice, straddling the active and passive, while the mysteries of undecidable agency serve what Wallace Stevens called "the poem of the act of the mind" and the transformative awareness its operations afford.[29] The ecoproprioceptual practice of reflecting on language even as we are using it accesses a built-in grammatical flexibility. In poiesis this can be tantamount to letting the language do the thinking. There's a long history of displaced agency right back to Pythia, the Oracle of Delphi. In this "inner tradition," Blake channeled his deceased brother for lines of *Jerusalem*, Surrealist Robert Desnos performed automatic writing, and Rimbaud wrote *"Je est un autre."*[30] And Wolfond seems to discover this self-generating *languaging* through an urgent energetics of his own body, excited into space awareness by body's raw relational powers. Indeed, his title *S/Pace* suggests that the rhythmic pacing realigns body and space, the self-composing field, and he can say there: "Good thought moves/like fluid water/and the way of water is raining/really into/the seething good cracks/of wanting thought." Where is the agency, and where the cherished boundaries of self?

In Merlin Sheldrake's *Entangled Life: How Fungi Make Our Worlds, Change Our Minds and Shape Our Futures*, we encounter the vast underpinnings of biological life and the seemingly limitless complexity of what constitutes individual being, indeed, *self, subject*.[31] It seems that we are networked through and through and mind is everywhere and thinking is "non-local." How do you language such complexity, or rather how is nature, with which we are coterminous, languaging itself? Marjolein Oele sees the grammar of middle voice as offering "an alternative way to grasping the relationship between subject and process."[32] She goes further to "propose that the middle voice is *materialized* in the ambiguous inner-outer potentiality of the plant [...] [and] plants *live and embody* the open, futural, participatory existence of the middle voice."[33] Language is not just an intellectual human construct and instrument; it's embodied, made flesh, alive, intersubjective, interdependent, co-creative—in short, ecoproprioceptual.

The Body of Language Singularity

"I start in the middle of a sentence," said John Coltrane, "and move both directions at once."[34]

When a dancer is imaginally moving in and with the forest and its inhabitants, the level of identification may seriously confuse the boundaries of self, space, and alien entitativeness. The body has inherent rhythm and syntax potential like anything that communicates using biological and neurological feedback loops. And somatic practices and dance approaches often develop ways of *tuning into* inherent rhythm and intuitive ordering like the previously mentioned rhythms of nature (*Jinen*). If every place has a vibe, it also has a music and a "tongue." Sondra Fraleigh speaks of "intrinsic dancing"—"predicated on pleasure rather than perfection"— reflections of inherent value, which are amplified and subtly discriminated in body and language poiesis.[35] The butoh dancer reads the ground and aligns through energic feedback; likewise the poet *listens in* on the unfolding syntax for its subtle directions. In "The Structure of Rime I," Robert Duncan writes:

> I ask the unyielding Sentence that shows Itself forth in the
> language as I make it,
>
> Speak! For I name myself your master, who come to serve,
> Writing is first a search in obedience.[36]

Agency is axial, interactive, and enactive, and language itself is the ground, the shifting source of feedback (Duncan speaks of "the tone-leading of vowels"), the contraries and energies with which the poet is in continuous oscillation.[37] It's an art of *tuning*.

Somatic disciplines, going back to Gerda Alexander (1908–1994) and Eutony, discussed by Fraleigh as an important source for health-conscious innovations in dance, teach movement awareness that *tones* the moving body by finding what accords with one's own biorhythms.[38] Over the years my own experience of dancers like Simone Forti, Trisha Brown, Eva Karczag, and Nancy Stark Smith contributed to a theory and practice of *axial poiesis* because these dancers—or what I call movement poets— bring out a sense of *singular body syntax*. Their somapoiesis imparts awareness of...—here I have to give it a place-holding name: *imaginational performativity in precarity, working the edge*.[39]

We are learning more and more through what our poiesis, our "search in obedience," teaches us; namely, that the natural world enacts and co-enacts the principles that guide us. As mentioned, Marjolein Oele explores middle voice as the appropriate grammatical mood to open up "an alternative and promising stance to contemplate the affectivity at the heart of plant life":

> If it is the case that middle voice offers us a subject that is "inside the
> process of which it is the agent," and provides a vision of life that grows

from the middle, without origin or end, then plant life *par excellence* offers us an instance of middle-voiced life [...]. Plants emerge and become who they are as products of a polyphonic, communal space. In this space, they result from hybrid, trans-species and trans-generational partnerships, which both sensitize and protect plants in numerous assemblages and through numerous means [...].[40]

If for the word "plant(s)" in the above we substitute "poem(s)" or "dance(s)" or "musical composition(s)," we get further insight into our connectedness, our entangled placing, what Edward Casey has called "co-implacement: body-in-place and place-for-body."[41] One kind of insight is how this connection with Earth life is the occasion of somapoiesis, the further languaging in conjunction with our place here. The action is both ways at once, and the possibility of mutual healing—an event from the middle of the sentence underway, as Coltrane said his music is.

We use "body language" to bring us closer to sensory experience of the body. We also need to cultivate Whitman's awareness that language is *of* the body as well as the psyche/mind and therefore itself may be said to *have* body. The different nuances of the word *body*, even in how we're using it here, are suggestive of the many senses in which both body and language *can* be used—a multiplicity calling up the subtly different ways we *experience* body, reflect on it, and transform it. Ambiguity and variability become opportunities for further nuance and avoidance of "misplaced concreteness."[42] What arises into view in any particular instance of close somatic experience is its singular syntax. In the case of axial poiesis there's a similar phenomenon: each spiraling breath in its distinctive particularity impels words forward while connecting them with other words by somatic-psychical attraction. Spiraling breath issues at the core of deep body ecoproprioception. It involves a curve of connectivity according to the enacted psychical biorhythms that issue forth. This unprecedented syntactic event is a matrix of new and indeed *axial* thinking, happening on the spot, true in alignment with the axis of the moment.

Reading and hearing the axis, the center of balance in syntax or in relational events—indeed one's own spine or another person's or a plant's stem—may induce a feeling of being touched, a haptic eye-ear reading-listening that gives rise to singular non-reducible meaning. It can have the power like that, for instance, of hearing the recorded voice of a missed friend. Such a unique significance brings to the surface a now amplified awareness of our implicit singular being, and its paradoxical link with our connectedness. It transmits a state of free being, as perceptibly true on the outside as on the inside; it *ecoproprioceives*. How could this not be a deep intention behind the artistic goal of *mimesis*—life and art inter-imitating, co-entrainment, and co-embodiment? That would be functionally a principle of ecopoiesis with free play in the somatic.

A seldom discussed intention of poiesis (relating to a core principle emphasized in axial poiesis) is a work's communication of its own *state* of creative being. It projects an attractive force of its *level* of somatic and psychical realization. This points to a vital social as well as environmental aspect of ecoproprioception: the invitation implicit in an artistic process for the receiver to take on its state of singular realization—to entrain to an *other possibility*. I call this *mirroring by alterity*. Its embrace of performativity qualifies the more traditional aesthetic focus of appreciating the formal qualities and "perfections" of a work in favor of valorizing its transformative impact and transmission of enactive value. That value is resolutely non-coercive. It invites. As Wolfond says: "Each wave art moves/ is like seconds/that rally movement." Such performative work projects outward its self-regulating, self-directing action of field awareness—its state of singular possibility. The action is not so much forward or toward as spiraling and radial. Its performative expression creates a new *linguality*— a vehicular languaging able to transform what we call reality.[43]

Practice Makes Perfect Variability in Non-Perfective Thinking

How to speak in the middle voice? You can only try: language happens "inside" us even as we happen through its agency "outside"—the bodily act of poiesis at large. Did I say *try*? That may be misleading. Ecoproprioception is not helpfully viewed as a skill, the result of repetitive action, calling for a new method, evaluated by codified standards. There's no mastery in the relational and the psychonautic. (And there's no sure path to the *'lectric*.)

In my poiesis, "preverbs" effect middle voice:

Let's be in touch through it all is the cry of the field.
Body talks non-stop.
An action by field lets the whole thing die in your arms.[44]

We're aiming to trap an elusive sense here: it's not motivated by the wish to improve but to *know further and other* (more *connaître* than *savoir*, more like knowing biblically as it were than abstractly). Its evolutionary principle is less getting fit to survive than erotic and playful to enjoy, and it's far more collaborative than competitive. The "-ception" in *ecoproprioception* may be closer to *conception*, toward birthing, a kind of bodymind adjustment that puts a pressure on thinking to be something *more*, some further state of connectedness it doesn't know just how to be. But as *prehensive* it's open to sensing.

Ecoproprioception responds to a *discipline of alert release* wherein space opens for connective thinking to occur of its own accord, or an *open accordance*, accommodating insight, a mindbody event of unaccountable origin and projective force. In this practice frame, discipline occurs receptively and without coercion and retains the possibility of pleasure, indeed a somatic sense of ecstatics. There are precedents to help guide us in the West with, say, an idea like *Gelassenheit*—"releasement, state of letting go into what is"—from the medieval Meister Eckhart to Heidegger.[45] More commonly now we have t'ai chi and the "soft" martial arts and related therapies, including a range of somatic practices already mentioned. On the philosophical plane there's the process philosophy of Whitehead, Bergson, and Deleuze; our native pragmatisms of William James, Charles Sanders Peirce, and John Dewey; and "enactivism" in the cognitive science of Francisco Varela and Evan Thompson; among other great intellectual adventures.[46] Yet note that, unlike cognitive science, ecoproprioceptual awareness is not focused on causal explanation or physiological mapping as such but on participatory, indeed performative, and relational enactment. Thinking-writing ecoproprioceptually, and even reading such writing, can optimally serve as induction into a further state of ecoproprioception; this would embody the root principle beyond mere conceptualization, so that one's processual engagement awakens an actual practice to be integrated in the course of life itself. These approaches and many current practice disciplines—somatic, noetic, gnoetic—offer guidance in developing an extended orientation by field. Ecoproprioceptive events are not so much planned as invited, or called home, to find ways to let nature, including our own nature, do the talking.

That talking is drawn to our poiesis, the immediate energy of language renewing itself, revitalizing ecosomatic thinking, in training to face the crisis that is the new life on Earth.

And despite the utter seriousness of the crisis facing us we must not lose the sense of play, always the first thing to go in "serious" environmental discourse, because play—enjoyment—is the attractor of inspired making which is critical to new thinking—the willing disorientation that conjures new orientation; the precariousness and being-on-edge that brings leaping to thought. Otherwise, we lose motivation through broken connection with "the force that through the green fuse drives the flower."[47]

Notes

1 David Bohm, *On Dialogue* (London and New York: Taylor & Francis e-Library, 2003), 25.
2 Italics in original. Ibid., 79.

3 The force of this usage of "edge" is explored in depth in Edward S. Casey, *The World on Edge* (Bloomington: Indiana University Press, 2017).

4 Gregory Bateson, *Steps to an Ecology of Mind: Collected Essays in Anthropology, Psychiatry, Evolution, and Epistemology* (San Francisco: Chandler Pub. Co., 1972); Gregory Bateson, *Mind and Nature: A Necessary Unity* (New York: Dutton, 1979). Bateson's work goes beyond the interdisciplinary to the trans-categorial for a further rationality.

5 The neologism "somapoeia" adds a somatic dimension to Ezra Pound's influential usage (1918) for three dominant tendencies in poetry: *phanopoeia, melopoeia,* and *logopoeia. See* T. S. Eliot, ed., *Literary Essays of Ezra Pound* (New York: New Directions Publishing, 1954), 25. *See also* my serial poem referencing *soma*'s double origin as "body" (Gk.) and the Vedic god and ritual drink "Soma" (Sk.): George Quasha, *Somapoetics* (Fremont: Sumac Press, 1973).

6 From the poem, "Dance Concentric Waves," in Cheryl Pallant, *Her Body Listening* (Kenmore: BlazeVOX [books], 2017), 19. Pallant is a poet, writer, somatics specialist, energy healer, dancer, and professor whose writing pervasively embodies ecoproprioception.

7 From the poem, "Since You Asked," in Billie Chernicoff, *The Red Dress: Charms Troubled & Amorous* (New York: Dr. Cicero Books, 2015), 69. Chernicoff's work brings out the singular intimacy of ecopoiesis.

8 Bateson, *Steps to an Ecology of Mind: Collected Essays in Anthropology, Psychiatry, Evolution, and Epistemology,* 1.

9 *See* Charles Olson, "Proprioception," in *Collected Prose,* eds. Donald Allen and Benjamin Friedlander (Berkeley: University of California Press, 1997); Olson, "Projective Verse." A rare instance of body-centered thinking about poiesis previous to Olson's influence was Francis Berry, *Poetry and the Physical Voice* (New York: Oxford University Press, 1962). Both the Olson and Berry works influenced the present writer in his mid-20s.

10 Walt Whitman, "A Song of the Rolling Earth," in *Leaves of Grass* (Philadelphia: David McKay, Publisher, 1891–1892), 176.

11 Albert Bates Lord, *The Singer of Tales* (Cambridge: Harvard University Press, 1960).

12 Sondra Fraleigh, *Dancing Identity: Metaphysics in Motion* (Pittsburgh: University of Pittsburgh Press, 2004), 52. Fraleigh's book is implicitly one of the major expositions of ecoproprioception.

13 In a letter to Dr. Trusler dated August 23, 1799. William Blake, David V. Erdman, and Harold Bloom, *The Complete Poetry and Prose of William Blake,* newly revised ed. (New York: Anchor Books, 1988), 702.

14 Estée Klar and Adam Wolfond, "S/pace: Neurodiversity in Relation," *Ought: The Journal of Autistic Culture* 2, no. 1 (2020), https://scholarworks. gvsu.edu/ought/vol2/iss1/7. *See also* Estée Klar-Wolfond, "Neurodiversity in Relation: An Artistic Intraethnography" (Ph.D. York University, 2020), 3.

15 Qtd. in Eva Kolcze, Adam Wolfond, and Estée Klar, *S/Pace* (2019), film. *See* https://vimeo.com/363876394. Wolfond is the author of *The Wanting Way: Poems* (Minneapolis: Milkweed Editions, 2022) and a founding member of *dis assembly,* a neurodiverse artist collective based in Toronto. On the

importance of collaboration, *see* Jeevika Verma, "For Neurodivergent, Non-speaking Poets, Collaboration is the Basis of Language," *NPR* (April 29, 2022). www.npr.org/2022/04/29/1095206261/for-neurodivergent-non-speaking-poets-collaboration-is-the-basis-of-language.

16 Kolcze, Wolfond, and Klar, *S/Pace*. Regarding atmosphere as conveyor of affect, *see* Edward S. Casey, *Turning Emotion Inside Out* (Chicago: Northwestern University Press, 2022).

17 For a far-reaching, cross-disciplinary study of successful intentional encounters with non-human language, which suggest that human beings have the intrinsic ability to learn non-natural language ("*xeno*proprioception"), *see* Diana Reed Slattery, *Xenolinguistics: Psychedelics, Language, and the Evolution of Consciousness* (Berkeley: Evolver Editions, an imprint of North Atlantic Books, 2015).

18 Estée Klar and Adam Wolfond, *Painting on Film with Adam Wolfond* (2019), film. See www.esteerelation.com/media.

19 Wolfond qtd. in Klar-Wolfond, "Neurodiversity in Relation: An Artistic Intraethnography," 185; 199.

20 Ibid., 8.

21 Ibid., 186, 48.

22 On the Eigner-Olson relationship *see* Larry Eigner and George Hart, "Six Letters," *Poetry* 205, no. 3 (2014).

23 Larry Eigner, *Air the Trees* (Los Angeles: Black Sparrow Press, 1967).

24 From Robert Duncan, "Pages from a Notebook [1953]," in *The New American Poetry, 1945–1960*, ed. Donald Allen (New York: Grove Press, 1960), 406.

25 For how far science has come since intelligence and language in animals and plants were viewed as "New Age nonsense," *see* Jeremy Narby, *Intelligence in Nature: An Inquiry into Knowledge* (New York: Jeremy P. Tarcher/Penguin, 2005); Erna Buffie, "What Plants Talk About," *PBS Nature* (2013).

26 Rebecca R. Burrill, "Languages of Nature, Languages of Art," *Humans and Nature* (October 27, 2020). https://humansandnature.org/languages-of-nature-languages-of-art/.

27 Laynie Browne, *You Envelop Me* (Oakland: Omnidawn Publishing, 2017), 28.

28 Manning and Massumi's innovative discourse with its clear poiesis of transformative thinking directly addresses our issues here (e.g., neurodivergence) in Erin Manning and Brian Massumi, *Thought in the Act: Passages in the Ecology of Experience* (Minneapolis: University of Minnesota Press, 2014). *See also* Erin Manning, *The Minor Gesture* (Durham: Duke University Press, 2016). Her work with the SenseLab at Concordia University and in "relational art" is itself a laboratory in the ecoproprioceptive. Adam Wolfond has studied with her in and out of classes since age 12, as non-speaking children do not have access to education in Canada.

29 Wallace Stevens, "Of Modern Poetry," in *The Collected Poems of Wallace Stevens* (New York: Vintage Books: A Division of Random House, 1990), 239.

30 Letter to Paul Demeny (Charleville, May 15, 1871) in Arthur Rimbaud, Wallace Fowlie, and Seth Adam Whidden, *Rimbaud: Complete Works, Selected Letters: A Bilingual Edition* (Chicago: University of Chicago Press, 2005), 374.

31 Merlin Sheldrake, *Entangled Life: How Fungi Make Our Worlds, Change Our Minds and Shape Our Futures* (New York: Random House, 2020).

32 Marjolein Oele, *E-Co-Affectivity: Exploring Pathos at Life's Material Interfaces* (Albany: State University of New York Press, 2020), 21.

33 Italics in original. Ibid., 20.

34 Dan Webb, "John Coltrane—Both Directions at Once: The Lost Album," *Sungenre*, 2018, https://sungenre.com/review/john-coltrane-both-directions-at-once-the-lost-album/.

35 Fraleigh, *Dancing Identity*, 168.

36 Robert Duncan, *The Collected Later Poems and Plays*, ed. Peter Quartermain (Berkeley: University of California Press, 2014), 8.

37 Robert Duncan, "The Lasting Contribution of Ezra Pound," in *Collected Essays and other Prose*, ed. James Maynard (Berkeley: University of California Press, 2014), 101.

38 Fraleigh, *Dancing Identity*, 167.

39 My own celebration of the underlying principle ("beauty=optimal x precarious") is in George Quasha and Carter Ratcliff, *Axial Stones: An Art of Precarious Balance* (Berkeley: North Atlantic Books, 2006). *See also* Edward S. Casey, "Quasha at the Edge," in *Zero Point Poiesis: George Quasha's Axial Art*, ed. Burt Kimmelman (New York: Aporeia, Marsh Hawk Press, 2022).

40 Oele, *E-Co-Affectivity*, 46.

41 Email from Edward S. Casey to the author, July 24, 2022.

42 On the "fallacy of misplaced concreteness," *see* Alfred North Whitehead, *Science and the Modern World [1925]*, Lowell lectures (New York: Free Press [Simon & Schuster], 1967), 52.

43 I focus on the healing potential of this transformatively performative event in "Healing Poetics" in George Quasha and Edward S. Casey, *Poetry in Principle* (New York: Spuyten Duyvil, 2019), 29–63.

44 Preverbs, my invented genre for over 20 years, are axial acts of language. *See also* George Quasha, "Preverbs," (2011). https://quasha.com/poetry/preverbs-and-axial-poems/. The first two lines are from "Gnostalgia for the Present," unpublished. The third line is from "Rippling Scales," in George Quasha, *Waking from Myself (preverbs)* (Barrytown: Station Hill Press, 2022), 116.

45 Martin Heidegger, *Country Path Conversations*, English ed. (Bloomington: Indiana University Press, 2010), 13 passim.

46 For the core enactivist position that cognition arises through dynamic interaction between active agent and environment, *see* Evan Thompson, *Mind in Life: Biology, Phenomenology, and the Sciences of Mind* (Cambridge, MA: Belknap Press of Harvard University Press, 2007). A further development in social neuroscience, the Interactive Brain Hypothesis (IBH), offers a likely supportive hypothesis to the thinking involved in ecoproprioception, one that shows an evolution within cognitive science important to ecology-of-mind inquiry. *See* Ezequiel Di Paolo and Hanne De Jaegher, "The Interactive Brain Hypothesis," *Frontiers in Human Neuroscience* 6, no. 163 (June 7, 2012). https://doi.org/doi:10.3389/fnhum.2012.00163.

47 Dylan Thomas and Daniel Jones, "The Force that Through the Green Fuse Drives the Flower," in *The Poems of Dylan Thomas* (New York: New Directions, 2003), 90.

Bibliography

Bateson, Gregory. *Steps to an Ecology of Mind: Collected Essays in Anthropology, Psychiatry, Evolution, and Epistemology.* San Francisco: Chandler Pub. Co., 1972.

———. *Mind and Nature: A Necessary Unity.* New York: Dutton, 1979.

Berry, Francis. *Poetry and the Physical Voice.* New York: Oxford University Press, 1962.

Blake, William, David V Erdman, and Harold Bloom. *The Complete Poetry and Prose of William Blake,* newly revised ed. New York: Anchor Books, 1988.

Bohm, David. *On Dialogue.* London; New York: Taylor & Francis e-Library, 2003.

Browne, Laynie. *You Envelop Me.* Oakland: Omnidawn Publishing, 2017.

Buffie, Erna. "What Plants Talk About," season 31, episode 9 *PBS Nature,* 2013, video, 30 min. Produced by Merit Jensen Carr. https://www.pbs.org/wnet/nature/what-plants-talk-about-introduction/8228/.

Burrill, Rebecca R. "Languages of Nature, Languages of Art." *Humans and Nature* (October 27, 2020). https://humansandnature.org/languages-of-nature-languages-of-art/.

Casey, Edward S. *The World on Edge.* Bloomington: Indiana University Press, 2017.

———. "Quasha at the Edge." In *Zero Point Poiesis: George Quasha's Axial Art,* edited by Burt Kimmelman, 69–66. New York: Aporeia, Marsh Hawk Press, 2022.

Chernicoff, Billie. *The Red Dress: Charms Troubled & Amorous.* New York: Dr. Cicero Books, 2015.

Di Paolo, Ezequiel, and Hanne De Jaegher. "The Interactive Brain Hypothesis." *Frontiers in Human Neuroscience* 6, no. 163 (June 7, 2012). https://doi.org/doi:10.3389/fnhum.2012.00163.

Duncan, Robert. "Pages from a Notebook [1953]." In *The New American Poetry, 1945–1960,* edited by Donald Allen, 400–407. New York: Grove Press, 1960.

———. *The Collected Later Poems and Plays.* Edited by Peter Quartermain. Berkeley: University of California Press, 2014.

———. "The Lasting Contribution of Ezra Pound." In *Collected Essays and Other Prose,* edited by James Maynard, 99–102. Berkeley: University of California Press, 2014.

Eigner, Larry. *Air the Trees.* Los Angeles: Black Sparrow Press, 1967.

Eigner, Larry, and George Hart. "Six Letters." *Poetry* 205, no. 3 (2014): 250–268.

Eliot, T. S., ed. *Literary Essays of Ezra Pound.* New York: New Directions Publishing, 1954.

Fraleigh, Sondra. *Dancing Identity: Metaphysics in Motion.* Pittsburgh: University of Pittsburgh Press, 2004.

Heidegger, Martin. *Country Path Conversations.* English ed. Bloomington: Indiana University Press, 2010.

Klar, Estée, and Adam Wolfond. *Painting on Film with Adam Wolfond.* 2019. film.

———. "S/pace: Neurodiversity in Relation." *Ought: The Journal of Autistic Culture* 2, no. 1 (2020). https://scholarworks.gvsu.edu/ought/vol2/iss1/7.

Klar-Wolfond, Estée. "Neurodiversity in Relation: An Artistic Intraethnography." Ph.D., York University, 2020.

Kolcze, Eva, Adam Wolfond, and Estée Klar. *S/Pace*. 2019. film.

Lord, Albert Bates. *The Singer of Tales*. Cambridge: Harvard University Press, 1960.

Manning, Erin. *The Minor Gesture*. Durham: Duke University Press, 2016.

Manning, Erin, and Brian Massumi. *Thought in the Act: Passages in the Ecology of Experience*. Minneapolis: University of Minnesota Press, 2014.

Narby, Jeremy. *Intelligence in Nature: An Inquiry into Knowledge*. New York: Jeremy P. Tarcher/Penguin, 2005.

Oele, Marjolein. *E-Co-Affectivity: Exploring Pathos at Life's Material Interfaces*. Albany: State University of New York Press, 2020.

Olson, Charles. "Projective Verse." In *Collected Prose*, edited by Donald Allen, Benjamin Friedlander and with an introduction by Robert Creeley, 239–249. Berkeley: University of California Press, 1997.

———. "Proprioception." In *Collected Prose*, edited by Donald Allen and Benjamin Friedlander, 181–183. Berkeley: University of California Press, 1997.

Pallant, Cheryl. *Her Body Listening*. Kenmore: BlazeVOX [books], 2017.

Quasha, George. *Somapoetics*. Fremont: Sumac Press, 1973.

———. "Preverbs." (2011). https://quasha.com/poetry/preverbs-and-axial-poems/

———. *Waking from Myself (Preverbs)*. Barrytown: Station Hill Press, 2022.

Quasha, George, and Edward S. Casey. *Poetry in Principle*. New York: Spuyten Duyvil, 2019.

Quasha, George, and Carter Ratcliff. *Axial Stones: An Art of Precarious Balance*. Berkeley: North Atlantic Books, 2006.

Rimbaud, Arthur, Wallace Fowlie, and Seth Adam Whidden. *Rimbaud: Complete Works, Selected Letters: A Bilingual Edition*. Chicago: University of Chicago Press, 2005.

Sheldrake, Merlin. *Entangled Life: How Fungi Make Our Worlds, Change Our Minds and Shape Our Futures*. New York: Random House, 2020.

Slattery, Diana Reed. *Xenolinguistics: Psychedelics, Language, and the Evolution of Consciousness*. Berkeley: Evolver Editions, an imprint of North Atlantic Books, 2015.

Stevens, Wallace. "Of Modern Poetry." In *The Collected Poems of Wallace Stevens*, 239. New York: Vintage Books; A Division of Random House, 1990.

Thomas, Dylan, and Daniel Jones. "The Force that through the Green Fuse Drives the Flower." In *The Poems of Dylan Thomas*, 90. New York: New Directions, 2003.

Thompson, Evan. *Mind in Life: Biology, Phenomenology, and the Sciences of Mind*. Cambridge, MA: Belknap Press of Harvard University Press, 2007.

Verma, Jeevika. "For Neurodivergent, Non-Speaking Poets, Collaboration Is the Basis of Language." *NPR* (April 29, 2022). www.npr.org/2022/04/29/109 5206261/for-neurodivergent-non-speaking-poets-collaboration-is-the-basis-of-language.

Webb, Dan, "John Coltrane—Both Directions at Once: The Lost Album." *Sungenre*, 2018, https://sungenre.com/review/john-coltrane-both-directions-at-once-the-lost-album/.

Whitehead, Alfred North. *Science and the Modern World [1925]*. Lowell lectures. New York: Free Press (Simon & Schuster), 1967.

Whitman, Walt. "A Song of the Rolling Earth." In *Leaves of Grass*, 176. Philadelphia: David McKay, Publisher, 1891–1892.

18

OUTDOOR DANCES

Meditations on Loss in the Finger Lakes and Beyond

Missy Pfohl Smith

Finger Lakes, New York, U.S.A.: 42.7238° N, 76.9297° W

The rapidly expanding changes in climate caused by humans and their lack of compassion for the Earth are making way for rising mental health concerns, and even worse, these impacts are disproportionately affecting impoverished and systemically oppressed groups.[1] An extensive literature review by Hurly and Walker shares convincing evidence that spending time in nature, or "nature relatedness," has psychological and physiological benefits and contributes to overall well-being.[2] Similarly, respondents in a study by Hansmann, Hug, and Seeland expressed a statistically significant reduction in stress level and headaches, and increase in feelings of well-being after walking in the forest and/or spending time in a park, in surveys taken in Zurich, Switzerland by 164 people aged 15 and over (with 49% over age 50).[3] Further, the higher the stress levels that were reported, the greater the reduction in stress was acknowledged, and those participating in sports within the forest or park reported significantly better benefits and stress reduction than those doing activities that were less physical.[4]

In 2020 and 2021, during the first and only global pandemic that I and my contemporaries have experienced, I created three outdoor dances. These creative practice meditations on environmental loss emerged simultaneously with the entrance to a personal grieving process from the loss of my father in 2020. The first dance, *Pilgrimage*, was welcomed by a forgotten wooded landscape in the Finger Lakes region near Keuka Lake, a place that my family has summered for decades. The site had been at one distant time a forest gallery for human-made and nature-inspired sculptures and structures. Bringing the dancers of BIODANCE together after a several-month hiatus, after failed attempts at remote online rehearsals, our joy of

DOI: 10.4324/9781003390985-23

dancing together again and immersing in nature was palpable, if balanced by the sense of collective bereavement we were experiencing. It was June of 2020, and we gathered as artists in this remote private property, open only to hikers, to contemplate our experiences of recent loss—of connection, of loved ones, of life as we had always known it. A weathered staircase to the sky that loomed above an overgrown labyrinth pieced together with stone many years prior became a home to this collective and meandering journey.

The dance wove together our individual loss with a collective environmental loss that had taken place in the area, most recently in 2014 when a flood wreaked havoc in the small neighboring towns of Branchport and Penn Yan. Though the towns still stand currently, the collapsed community center and washed-out roads, businesses, and homes would never return to their beloved origins.[5] But of course this is only the land's recent history, which compounds the great devastation of the Haudenosaunee peoples, otherwise known as the Seneca Nation, and their Earth-friendly ways of living.[6] A memorial for the mother of Red Jacket stands a stone's throw away from the lakefront cottages occupied by my husband's family since 1970.[7]

I was compelled to create a second dance, *Root to Leaf* (2021), an immersive solo in the woods of Powder Mills Park in New York State's Monroe County, with a sound score created by the wind, the woods, and its inhabitants, and informed by the writing of Mara Ahmed who also recognized the wisdom and comfort of the landscape in times of loss.[8] Lying on the earth in a bed of mayflowers towering over my prone body like tiny green umbrellas giving me shelter was at once settling and playful. I felt small, protected, comforted, and delighted.

I titled the third dance *Numbered Days*, inspired by the last line of a poem by Lauren K. Alleyne, who shares my sense of dread for our misused and disrespected Earth and its people of color: "For now, the beauty of it fills my heart to breaking as I head out into our numbered days."[9] *Numbered Days* (2021) became the seed for a staged dance *Wind*, part of a recent collaborative show with media artist W. Michelle Harris and the Dave Rivello Ensemble, *Elemental Forces Redux* (2022). In making *Wind*, in which projected site-specific video bathed the live dancers and brought the outdoors in, my collaborators and I considered human destruction of the Earth and catastrophic climate change as metaphors for human destruction of one another. Two weeks after we captured the video, I returned to the location for a hike, and in the clearing amid trees in Powder Mills Park, I discovered the devastation of the landscape from a violent windstorm. It was as if mother and father nature themselves became collaborators, warning us all to care for the Earth, and for one another. Glenn Albrecht proposed the concept of solastalgia in 2003, defined as "place-based distress" resulting from loss due to climate change in the places we call home.[10] I felt this sense of solastalgia in my constricted chest, and in my clenched fists during restless nights. Later, the collaborative show, *Elemental Forces Redux* (2022), connected recent stories of local and global climate events to my choreographic works, *Wind, Earth, Water, Fire*, and *Air*. *Air* tells the story of a small Italian village banding together to save their oldest tree from fire damage, imagining a world in which we care.[11] A recent review of climate change studies reported that a wide array of researchers agree that making conscious the connections between extreme weather in local communities and climate change is helping to raise the belief that climate change is real.[12]

Deeply feeling the care that nature was providing, I started to experiment with some workshops to give participants somatic opportunities to connect movement choices with inspiration designed by nature itself. Through collaboration and creative process, I wanted to encourage and facilitate opportunities to immerse, notice, listen, and respond to our environment. Not surprisingly, the noticing of self is occurring at the same time as the noticing of nature (*see* Figure 18.1). While I wanted to draw attention to the fascinating and complex details that often go unnoticed and unappreciated

FIGURE 18.1 Smith's body bridging water and land, Keuka Lake, NY. Photo by Ella S. Smith.

in the natural world, I also wanted to provide an opportunity for embodied somatic experiences. By translating the discoveries of natural design into both internal dances of nature and performative choreographies, I hope to find a means for not only the healing of our relationship with nature, but also our relationship with self and with one another, and in doing so helping to heal the Earth itself.

My first foray was a workshop I created entitled, *Earth and Snow: An Outdoor Performance Adventure.* Designed for a conference in northern New York in March, I endeavored to invite participants to notice the still beauty of the winter season, to draw from nature's brilliant designs and create dances that might help us appreciate and protect our planet. Being in "less than ideal" weather conditions such as rain or snow invited a heightened sense of awareness and unexpected appreciation. Encouraged, I continued to experiment with workshops in nature, in the sun and the rain, the snow and the wind, the mud and the sand, the warmth and the cold. By choosing to embrace the elements, we open a door for becoming more aware of our own habitual responses and our bracing in relation to our potential for yielding (*see* Figure 18.2). Peggy Hackney takes this concept further in discussing how "yield and push" both connects and grounds us to the Earth and gravity, and "reach and pull" connects us with the surrounding world.[13]

FIGURE 18.2   Smith's body blends with the roots, branches and water of Keuka Lake, NY. Photo by Ella S. Smith.

Practice

I invite you to take in your surroundings, and to take your time to find a location where you can begin to be still. This can be standing, seated, prone, leaning by a tree, tucked into a wooded nook, or gazing over the land (*see* Figure 18.3). I invite you to choose a position that allows you to feel at home in your body and breath. Then if you are comfortable, I invite you to close your eyes or soften your gaze to heighten your other senses.

Let's start with the sense of touch. What parts of your body are in contact with the earth or your natural surroundings? What is the quality of that contact? Is it completely comfortable? How can you invite yielding into those points of contact and your breath? Do you notice any places of bracing in your body and/or your breath? What do the surfaces you are in contact with feel like? Do you feel you need to make adjustments in response to the invitation to cradle into the earth? Explore....

What are you hearing, or if hearing is not in reach, what are you feeling and sensing more deeply? Are you alone in nature? Who or what are you sharing this arena with? Do you hear or sense the rustling of leaves or branches, or the chirping of birds, or the shuffling of small animals, or

FIGURE 18.3   Missy Pfohl Smith sits nestled in a tangle of submerged roots, Keuka Lake, NY. Photo by Ella S. Smith.

the buzzing of insects sharing the space? Do you sense the sounds and movements of your own breath and how they interplay with the orchestra of nature? Do you hear/sense your own heartbeat, or the fluids of your swallow or your pulse? Do you notice different speeds and time signatures in your environment? I invite you to be responsive to these sounds/ movements both within and surrounding, moving with the fluids, with the power, and with the subtlety of the sounds in all directions, including within, above, below, and around you. Explore ....

I invite you to take in your visual surroundings, and if that is not accessible to you, what shapes and textures do you feel in your near surroundings? What repetitions do you notice? What lines and shapes and pathways? What large expanses do you notice and what in your own self feels expansive and large? How can you allow your expansiveness to converse with the expanses around you both near and far, seen and beyond seen, real and imagined? What tiny details can you detect very close to you. Are they organic, geometric, meandering, organized, or random? Explore....

What do you smell, and what is inviting you to breathe it in? What causes you to draw back? What do these multiple and dynamic scents feel like in your body? How do they move? How do they shift? Are they strong, or subtle and light? And what do you taste in this moment? Is it a real taste or an imagined taste? If you could taste that pathway, or taste that distant rustle, what would that be like in your body? Explore ....

Let's return to a stillness that has subtle motion as a constant. Let's return to a simple attention to breath and our sense of connection, and yield to the surfaces with which we are connected.

Do you notice any shifts:

In your "self?"
In your state of mind or state of being?
In your connection to your environment or any of the beings with
    which you are sharing your environment?

Evidence suggests that climate change is contributing to feelings of deep frustration, mourning, and hopelessness, particularly for women and for those whose awareness is acute.[14] By witnessing and discussing explorations or choreographic creations of others inspired by nature, I hope that we can invite conversation and consideration of the details of the environment and its inhabitants, including nature as partner. Witnessing can acknowledge the complexities, gifts, and surprises that our Earth and sky and all beings have to share. We can support the well-being of nature just as nature promotes ours.

Notes

1 Katie Hayes et al., "Climate Change and Mental Health: Risks, Impacts and Priority Actions," *International Journal of Mental Health Systems* 12, no. 1 (2018): 28, https://doi.org/10.1186/s13033-018-0210-6.

2 Jane Hurly and Gordon J. Walker, "Nature in Our Lives: Examining the Human Need for Nature Relatedness as a Basic Psychological Need," *Journal of Leisure Research* 50, no. 4 (2019): 303, https://doi.org/10.1080/00222 216.2019.1578939.

3 Ralf Hansmann, Stella-Maria Hug, and Klaus Seeland, "Restoration and Stress Relief through Physical Activities in Forests and Parks," *Urban Forestry & Urban Greening* 6, no. 4 (2007): 216, 219, https://doi.org/10.1016/ j.ufug.2007.08.004.

4 Ibid., 220, 222.

5 Mike Hibbard, "Hard to Believe: Flood Waters Cause Heavy Damage in Penn Yan, Parts of Yates," *Finger Lakes Times*, May 14, 2014, www.fltimes.com/ news/hard-to-believe-flood-waters-cause-heavy-damage-in-penn-yan-parts-of-yates/article_446c0090-db71-11e3-9783-001a4bcf887a.html.

6 Seneca Nation of Indians. "Seneca Nation of Indians Culture," Seneca Nation of Indians, accessed February 1, 2023, https://sni.org/culture/history/.

7 Red Jacket (c. 1758–1830), later named Sagoyewatha, was a skilled orator and a Seneca chief of the Six Nations of the Haudenosaunee. *See* Christopher Densmore, *Red Jacket: Iroquois Diplomat and Orator* (Syracuse: Syracuse University Press, 1999); Arthur C. Parker, *Red Jacket, Seneca Chief* (Lincoln: University of Nebraska Press, 1998). Parker was a Native American ethnographer born on the Cattaraugus Seneca Reservation.

8 *See* Mara Ahmed, "The Warp & Weft," Mara Ahmed Studio, May 2021, www. maraahmedstudio.com/category/warpweft/.

9 Lauren K. Alleyne, *Honeyfish*, 1st ed. (Kalamazoo: New Issues/Western Michigan University, 2019), 11.

10 Glenn Albrecht et al., "Solastalgia: The Distress Caused by Environmental Change," *Australasian Psychiatry* 15, no. S1 (2007): S96, https://doi.org/ 10.1080/10398560701701288.

11 Gaia Pianigiani, "Sardinian Village Tries to Save an Ancient Tree Scorched by Fire," *The New York Times*, July 26, 2021, www.nytimes.com/2021/08/22/ world/europe/italy-sardinia-fire-tree.html.

12 *See* Joseph P. Reser, Graham L. Bradley, and Michelle C. Ellul, "Encountering Climate Change: 'Seeing' Is More than 'Believing'," *WIREs Climate Change* 5, no. 4 (2014), https://doi.org/10.1002/wcc.286.

13 Peggy Hackney, *Making Connections: Total Body Integration through Bartenieff Fundamentals* (London; New York: Routledge, 2002), 97.

14 Julia Rothschild and Elizabeth Haase, "Women's Mental Health and Climate Change Part II: Socioeconomic Stresses of Climate Change and Eco-anxiety for Women and their Children," *International Journal of Gynecology and Obstetrics* 160, no. 2 (2023), https://doi.org/10.1002/ijgo.14514; Hayes et al., "Climate Change and Mental Health: Risks, Impacts and Priority Actions."

Bibliography

Albrecht, Glenn, Gina-Maree Sartore, Linda Connor, Nick Higginbotham, Sonia Freeman, Brian Kelly, Helen Stain, Anne Tonna, and Georgia Pollard. "Solastalgia: The Distress Caused by Environmental Change." *Australasian Psychiatry* 15, no. S1 (2007): S95–S98. https://doi.org/10.1080/1039856070 1701288.

Alleyne, Lauren K. *Honeyfish*. 1st ed. Kalamazoo: New Issues/Western Michigan University, 2019.

Densmore, Christopher. *Red Jacket: Iroquois Diplomat and Orator*. Syracuse: Syracuse University Press, 1999.

Hackney, Peggy. *Making Connections: Total Body Integration through Bartenieff Fundamentals*. London; New York: Routledge, 2002.

Hansmann, Ralf, Stella-Maria Hug, and Klaus Seeland. "Restoration and Stress Relief through Physical Activities in Forests and Parks." *Urban Forestry & Urban Greening* 6, no. 4 (2007): 213–225. https://doi.org/10.1016/j.ufug.2007.08.004.

Hayes, Katie, Grant Blashki, John Wiseman, Susie Burke, and Lennart Reifels. "Climate Change and Mental Health: Risks, Impacts and Priority Actions." *International Journal of Mental Health Systems* 12, no. 1 (2018): 28–12. https://doi.org/10.1186/s13033-018-0210-6.

Hibbard, Mike. "Hard to Believe: Flood Waters Cause Heavy Damage in Penn Yan, Parts of Yates." *Finger Lakes Times*, May 14, 2014. www.fltimes.com/news/hard-to-believe-flood-waters-cause-heavy-damage-in-penn-yan-parts-of-yates/article_446c0090-db71-11e3-9783-001a4bcf887a.html.

Hurly, Jane, and Gordon J. Walker. "Nature in Our Lives: Examining the Human Need for Nature Relatedness as a Basic Psychological Need." *Journal of Leisure Research* 50, no. 4 (2019): 290–310. https://doi.org/10.1080/00222 216.2019.1578939.

Parker, Arthur C. *Red Jacket, Seneca Chief*. Lincoln: University of Nebraska Press, 1998.

Pianigiani, Gaia. "Sardinian Village Tries to Save an Ancient Tree Scorched by Fire." *The New York Times*, July 26, 2021. www.nytimes.com/2021/08/22/world/europe/italy-sardinia-fire-tree.html.

Reser, Joseph P., Graham L. Bradley, and Michelle C. Ellul. "Encountering Climate Change: 'Seeing' Is More than 'Believing'." *WIREs Climate Change* 5, no. 4 (2014): 521–537. https://doi.org/10.1002/wcc.286.

Rothschild, Julia, and Elizabeth Haase. "Women's Mental Health and Climate Change Part II: Socioeconomic Stresses of Climate Change and Eco-Anxiety for Women and their Children." *International Journal of Gynecology and Obstetrics* 160, no. 2 (2023): 414–420. https://doi.org/10.1002/ijgo.14514.

Seneca Nation of Indians. n.d. "Seneca Nation of Indians Culture." Seneca Nation of Indians, accessed February 1, 2023, https://sni.org/culture/history/.

19

AWE AND EMPATHY

Edward S. Casey

Stony Brook, New York, U.S.A.: 40.9027° N, 73.1338° W

I here explore two aspects of our relationship with the wild world, each of which has important ecosomatic dimensions and both of which relate intimately to ecoproprioception. These are awe and empathy. Awe provides the specific emotionality of this relationship, while empathy offers an ethical dimension. Neither has exclusive domain. Awe is supplemented by other emotionalities such as intrigue and wonder; empathy is often embedded in a complex whose other members include concern and sympathy. In all these variations, the lived body is the animating factor, providing at once an exploratory edge and a sustaining force.

All too often we think of emotion as self-expressive: a matter of what *I* feel Now—something I want to "get off my chest." And to back this up, many theories of emotion emphasize its roots within ourselves as singular isolated persons—roots in the heart, the mind, and (assiduously pursued these days) the brain. This is to think of emotion as *mine*, something I generate on my own, something I can claim as belonging to my *propria persona*, my own personal being. But what if things are different? What if emotion may come upon us, stemming from *out there*—out there in the world around and beyond me? What then?

Today I want to pursue two cases in point, two instances in which emotion comes to us from elsewhere first of all and primarily rather than from inside me. Once I take it over, it can very well become mine, part of my personal history, and I can give to it my own signature and stamp. But in the beginning it arrives from somewhere else—from *out there* rather than from *in here*. In one case emotion emerges from my interface with the natural world, the other from my relationships with other humans—and

DOI: 10.4324/9781003390985-24

with animals as well. The first is the awe I experience when confronted with a striking scene from nature—paradigmatically, an open ocean, a cluster of mountains, a rainbow, a violent storm. The other is my empathic connection with someone I know to be suffering—to whom, as we say revealingly, "my heart goes out." This empathic bond also obtains among non-human animals.

In awe we *stand back*—back before the sheer spectacle that nature presents to us in its more dramatic moments. We are admiring bystanders in such situations, preoccupied with taking the spectacle in from without. In empathy we *step forward* in feeling the gravity of another's stressful situation and are moved to do something about it: to intervene in some constructive way. Different as they are, in both situations the emotionality exceeds us—whether it stems from a mountain range we are watching, or from witnessing young people on the streets of Iran being attacked by the police. Whatever happens subsequently, the awe and the empathy come first of all *to us* rather than *from us*.

Awe and empathy alike each bear unexpected dimensions. Even if awe is typically felt in the presence of certain massive phenomena of nature—a roaring waterfall, a colossal mountain range—we can also be awed by certain miniscule phenomena such as hummingbirds or the tiny buds of various plants. We think of empathy as confined to the human realm, but elephants and whales are able to engage in empathic relations.

In awe and empathy, we have two forms of expanded emotion—one largely (but not exclusively) concerned with what is experienced in the outside natural world, the other mostly (but not entirely) happening in the intersubjective context of human and more-than-human interaction. Taken together, awe and empathy represent two primary ways in which we take emotion in as it presents itself before us: as it is *given to us* in a vast variety of settings.

What follows are descriptive forays into these two emotionalities taken separately, followed by a discernment of a common bonding.

Awe

First Impression

At the Santa Barbara Sea Center recently, I was struck by a film that was showing on an enormous screen—at least 12'X12'—and that took up an entire wall at the end of the museum. In the film itself, gigantic swaying plants rooted at the bottom of the ocean were pervaded by numerous fish of many sizes and kinds that threaded their way through these same plants. Sometimes a fish, or a stray seal, would come straight toward me, as if

trying to establish communication—only to turn away abruptly at the last
moment. I was not just captivated by this spectacle but carried away by
it: I was awed by it.

Where was such a sublime spectacle situated? Certainly not *in me*,
though I could feel its resonance all around me. What induced my awe was
out there in the marine world that was captured in the film I encountered.
I may claim the awe as mine, but what moved me was something awesome
in the undersea world to which I was given momentary access at the Sea
Center.

Three key factors in the experience of the awesome are force, scale,
and image.

Force names the way that what induces awe bears down upon us, often
without full warning. The very film that occasioned these reflections was
thrust upon me as I entered the final room of the Sea Center's first floor.
I wasn't expecting it, much less looking for it. It is as if it *asserted itself*,
coming forward to confront me with its full force. This force was that
of another world than my own—an undersea world exhibiting a special
vitality that enlivened it before my very eyes, drawing my perceiving body
into that world as if I were there in person.

Scale bears upon the comparative size of this same world. I grasped
immediately that it is enormous—indeed, that it has virtually no limits
and that whatever limits it does possess (e.g., the sea bottom, the edges
of islands, and whole continents) are remotely located vis-à-vis the scenes
I was witnessing in this film. In short, the scale was *immense*—literally im-
measurable. But the awesome can also come forward on a much smaller
scale, as with the collection of small sea creatures in open-top tanks in
the next room over. What is awesome here is the presence of the intricate
structures of living beings who live within shells measuring only a few
centimeters.

Image. None of this would have been accessible without the explicit
imagistic content contained both in the film and in the nearby water tanks.
At each extreme, I was presented with parts of a much larger seascape that
I could sense through and in an imagistic format, whether that given to me
by filmic imagery or by the directly perceived bodies of the small entities
on display in the tanks. Different as were these two situations—in one case,
a matter of seeing things directly presented to my sensory awareness; in
the other, contents mediated by a moving picture camera—each conveyed
a vivid sense of the dynamics of living inhabitants of an encompassing
sea world. Such imagery renders coherent what, without imagery, would
likely be experienced as incoherent: items tossed around in blind con-
fusion: creatures with no place to be and no distinctive marks by which
to identify them. With perceptual imagery, what would otherwise be very

murky or too small to discern is set before us, and on this basis we can witness an entire world at a glance.[1]

Second Impression

Underlying force, scale, and image as they figure into experiences of the awesome is *place*. Land- and seascape painters often seek to convey their experience of awe felt before or *in a given place*: whether a valley or a sea channel, a swamp or a forest. The awesomeness of such places may well defy the application of descriptive signifiers indicating exact extent, palpability, manipulability, etc., but they exude a special power of implacement. A film of underwater life *puts us in the place* where animate bodies are shown swimming around or swaying in deep sea water. The aqueous place afforded by the sea not only surrounds them; it *situates* them. From the standpoint of the wild life it literally supports, it gives them a place-to-be; whereas from our human point of view, it affords them a place-to-be-witnessed in their own elemental presence.

This place is extraordinary not just in this doubly situational character, but also insofar as it defies any effort to measure its extent: to quantify it. As witnessed in the film, it is a complex qualitative whole whose constituents are continually changing even if the view offered by the film is recognizably the same thanks to the steady stationing of the camera and of my own position in the room where it is being shown. A delicate balance is struck between these two placial parameters, which combine dynamics of movement with continuing recognizability of content. Such a place is unique and cannot be exchanged with any other.[2]

Place in the natural realm is never entirely isolated. It exists always and only in relation to other places—a circumstance we can designate as *co-implacement*, the composition made up of a set of contiguous places. The natural world—whether on land or under the sea—is a scene of continual (and continually changing) co-implacements. It is comparable to a gigantic quilt of places whose con-figuring presences make up a unique pattern for any given stretch of land or sea. The full pattern is best seen from a distance though we can catch a glimpse of it from close-up as well.

So far from being a passive constituent of our experience of the natural world, place can dominate the field of vision almost defiantly, rendering our sensing selves subordinate to it as its sheer witnesses. When we are in awesome situations in particular, the factor of place contributes directly to its overwhelming force. What strikes us with awe need not be a singular formation—as with the Pitons in St. Lucia, two mountains that rise together in a radical verticality at the very edge of the sea—but is often an entire scene, a single but complex place: as when viewing the Grand Canyon from its southern edge. And the two scenarios may combine as well, as at Monument Valley in Utah, where towering sandstone rocks form part of a much larger landscape. I find myself agape at such a scene: my mouth falls open, my retinas expand, and I find myself out of breath in its presence. I have become a witness of its massive presence, awestruck by it. And not only a witness but actively and affectively moved as I gaze upon it. Its emotionality—the way it moves me—is a combination of the placiality of its sheer appearance, its being-*there*, with my lived body's being-*here*: here where I experience it in first person.

Third Impression

When I am struck by awe, I am moved in my bodily being. The primary phenomenon may be visual or auditory—hence situated outside me—but my lived body is an integral part of all experience of awe. Such experience takes the form of resonances that are often subtle—so subtle I cannot link them to a given organ or portion of my body. When we say that we are "moved" in awe, this is not only psychical or spiritual (though these are also often implicated) but also refers to ways in which my whole physical being is drawn out. The emotionality spreads through my entire lived body: as we imply when we say that we are "carried away" by awe. Then we feel the full force of the awesome in various parts of our experienced body: in the arms, or legs, or spine, or face. Our body, rather than being an intentional agent, has become a resonating field of diverse forces—some of which are easily locatable (as in a beating heart, or drawing in one's breath) but others are not so, given their subtlety: as with tremors or tinglings that we cannot simply locate.

The body in awe is the corporeal correlate of ecoproprioception in George Quasha's sense of the term: a form of field awareness that emerges at the edges of our familiar focal areas. In contrast with the focalized nature of proprioception (by which we situate ourselves in a determinate way), ecoproprioception is a matter of subtle peripheral consciousness that is often captured best by poetry that takes us out into the perimeters of ongoing experience. As Wallace Stevens puts it in his poem "Of Mere Being":

The palm at the end of the mind,
Beyond the last thought, rises
In the bronze décor ...

The palm stands on the edge of space.
The wind moves slowly in the branches.
The bird's fire-fangled feathers dangle down.[3]

Awe emanates from an entire field of presentation even if certain parts are more striking than others, such as a palm tree perceived at the horizon even as the sun sinks under the ocean.

Just as it is a matter of expanding the usual range of emotion when we attend to awe, so the setting of awe—its field, its place of presentation—is extended beyond the delimited focus of the Husserlian intentional object. The awesome is no object; it is an entire spectacle. And as such, a spectacle calls for a correlative expansion of the ecoproprioceptive field of awareness, so it implicates an extension of the felt range of lived bodily awareness into subtle domains rarely experienced in ordinary experience. Awe draws us out in every way we can imagine—emotionally, bodily, and by way of various modalities of pre-reflective consciousness. It draws us out into the peripheries of our experience—which is just where ecoproprioceptive awareness already takes us: into the larger landscape, a world beyond the proximal edges we are accustomed to in ordinary perception.[4]

Fourth Impression

Let us be clear: awe is not a single determinate experience but a complex collocation of different kinds of experience, ranging from fear and terror to reverence and respect as well as many variations of these basic states. Most of these experiences are covered in this single formulation: *Awe is a state of wonderment whose expressions range from extreme surprise to fear (including terror as an extreme) to reverent appreciation (whether the reverence bears on a deity or on a feature of the natural world).*

In other words, awe is not a single emotion but an epitome of what I have come to call "emotionality." By this I mean a complex combination of affective states, each of which counts as what we conventionally call "an emotion." When several such states combine, we have emotionality. A person who is caught up in emotionality can move back and forth between the various constituent affective modes—from fear to reverence and back again, or from being transported (carried away, inspired) to being focused on one feature of a moving spectacle. Unlike a single emotion such as fear—which has a determinate target as it were—emotionality allows the experiencer to oscillate between very different affective states and yet to feel caught up in a single experience termed "awe" in English. It is this versatility that, in the specific case of awe, allows the subject to range broadly within certain generous limits—and still to be able to say that "I am awed." In this case, the intentional object is no object at all but various transitional states, each with its own peculiar character—yet all of which configure in generating awe. This is not a case of mere ambiguity or outright ambivalence but of a polymorphic assemblage of affective states that remain alternative animations of a single event of emotionality. Awe is just such a multiplex happening.

Fifth Impression

I have left out one very significant factor in my analysis of awe: *wildness*. We rarely have awe before what is civilized, indeed anything that is well ordered and organized in accordance with all too well-known criteria with which we are fully familiar. But the wildness before which we experience awe is also not sheer disarray or utter chaos either. In its minimal acceptation, wild means uncultivated, as with wild plants and wild animals. But in the case of human beings, it also refers to such things as licentious behavior; passion; being reckless or careless; ferocious, destructive, cruel. "Going wild" means getting out of control, radically unruly. In the larger context of the natural world, it means getting out of control, tempestuous, violent. As Richard Powers puts it: "*That is so wild.* 'That's the word for it.'"[5]

For our purposes, "wild" signifies an essential dimension of that which occasions awe in us. Instead of looking for a single trait that encompasses all the ways in which what is wild occurs, our focus should be on ways of *going wild*. Examples include a storm that starts modestly but that gets out of control by turning into a tempest; an oceanic vista that looks like a simple sequence of regular waves but that builds into a raging surf that crashes against rocks on the shore, creating blasts of water at irregular intervals; an earthquake that moves with its own undermining rhythm; a child who at first seems calm and even timid who suddenly gets out of control, thrashing at his playmates for no apparent reason. What all these otherwise so different examples have in common is that they move from a comparatively sedate state into a volatility in which we can no longer be sure of what is coming next. Our awe is activated by, and focused on, this very becoming-disfigured. It is the very going-wild that occasions our awe as much as any resultant state.

In other words, our awe bears on the sheer *becoming* of what is happening, its radically protean character. We are witnessing what we not only cannot control but also cannot foretell. The unfolding need not be rapid; the way that certain plants develop are also instances of going wild; even if we possess considerable botanical knowledge, the exact form of the final plant whose development we have been following may well escape our predictive powers. Even the most gentle plant can go wild in its own way.

Sometimes we are awed by particular moments of the process of going wild: the sunset that seems to linger as it hovers over the horizon of an ocean onto which we are looking. But we know that the sun's brilliant display will diminish and eventually disappear. This, too, is part of its *élan vital* as a natural process.

What counts as wild is not only found in the woods or on mountains or at the horizon. Our very existence goes wild in its own ways: the growth of a tumor, the spread of a virus: so much so is this the case, and in so many ways, that we can be said to be ourselves a process of going wild: we *are* wild thanks to continual and often non-conscious *wilderizing* as we can call it. The natural world is *wild all over*. Wildness is found in Central Park (which attracts more species of birds than any other park in the Northeast United States) as well as in the Appalachian mountains (where millions of bison once roamed) but it is also *in us*: indeed, it *is us*. Wildness is all over—wherever we look, wherever we walk, wherever we go, wherever we are. Wildness, as Gary Snyder has it, "is *everywhere*."[6] It is found not only in the high drama that Kant, following an ancient tradition, calls "sublime." It also emerges in much more modest settings such as the small open tanks in the Sea Center in Santa Barbara.

Sixth Impression

Answering to all the wildness we experience in very different settings is the emotionality of awe itself. Awe ranges from sudden surprise to a state of being astounded. Just as wildness is found all over, so is the awe occasioned by it that arises in many different affective forms. The more we acknowledge the extent of awe, the more attuned we become to our environing world in all its variety. Awe, dramatic or modest, takes us into this world and allows us to see it with fresh eyes.

It follows that we should be open to moving freely back and forth between various forms of awe, each of which represents a form of openness to wildness of a particular stripe: a form of *bewilderment* in Powers' term. Awe is indispensable to our connecting with diverse forms of wildness—which means different ways of existing on terms that are not dictated by human interests and concerns. Trusting the emotionalities that ensue from the awesome in its diverse forms, we can enter into horizons in ways that will instruct us about what it is like to exist in worlds not of our own devising or direction. We will be increasingly expansive in the ways we take in the natural world, including its quite mundane manifestations—instead of always looking for new ways to exploit this world in an extractive and dominating spirit.

Aristotle said that philosophy begins in wonder.[7] One way of honoring this proto-axiom is to appreciate and cultivate the many modes of awe which the natural world occasions in us. Experiencing awe in these multiple ways can be an effective way of avoiding the kind of paralysis that has been happening all too frequently on this planet in the last several centuries as we face radical climate change but fail to face it down.

It follows that we must more fully recognize the importance and pervasiveness of awe in its full scope instead of limiting it to our reaction to hyper-dramatic situations. Otherwise put, we have to learn to appreciate the diversity of ways in which we humans, and all living things, can go wild.

Empathy

First Impression

Empathy is the emotionality that animates caring for others. In empathic caring, several criteria are at stake: attentiveness, persistence, generosity. In an emergency, our caring for someone may be momentary; but to continue to care for that same person takes a longer commitment on our part; it means caring for them through thick and thin, not just for now but indefinitely, with no effective limits. For we are caring for their *well-being*, which is not just something momentary but also a lasting state.

I here think of the way in which Dr. Bruce Anderson cared for my sister Connie over a period of 27 years at the St. Helena State Hospital. During that period, she was in and out of the hospital many times, but each time she was admitted, Dr. Anderson agreed to care for her without hesitation. This was not always pleasant or rewarding, far from it; my sister's mental troubles meant that she was very difficult to relate to and rarely grateful for Dr. Anderson's arduous interventions. But he persisted, employing both traditional medications and ECT at a time when the latter was a largely untested procedure. He never wavered in his care. This was more than attending to someone in a very demanding state; it was an extraordinary exercise of empathy with her troubled existence. He *felt his way* into her deeply disturbed state and came to understand it better than any other professional caretaker was ever able to do.

To care for an asylum seeker who is not familiar with American ways of life, and who may know very little English, is to take on a commitment that can last for decades. Members of the Stony Brook Asylee Advocate group have learned this in spades, looking after those who have been granted asylum after being held for years in detention, and then suddenly released with no money, no housing, and no work in sight. Former detainees need help in getting off the ground, but they may also call for continuing support of many kinds: getting a green card, obtaining medical insurance, paying taxes, and much more. And if other members of an asylee's family join him or her, the complexities compound and the need for ongoing care only increases. (I speak here from eight years of experience helping my Nigerian asylee friend, Chris Ebele, enter American life with some genuine success after many daunting challenges.) Caring for immigrants teaches us that it is one thing to gain access to the United States—this is often already a major struggle as we witness presently at La Frontera, the U.S.-México border— and something else again to get settled in this country: for which many kinds of dedicated care and an abundance of empathy are often required.

All these forms of helping others in distress require empathy. Empathy has many modalities, indicating that it is not a single action or kind of action. Empathy in German—where the word was developed by Theodore Lipps in the late nineteenth and early twentieth centuries—is *Einfühlung*.[8] The etymology suggests that in empathy I *feel at one with* another person: *Einfühlung*. Thanks to empathy, I enter into their suffering. I become one with them not ontologically—that would require an act of thaumaturgy— but emotionally: *my* emotion is not the same as *theirs* but responds to it without corresponding with it. Empathy is a two-way dialectic between my feelings and those of another. It is my way of dealing emotively with a situation in which the suffering of someone else "speaks to me" from my being present at, hearing out, or just noticing signs of the other's distress.

Empathy is not only psychological. Crucial to it is *bodily enactment*. Empathy is not just something I *think* I should do but something I *feel* I must do, even to the point of physically sensing it in my bones. My body is telling me: help this person and do it right here and now in a fully embodied action. I feel I have to take action, and this action is bodily. I come to the assistance of the suffering other, and I do so by moving my body into a place that will indeed help the other in their suffering.

By way of embodiment, I am *here for the other*: right here and now, with my own lived body. This means that I am also *there for the other*: where "there" means putting myself in the place of the person with whom I empathize, actively imagining what that place is like. When I say to another in such a situation that "I am there for you," it means that I am willing to go to some considerable lengths to help you recover from, or least to lighten the burden of, your affliction.

I link into the other's affective life in order to do two things: to grasp better what they are suffering from; and to relieve them from this suffering if I am in a position to do so. This is only rarely an aggressive intervention except in a scene of violence where my coming forward might make a difference, as when in Chicago I once moved in a menacing way toward a couple whose male member was threatening a woman—and the male ran away. More often, it is a delicate operation that requires nuanced expressions and gestures on my part. If I have the time, I seek to understand the source of the other's distress and, as a consequence, to act to diminish the distress. Length of time is not here a primary parameter. An empathic response can come all at once, in the present moment (as when a brief remark relieves the other of their current suffering) or can be spread out over a lifetime of caring concern (as when my partner Mary continues to care for a daughter who suffers from premature macular degeneration: a caring that has been there since she was first adopted as a baby in India).

Another dimension of empathy emerges when I *find myself through the other*—through my empathic connection with them. This is not always the case, and it rarely happens with just a few brief encounters with the other. Rather, it is by engaging in a regular practice of empathizing with suffering others that I *become myself* as a caring person. This is to take on an enhanced new identity. This identity need not be incompatible with that of my prior self; it is likely to have deep roots here; but it can also be something radically new in my life. Whichever is the case, my personal identity alters in keeping with my active empathic life. I become who I am as a direct reflection of my empathic bonding with others. All conscientious teachers experience this, even if they may not employ the specific language of "empathy." I become who I am—emerging in the present and the future—through my empathic outreach to others who call, implicitly if not explicitly, to be cared for: whether students or family members or a homeless person we have recently befriended.

Second Impression

Empathy, like awe, extends beyond the human realm. Indeed, these two emotionalities, seemingly so different when we think of awe as situated squarely in nature and empathy as happening between humans, converge when we come to care for living occupants of the natural world before which we find ourselves in awe but which are in deep distress today due to the ravages of climate change and other situations of deprivation and disaster. Thanks to our empathic powers, we are drawn to find ways to care for endangered species as well as the devastated landscapes in which they live. Our empathic interventions can address an entire class of suffering creatures as well as a single injured animal or plant. In my Kansas youth, there was a major movement to restore bison to the Flint Hills. The empathy that lay behind this public outreach came from the growing sense that the entire species was in danger of extinction—whereas in the nineteenth century, they had existed in the millions all the way from Tennessee to Montana. Contemporary Midwesterners became increasingly alert to the difficult living conditions of the bison and responded with a widespread public effort that succeeded in drawing upon major federal support and bringing back large numbers of these creatures.

Once we open the empathic field in this way, there is little limit to its potential extent: not only is any species of animal fair game for empathic support, but such support construed as caring-for can extend to birds and insects, indeed to trees and flowers. It includes creatures under the Earth's surface—snakes, insects, fungi, lichen, even bacteria. Here the expansion of emotion leads us to appreciate an enlarged field of care with which we, as fellow creatures, can empathize. The scope of this field of care includes the Earth's surface itself in the form of its poisoned and deteriorating soil, its heavily polluted oceans, indeed the sky itself as contaminated by excessive amounts of CO_2 and methane. Increasingly, we are called upon to extend the scope of empathy beyond the limits that humans themselves impose upon it. A basic awe for all these domains remains as a deeply motivating factor, felt in the very face of so much environmental destruction, indeed heightened there as what is awesome on its own terms is undermined by human machinations and calls for empathic intervention on our part to save its awesome aura.

In the context of understanding the mystery of emotion better, we have here direct and massive testimony to what I would underscore as the *outwardness* of certain emotions: their being precipitated by phenomena out and around us, and from there becoming something that we experience as happening *to us* rather than *in us*. Among these emotions, awe and empathy are leading but not exclusive cases in point: others include a special form of environmental compassion that is expressly ethical in its motivation, an exuberance that is keyed by a lively scene we are witnessing, a certain depression occasioned by films of environmental degradation, and a spirit of environmental activism as inspired by figures such as Winona LaDuke (Anishinaabekwe/Ojibwe), Naomi Klein, and Bill McKibben.[9]

Third Impression

Integral to the combination of awe and empathy to which I here point is a certain wildness on the part of those creatures and places that engage us. Here "wild" signifies nothing unruly or chaotic but rather what belongs intrinsically to the non-human natural world—thus to what cannot be construed in strictly human terms but which draws our attention and concern. As we come to know this world more fully, we realize that some things we had formerly considered escaping our experiential limits altogether we can come to understand in terms that make eminent sense to us. One such circumstance is found in the forms of communication that transpire among animals and among plants—and between both of these and human beings. At stake here is not just a sending of sounds regarded as signals according to a preexisting code. For the sounds are immediately *understood* by the recipients who hear or see them—just as much as young human children, before they acquire language, are able to comprehend their parents' communications with them—as conveyed in facial or hand or full body gestures.[10]

An especially striking case is communication between different members of a given animal species. The signifiers at stake here are not words but *affects*, with the result that in a communicative exchange between animals, it is a matter of the transmission of emotion—understandable by members of the same species who learn how to engage in such transmission from an early age. Naturalist Gay A. Bradshaw maintains a sizeable private animal park in Oregon, where she has observed how the death of a mother elephant has occasioned profound grief and rage among her surviving children—intense emotions which they communicated forcefully among themselves.[11] In many other species as well we can observe communication by means of the exchange of affect. Cynthia Willett has documented this extensively in her book *Interspecies Ethics*, where she maintains that "affect transmission" is an effective semiotic medium for intra- as well as interspecies communication.[12] Regarding the latter she asks: "Are there not multiple channels for biosocial exchange across diverse species? [...] multimodal flow[s] of affect attunement in mixed species societies?"[13] Willett's interest is not just in communication between a pair of animals but "affect-laden conversations" that "weave substantial threads of a communicative ethics across [entire] regions of the biosphere."[14] These communicative exchanges are based on "attunement through rhythms, smells, and affect-laden tones."[15] The affects that are the vital medium of these exchanges are not altogether discrete emotional units. Instead, they form what Willett—drawing on Brian Massumi's recent work—calls "affect clouds" and "atmospheric attunement."[16] These indicate that what is at stake is not just the transmission of one discrete affect or another but also whole *clusters of affects* that together signify entire states of emotional being on the part of the animals who communicate in this way and that become building blocks of entire animal societies.[17]

Other non-animal species also engage in close communication. On the one hand, trees have been shown to communicate to each other oncoming dangers. As Peter Wohlleben states, "trees communicate by means of olfactory, visual, and electrical signals."[18] On the other, fungi are known to form entire communities underground that communicate with other such communities nearby.[19] In neither case is affect transmission as such at stake; but there can be no question but that significant forms of intra-species and interspecies semiosis are here actively present.

All this has direct consequences for empathic caring. First, it shows that such caring does not require *language* to be realized. It can transpire entirely at the level of the exchange of affects or other non-affective signals. Second, such an exchange can take many forms and occur in many modalities: linguistic, affective, and semiotic signaling. Third, it is precisely their pluri-valence that gives them a flexibility that suits many situations of caring-for that do not follow a simple or predictable pattern: as when we are not sure just what kind of care we should give to a friend who has fallen ill with a so far undiagnosed disease. But the very fact of being concerned with her situation and offering our concrete assistance suffices to make it clear that we care for her well-being and wish to contribute to it in some concretely appropriate way by bodily actions we take that enhance her quality of life at the time. Otherwise put, I and other humans and animals are *affected* by the suffering of others, and our heartfelt response is likely to be received by those same others as a gesture of genuine caring.

Conclusion

Much as awe is occasioned by a vast variety of scenes we experience in very different places, so empathic caring is not restricted to human beings but is solicited in highly diverse populations of the natural world, human and non-human alike. Awe and empathy, seemingly at first glance such radically different emotionalities, show themselves to converge in unexpected but highly significant ways.

That which occasions awe, rather than being an altogether autonomous event as we are often tempted to think (e.g., as when we are "overwhelmed" by a stunning view of mountains), shows itself to be something we deeply care about, thus which induces active empathy—especially when it is a matter of situations of destruction or degeneration. We care for the well-being of those who are in danger and for the restoration of damage done to buildings and the larger landscape. At the same time, we humans come to care about the fate of other species, animal as well as plant, especially those in danger of extinction.

In the end, awe and empathy converge even if they are far from identical. They overlap at their edges. This happens above all in natural settings and is facilitated by the presence of wildness in one or another of its many modes of manifestation. The wildness gives each its own characteristic spontaneity and unpredictability. But such wildness is mitigated by place considered a coherent scene of appearance as well as by the intervention of the lived body—both of which help to make the wild make sense. But the sense thus made is never total, given that wildness has a life of its own that is never entirely congruent with human life, much as we might wish it were otherwise.

To engage in awe and empathy is at once to acknowledge the abiding factor of the wild and to offer to us ways to cope with it effectively as well as creatively. Each offers us *ways into the wild*—awe more expressly so than empathy—and for this we should be grateful. They may not be salvational, but without them our lives would be very considerably diminished.

Notes

1 *See* Edward S. Casey, *The World at a Glance* (Bloomington: Indiana University Press, 2007). Part of my claim in this book is that a bare glance can take in entire environments and placial complexes.
2 Exchangeability obtains only for what Whitehead terms "sites," which are reducible to measurable and unchanging volumes of space that are indifferent as to their contents. *See* Alfred North Whitehead, David Ray Griffin, and Donald W. Sherburne, *Process and Reality: An Essay in Cosmology* (New York: Free Press, 1978), 72, 92, 105.
3 Wallace Stevens, "Of Mere Being," in *The Palm at the End of the Mind: Selected Poems and a Play by Wallace Stevens*, ed. Holly Stevens (New York: Vintage Books, 1990), 398.
4 On ecoproprioception, *see* Chapter 17 by George Quasha, "Languaging Body by Field: Ecoproprioception" as well as Quasha's earlier treatment in George Quasha and Edward S. Casey, *Poetry in Principle* (New York: Spuyten Duyvil, 2019), 67–84.
5 Richard Powers, *Bewilderment: A Novel* (New York: Norton, 2021), 17–18. His italics.
6 Gary Snyder, *The Practice of the Wild* (Washington: Shoemaker Hoard, 1990), 15. My italics.
7 Aristotle, *Metaphysics*, Book 1, section 982b, 1.
8 Theodor Lipps, *Zur Einfühlung* (Leipzig: Verlag von Wilhelm Engelmann, 1913).
9 *See* the description of one such scene at Jalama Beach I give in Edward S. Casey, *Turning Emotion Inside Out: Affective Life beyond the Subject* (Evanston: Northwestern University Press, 2022), 186–187, 205. *See also* Winona LaDuke, *All Our Relations: Native Struggles for Land and Life* (Cambridge, MA: South End Press, 1999).
10 This is what Julia Kristeva calls the "semiotic stage" in human infancy: "a psychosomatic modality of the signifying process." *See* Julia Kristeva, *Revolution in Poetic Language* as reprinted (in part) in Julia Kristeva and Kelly Oliver, *The Portable Kristeva*, European Perspectives (New York: Columbia University Press, 2002), 38.
11 *See* G. A. Bradshaw, *Elephants on the Edge: What Animals Teach Us about Humanity* (New Haven: Yale University Press, 2009).
12 Cynthia Willett, "Affect Attunement: Discourse Ethics across Species," in *Interspecies Ethics* (New York: Columbia University Press, 2014).
13 Ibid., 80–81.
14 Ibid., 82.
15 Ibid., 83.

16 Brian Massumi, *Politics of Affect* (Cambridge; Malden: Polity, 2015). I develop my own view of emotion as something given to us from without rather than altogether generated within us in my *Turning Emotion Inside Out*.

17 It is a matter of "social bonds" between communicating animals who through such attunement form their own versions of "intimate and political solidarities" that constitute genuine political *communities* (Willett, "Affect Attunement," 82). These communities can be constituted by animals of the same species (as with the elephants discussed by Bradshaw) but also, though more rarely, between members of different species who inhabit the same geographical region and thus share certain places in common.

18 Peter Wohlleben, *The Hidden Life of Trees: What They Feel, How They Communicate—Discoveries from a Secret World* (Vancouver; Berkeley: Greystone Books, 2015), 12.

19 Merlin Sheldrake, *Entangled Life: How Fungi Make Our Worlds, Change Our Minds and Shape Our Futures* (New York: Random House, 2020). For a further study of such communication, *see* Michael Marder and Edward S. Casey, "The Shared Sociality of Trees, with Implications for Place," to appear in a volume edited by David McCullough.

Bibliography

Aristotle. *Metaphysics*, edited by W.D. Ross. Oxford: Clarendon Press. 1924.

Bradshaw, G. A. *Elephants on the Edge: What Animals Teach Us about Humanity.* New Haven: Yale University Press, 2009.

Casey, Edward S. *The World at a Glance.* Bloomington: Indiana University Press, 2007.

———. *Turning Emotion Inside Out: Affective Life beyond the Subject.* Evanston: Northwestern University Press, 2022.

Kristeva, Julia, and Kelly Oliver. *The Portable Kristeva.* European Perspectives. New York: Columbia University Press, 2002.

LaDuke, Winona. *All Our Relations: Native Struggles for Land and Life.* Cambridge, MA: South End Press, 1999.

Lipps, Theodor. *Zur Einfühlung.* Leipzig: Verlag von Wilhelm Engelmann, 1913.

Massumi, Brian. *Politics of Affect.* Cambridge; Malden: Polity, 2015.

Powers, Richard. *Bewilderment: A Novel.* New York: Norton, 2021.

Quasha, George, and Edward S. Casey. *Poetry in Principle.* New York: Spuyten Duyvil, 2019.

Sheldrake, Merlin. *Entangled Life: How Fungi Make Our Worlds, Change Our Minds and Shape Our Futures.* New York: Random House, 2020.

Snyder, Gary. *The Practice of the Wild.* Washington: Shoemaker Hoard, 1990.

Stevens, Wallace. "Of Mere Being." In *The Palm at the End of the Mind: Selected Poems and a Play by Wallace Stevens*, edited by Holly Stevens, 398. New York: Vintage Books, 1990.

Whitehead, Alfred North, David Ray Griffin, and Donald W. Sherburne. *Process and Reality: An Essay in Cosmology.* New York: Free Press, 1978.

Willett, Cynthia. "Affect Attunement: Discourse Ethics across Species." In *Interspecies Ethics*, 80–99. New York: Columbia University Press, 2014.

Wohlleben, Peter. *The Hidden Life of Trees: What They Feel, How They Communicate—Discoveries from a Secret World*. Translated by Jane Billinghurst. Vancouver and Berkeley: Greystone Books, 2015.

20

ENWORLDING PLACE DANCES AND POTATOES

Sondra Fraleigh

Snow Canyon, Utah, U.S.A.: 37.2145° N, 113.6402° W
Yokohama, Japan: 35.4437° N, 139.6380° E
Circleville, Utah, U.S.A.: 38.1688° N, 112.2696° W

> I wait at the start,
> pause, and hold my horses,
> til breath finds me.

Placescapes: Autoethnography Meets Phenomenology

My body is *spatializing*, tight or diffuse, giving me access to environments and mentalities I live and dance. My body lives its somatics of space and time, especially attuned when I pay attention to the poiesis of place in making and doing dances. I am drawn to places that encourage creative abundance in consciousness, "placescapes" both built and untended that challenge and decenter my ordinary self-orientations.[1]

Audrey Lane Ellis writes about corporeal activations of dance improvisation in descriptive tones of analytical phenomenology. She describes somatic, environmental affordances of dancing on the road. I appreciate the relational aspects of her description and practice:

> This body, as a tight place, practices desperate leaps out of itself; never transcending itself to achieve a particular end, but trans-forming its own idea. All of the impossibilities of my body-as-wolf are right there for my playful imaginary, activated by and through movement. I am both hopeful and fearful of what I might become; going down the road once again is the opportunity the improvisational frame offers. The road is not a progressive accumulation of corporeal knowledge or a material

DOI: 10.4324/9781003390985-25

transcendence, but a way of being in relationship to myself, others, and my environ-mentality.[2]

We are already emplaced in familiar places, in quiet places inviting us to stop chasing thoughts, noisy cityscapes, hardscapes, and ruralscapes. I don't forget that places are spaces, a key aspect of phenomenology; as Edward S. Casey writes: "There is no position outside of space... just as there is no space without a place."[3]

What Are Landscapes?

Landscapes are tints of countryside and places of appeal,
Bits of imagination bleeding through grass and rock.
Or they spill across a canvas stain, across some memory
forgotten in the landscapes of the mind.
Landscapes are sad lakes in the air,
romance and calm dark wine.

Ecosomatics locates experiential methods and practices that respect sensate knowledge of particular places. I use "placescape" through Casey's meaning, not simply a matter of relative position, not a *site*, and probably not a remote romanticized landscape. I welcome his view of "finding soul in place."[4] Places have souls, whether acknowledged or not. These arise through history, affective spirit, and tactile geography. Gaston Bachelard's classic book, *The Poetics of Space* (1964), articulates space and place through living narratives.[5] He focuses primarily on bodily lived emotional responses and evocative feeling tones of places like basement cellars, kitchens, and gardens. Integrating the experiential philosophy of John Dewey, Adesola Akinleye extends the work of somatic emplacement toward cityscapes and architecture.[6]

Placescapes have geographies; they are somewhere, not just anywhere. Likewise, nature is specific and somewhere—outside and within. In these autoethnographic practice pages, I dance in placescapes of nature to enhance my awareness of my own nature and remind myself of vastly ignored histories in places that suffer. I might fall in love with places through dance and performance practice or learn from them. Japan has been such a place for me. There I learned how to dance in new ways and meditate near pools, greenery, and raked rocks.[7] Concurrently, I studied America's tangled history with Japan; its atomic bombing of Hiroshima and Nagasaki and horrific firebombing of Tokyo.[8] Now rebuilt, Japan calls for bans of all nuclear testing as documented in letters to world leaders displayed on the Hiroshima Peace Memorial Museum walls.

Place Dances are performances of perceiving and remembering the soul and history of particular places. When I map (or plan) Place Dances beyond four walls, I affirm my human belonging to the naturalcultural world, whether its wild rhizomatic beauty and feral ecologies, or the perils of intervening weather.[9] Such dances recognize resonant places in promise of well-being and pay attention to problems of exploitation. Place Dances are performed for the experience of participants, audience engagement, and environmental activism. My earlier essay, "Canyon Consciousness," studies human affinities with nature and place in the rugged beauty of Utah's endangered canyonlands.[10] This is also a localized theme in my recent teaching, as shown in Figure 20.1.

FIGURE 20.1 Fraleigh's senior students in a Yoga Tree Circle in Snow Canyon, experiencing empathies with sandstone, lava rock, and each other in ecosomatic community. Photograph by Sondra Fraleigh.

Snow Canyon seems not to have been inhabited in the past. The harsh desert environment prohibited permanent settlement. Evidence of Native American use of the environment exists as far back as 500 B.C.E. in the form of arrowheads, pottery shards, grinding stones, fire pits, and petroglyphs. The Anasazi used the canyon for hunting and gathering from 200 to 1250, and the Paiute used the canyon from 1200 to the mid-1800s. They were using it still when Mormon settlers entered Snow Canyon in the 1850s while searching for lost cattle.[11]

In the 1950s, this timeless red sandstone canyon was dusted with nuclear fallout, enduring 17 years of tests blowing on the wind toward Southwest Utah from Frenchman's Flat testing ground in Nevada. The government officials waited for the wind to blow away from densely populated Los Angeles before they detonated the bombs. My family and I were part of the more or less 50,000 expendables in Nevada and Utah, but the fallout traveled more widely than expected. The nuclear clouds traveled as far as Troy, New York, and extended into Canada. I have studied this from extant sources and written more personally about my proximity to this testing, including the effects of nuclear fallout on my family of origin and me.[12] The detonations began when I was 12. I lived closer to the bombs in my early college years in Cedar City, and there I suffered long bouts of strep throat and unexplained seizures. These gradually subsided when I moved away.

Now in semi-retirement, I live once again in the vicinity and embrace my return with passion and healing. In the images below, dancers perform somatic affinities with Snow Canyon's immense swirling careens and the pooling brooks of Pine Valley Mountain (*see* Figures 20.2–20.4).[13] As a prelude to our dances, we ask permission of the colored canyon, its pools, and brooks to perform there, and we listen for responses in answers that come in feelings, not words. Asking shows respect in a gesture and reminds us that we are not tourists or trespassers. We have come to this place to fall in love, surrender to the earth, and wake its healing waters.

Networks of Nature and Culture

My perspective on Place Dances benefits from Watsuji Tetsurō's Japanese phenomenology of nature (in the wake of Heidegger), where being-in-the-world is relational and the *space of self* is a *network* of nature and culture.[14] One cannot have a philosophy of somatics without a philosophy of nature and culture. In this short autoethnography, I delineate nature in terms of character and present-centered consciousness. Nature is an interwoven dance of tendencies and potentials as much as anything. The founder of phenomenology, Edmund Husserl, describes the *emergent*

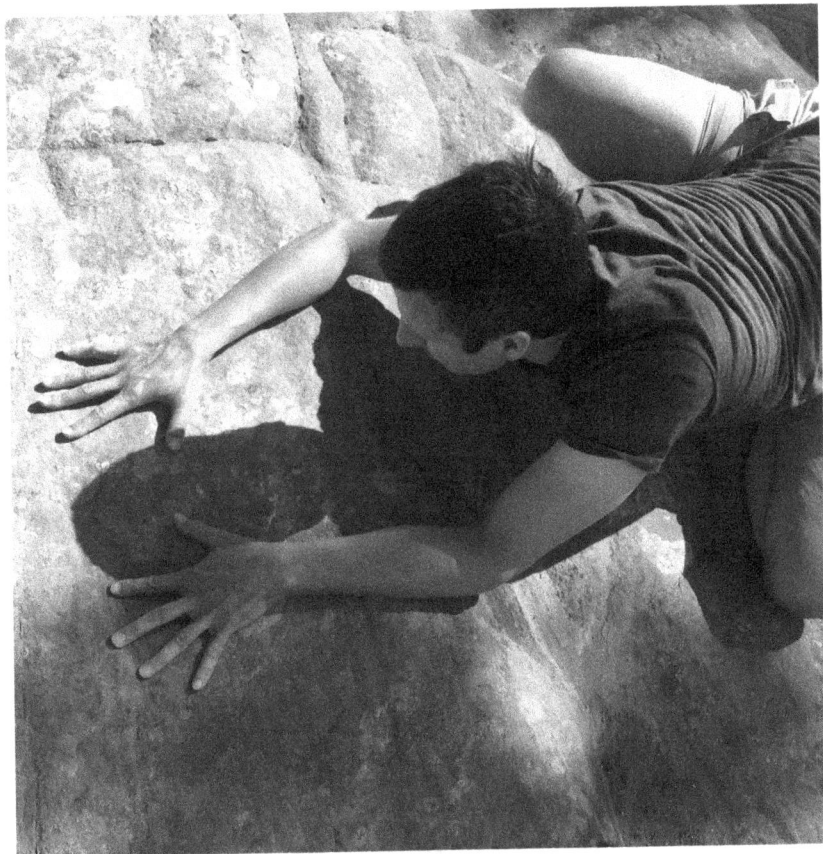

FIGURE 20.2 Robert Bingham on the red sandstone in *Sounding Earth* (2016). Photograph by Sondra Fraleigh.

kinetics of "the flowing live present" in bodily convergence with the environing world.[15] Movement guides his sense of world: "Consciousness of the world [...] is in constant motion."[16] Husserl's term *enworlding* (*verweltlichung*) indicates the ongoing nature of the world as we live and come to know it.[17] Consciousness is not abstract, except as a concept. It is we who develop and constitute our own sense of the world, and this sense is not finished; ontologically, we enworld as a "*being-tendency*."[18] This tendency leading out, *whither*, and returning, *wherefrom*, is connective and rooted (emphasis in original).[19] Grasped through an ontology of becoming, the world worlds in its tendencies as it emerges and unfolds. It dances and flows through us as we sense and allow its passage.

FIGURE 20.3   Robert Bingham merges with river, sun, and stone in *Pulsing Slumber* (2016). Photograph by Sondra Fraleigh.

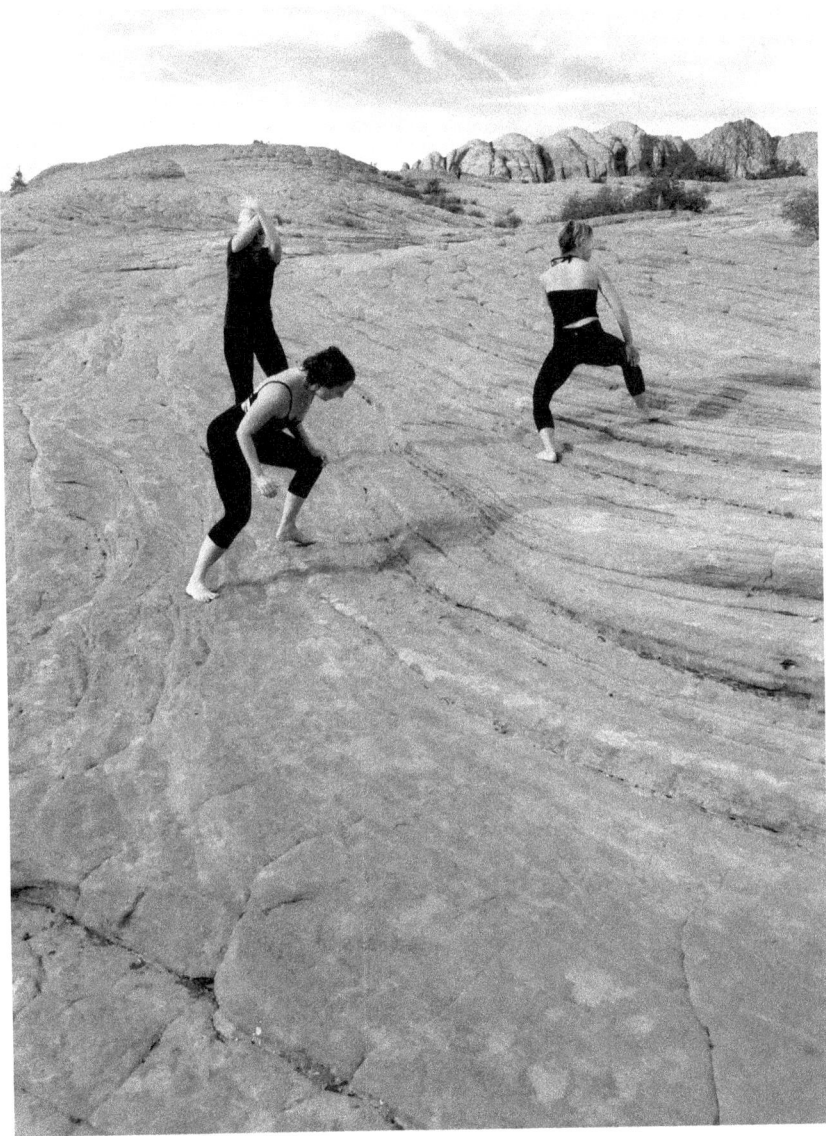

FIGURE 20.4 Dancers improvise with speed across the Navajo sandstone and basalt ridges in *Pulsing Slumber*, Snow Canyon and Pine Valley, Utah. Photograph by Tom Gallo © 2016.

In Practice
Dances in nature
world aplenty, whilst
entangling-culture envies.

Fields blow in the wind
rushing colors together
clusters at a time.

Yes, I chase words and dances
in a zone of communication,
moving past myself.

Grateful to be *Still Here* as choreographer
Bill T. Jones once put this,
Capturing his survival of AIDS
in an activist dance.

Now we cope with another disease on a grand scale,
side-stepping COVID's mutating variants.
Does this virus just want us to stay home?
Or push us toward the suchness of
our home in nature?

Not Wasting
Being grateful for being here now while enworlding like Jones.
Not wasting on wishing, resting "was" and "shall be."
Letting go expectations. Walking
to start over a being-tendency.

Enworlding the Lessons of Nature

Husserl speaks widely and inclusively of ontology. We *experience* and come to *know* the world through acting in it; thereby constituting and possessing a consciousness of the world, whether through the strife of politics or the soil of nature and history. We humans bear every sense of what is meant by *world*, and the lifeworld of nature has ontological being.[20] Husserl's student, Martin Heidegger, writes of nature as *seinlassen* (letting be) and critiques the exploitation of nature in treating it as a *resource*. He makes early predictions of environmental dangers inherent in technological worldviews.[21] Today, Michael Marder picks up this "onto-phenomenological 'letting-be'" in his philosophy of vegetal life.[22]

We can be romantic about nature because it is beautiful beyond imagination, even in its fires and floods. We can also envision nature productively outside of romance, particularly in its tough resilience and adaptive qualities of replenishment. Nature likes weeds as well as flowers, and it likes migrant species as well. Environmental scientist Fred Pearce writes of the importance of migration in nature.[23] Might we also take a lesson from nature in welcoming human immigrants? Nature doesn't favor one color or kind. Its lessons are about renewal, resilience, connectivity, and empathy.

These are lessons of somatic processes as well. I gain lessons in resilience from my Feldenkrais study, lessons that speak to me of an adaptive reserve, as I plum my body's healing potentials and replenish them by working with less effort. The Alexander Technique is also subtly adaptive and emergent. I have incorporated adaptivity from these practices into my somatics teaching, hopefully avoiding the sometimes-obvious patriarchy in the background. I also learn the lessons of replenishment by practicing conscious awareness and cultivating a love of the natural world through movement practices in nature.

Not going far to find nature
Walk outside in the morning
And perform a short gratitude ritual,
moving your arms in gestures that arise.
Practice wherever you are, at home or away.
Be well and create a well of wellness to draw from.
Mine is in transition and interruption.
Thus, I dance weakness,
and I don't push away the suchness
of frailty and infirmity—these
embellishments of my life. My dance
takes lessons from variance and suffering.

Attune to the suffering
Listen: Look: Smell and Touch
the brittle land.
Dance in the char and dirty snow,
deserving,
grieving and gruesome.
Trace your good actions and bare afflictions,
standing rudely muddled in the politics of place.

FIGURE 20.5 Sondra Fraleigh dances in winter for her Japanese mentor Ohno Kazuo-sensei's 100th birthday. Video still by Sondra Fraleigh, © 2006. Photograph by Alycia Bright Holland © 2006.

My butoh mentor Ohno Kazuo-sensei said, "When I dance, I carry all the dead with me." As a Japanese veteran of World War II, he didn't just indicate the Japanese war dead; he emphasized vast geographies of "all the dead." Ohno encouraged respect for those who have gone before us and all we have been given; and thus, I dance for him (*see* Figure 20.5). "You are not the be-all and end-all of life," he said in his Yokohama workshops. His dances and workshop words expressed a longing for origins as "source." Moving with death and morphic darkness, Ohno and Hijikata conjured the suffering, or "bleeding," of nature.[24] Many butohists since have translated this ethos in their own aesthetics of care.

Carry, lift, hold. Respect ....

Enworlding and Decentering Human Tendencies

Enworlding (verweltlichung), as we retrace this word, is Husserl's concept for the *emergence* (the "being-tendency") of consciousness relative to the environing lifeworld.[25] Some may feel threatened by enworlding as loss, that the fragility of the human might dissolve into the-more-than-human.

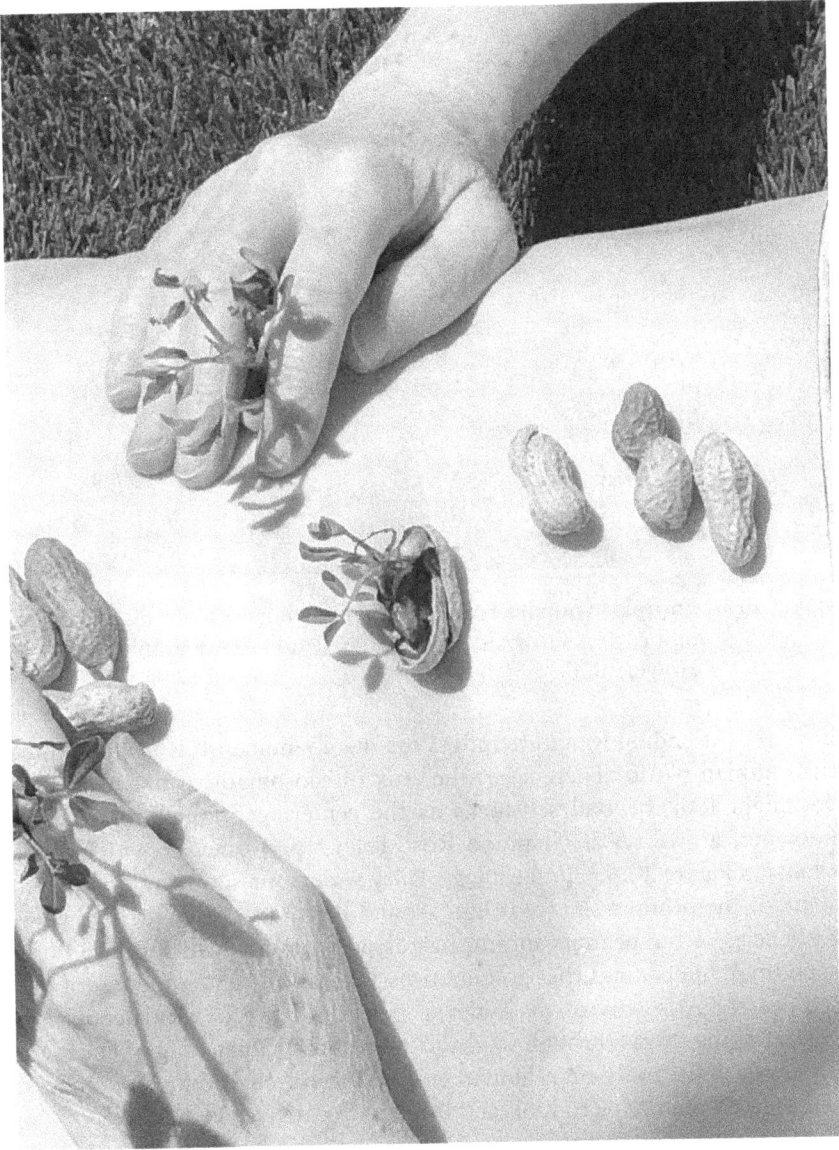

FIGURE 20.6 Shannon Rose Riley performing with sprouted peanuts in *Decentering the Human: A Story of Us*. Photograph by Shannon Rose Riley in a music video by Sondra Fraleigh.

Psychologically, human affects decenter when we focus our attention on the emergent potentials of nature, but some of us may have already sided with those who say there is no such thing as nature, as theorized in much sociology.[26] Human empathies with nature are not easily dismissed,

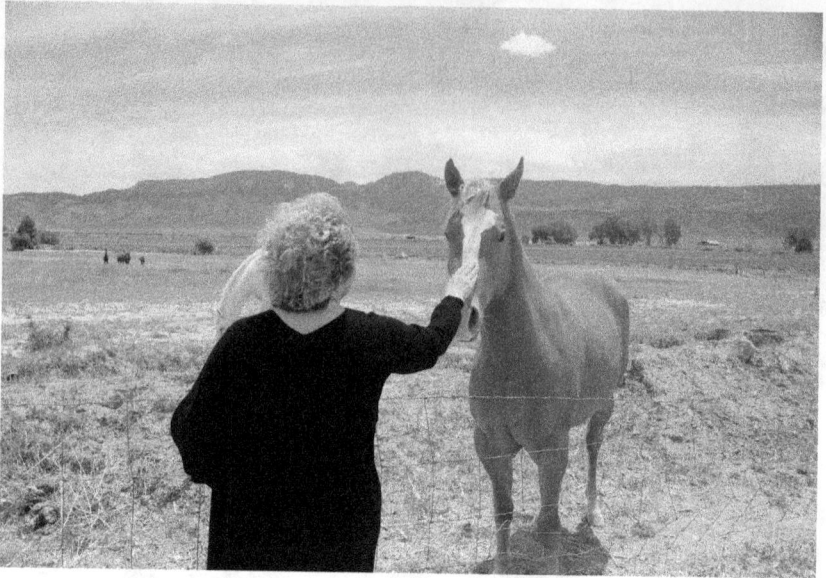

FIGURE 20.7   Portrait of Sondra Fraleigh stroking a horse's nose, place dancing in her hometown of Circleville, Utah. Photograph by Mark Howe, 2010.

however. Attending to nature moves the merely-human toward the more-than-human world. Therein lies the crux of ecosomatic attunement, that we might look beyond self-focus as the centering principle of somatic knowing, as we see in Shannon Rose Riley's performance with peanut shoots in Figure 20.6.[27] In the image, Riley waits for a scrub jay who plants some of the peanuts she feeds him around her house. In time, the being-tendencies of the peanuts sprout, just as potatoes too will sprout. Often, such small happenings that go unnoticed take us off-center and out of our self-centers. Arts activism magnifies such overlooked ecological details. My way of doing this is through workshops on somatic matching of emergent processes in the more-than-human world. These are not difficult practices, but they do require us to look and listen with care (*see* Figure 20.7).

> *Lean into falling*
> way off-center
> into the more-than-human world
> and link to all that is more.
>
> We become human through
> a continuum that is more;

being beside doesn't work.
Nature doesn't belong to us,
and culture will crop up.
Indigenous peoples teach
that we are simply stewards,
the problems and cures of continuities.

Enworlding the world stretches our human natures,
while our collective *geographies* wait at the finish,
pause, and hold their horses.

Home as Placescape and Picking Potatoes

Home lives in the heart, maybe not in a place or even a country. My sense of home spreads out like potato vines meandering underground. I have had many homes, but just one hometown. My first home is a hearty place with humor, many stories, and complicated histories. Circleville is where Butch Cassidy lived during his early life, and most of the tales about his subsequent exploits are not to be believed. I knew his youngest sister, Lula, who was always happy to tell and retell her favorite stories about Butch. She was a lot older than I and charming as ever when Paul Newman invited her to New York during the making of his movie about her brother. She returned home with more stories to tell about Butch, Paul, and N.Y.C. In this small town, people communicate through story and embellishment. My grandmother told me I would need to move away to find a husband because I was related to everyone in town. I traveled to San José, California, New York, and Detroit for this task. My husband's hometown is Detroit during the heydays of big-band jazz. He remembers hearing amazing music with student discounts in clubs throughout his youth, including his favorite singer, Billy Eckstine.

I loved jazz through radio and movies, but my hometown is poles apart from city life. Circleville, 135 miles north of Snow Canyon in southwest Utah, is part of the home of the original Paiute Nation, and the Navajo (Diné) lived in the vicinity when I was young; I remember attending school with Navajo children. The Mormons colonized this verdant valley home nestled in the Rocky Mountains in the mid-1800s. For generations, they lived there with a population of around 600—if you count the dogs and cats (and the two Catholics, as they like to say). Navajos were trucked into town to pick potatoes in the Fall. I scattered sacks for the pickers from my horse and then picked beside them. If you were strong and a good picker, you could earn $60 a day, and I made about $20 on a good day.

The whole town participated in the Fall harvest. Toward the end of a long day, I could be found in the fields sitting on my overturned basket. I couldn't "drag sacks" between my legs and throw the potatoes in behind, as most pickers did. I filled two baskets and poured them into a sack (with help). Gunny sacks (now vintage burlap) were hooked on nails in a belt around the waist, left to gape open, then filled by throwing the potatoes back between the legs as the sack filled up. Of course, this was all done in a bent-over position with the legs wide and staggering apart. I don't know how many pounds a filled sack of potatoes weighed, but it was a lot. This hard work broke backs, and I got used to it from childhood. The Navajo women worked in long velvet-tiered skirts dragging the ground. I sometimes asked them how they could do it as I sat exhausted in the warm dirt. High school was suspended for two weeks for the harvest (and we cheered). Now, most of the work is done by machinery, aided by migrant workers from México.

This little town has a soft and troubled soul. It was one of the poorest and most remote areas of the United States when I grew up there, and it is still poor. Circleville was incorporated as a township twice, first in 1864, but was abandoned two years later in the wake of mass murder. On April 21, 1866, during Utah's Black Hawk War, white settlers brutally murdered 24 Koosharem Paiute men, women, and children who were innocent of any wrongdoing.[28] Now a plaque in the town park remembers the horrible event with shame and apology. Soon after the massacre, Circleville was vacated entirely. It was reestablished in 1874. My ancestors, who had arrived earlier on wagon trains in the Salt Lake Valley, were sent by Brigham Young to the second establishment of Circleville. Some of my family were religious; all were survivors of harsh winters and privation. My great-great-grandmother had a cancerous breast removed in a covered wagon without an anesthetic. Mormon settlers in flight from religious persecution suffered, no doubt, but scholarship makes clear that settler colonialism enables terrible crimes.[29] Indeed, Utah's Black Hawk War consisted of over 150 bloody confrontations that resulted in the deaths of untold numbers of the Timpanogos Nation of the Wasatch, the Paiute, Ute, Navajo, and other tribes who lived on and nurtured the land when the Mormons arrived in 1847.[30]

If You Can Find It

I carry the wind and dusty placescapes of ranch life and the tragedies and meadows of Circleville in lasting tones of sorrow and gratitude. My aunt's crumbling grocery store is shuttered now but still sports a small hand-printed sign in the one remaining window that says, "we've got it if you can find it."

FIGURE 20.8   Sondra Fraleigh, high above Circleville in the hazy background, and halfway up the mountain to Timid Springs. Photograph by Mark Howe, 2010.

I like to visit Circleville (seen from surrounding mountains in Figure 20.8). The quiet potato fields slow me down—and sometimes I find a stray spud to hold in my hands. Near town, the pungent river mud is serene, and when I eye the patchwork geography from the Black Ridge and valley mountains, its ragged beauty makes me cry. As a child, I would ask my father to stop the truck on our bumpy way up the mountain so I could gaze back at the tiny village and the bright spot of my home in the distance, clutching it close to my wobbly body, as though it might dissolve in wonder. Circleville remains a town whose myths outlast its facts—whose geography slips through time with purple mountains still standing, starry nights and flickering sunsets unchanged. New York is approximately 3,000 miles to the east—its human capital awash with hopes of gold and art as we speak. Some things change. But what lasts? For me, it's potatoes. They love me. What have you plucked from the garden of life that edifies you? Can you find it?

For a Somatic Answer

I invite you to close your eyes:
not to remember your life, but to ask what you do not know
that you would like to know.
Pluck the gauze of failure and the heft of melodramatic hurt from
the dark

to ask what is reduced from your sustainable ancient life
that you have no idea is waiting.
Much more than you might expect, I suspect.
Have patience. Listen. Rest all expectations.
Take a deep breath, and clap once.

Notes

1 Edward Casey coined the term, "placescapes." Edward S. Casey, *Getting Back into Place: Toward a Renewed Understanding of the Place-world*, 2nd ed. (Bloomington: Indiana University Press, 2009), 25.
2 Audrey L. Ellis, "From Animation to Activation: Improvisational Dance as Invitation and as Interruption" (Ph.D. Doctoral, Stony Brook University, 2021), 204.
3 Edward S. Casey, *Spirit and Soul: Essays in Philosophical Psychology*, 2nd ed. (Putnam: Spring Publications, 2004), 231.
4 Ibid., 319–341.
5 Gaston Bachelard, *The Poetics of Space* (New York: Orion Press, 1964).
6 Adesola Akinleye, *Dance, Architecture, and Engineering* (London: Bloomsbury Publishing, 2021).
7 Sondra Horton Fraleigh, *Dancing into Darkness: Butoh, Zen, and Japan* (Pittsburgh: University of Pittsburgh Press, 1999); Sondra Fraleigh, *Butoh: Metamorphic Dance and Global Alchemy* (Urbana: University of Illinois Press,, 2010).
8 Sondra Fraleigh, *Dancing Identity: Metaphysics in Motion* (Pittsburgh: University of Pittsburgh Press, 2004), 174–179.
9 "Natureculture," a term that comes from cultural anthropology, primatology, and feminist science studies, refers to a "synthesis of nature and culture that recognizes their inseparability in ecological relationships that are both biophysically and socially formed." *See* Kathryn Ovenden and Nicholas Malone, "Natureculture," *The International Encyclopedia of Primatology* John Wiley & Sons, Inc. (2017), https://doi.org/10.1002/9781119179313.wbp rim0135. The term was originally employed by Fuentes (2010) and Haraway (2003). *See* Agustin Fuentes, "Naturalcultural Encounters in Bali: Monkeys, Temples, Tourists, and Ethnoprimatology," *Cultural Anthropology* 25 (2010), https://doi.org/DOI:10.1111/j.1548-1360.2010.01071.x; Donna Haraway, *The Companion Species Manifesto: Dogs, People, and Significant Otherness* (Chicago: Prickly Paradigm Press, 2003). "Feral ecologies" are ecologies that have been fostered through human-built infrastructures and that typically spread beyond human control. Anna L. Tsing, *Feral Atlas: The More-than-Human Anthropocene* (Stanford University Press, 2021). http://feralatlas.org/.
10 Sondra Fraleigh, "Canyon Consciousness," in *Dance and the Quality of Life*, ed. Karen Bond (Cham: Springer, 2019).
11 Geological history researched by Ruben Wadsworth: www.stgeorgeutah.com/news/archive/2020/11/29/raw-snow-canyon-day-a-breathtaking-land-of-stark-contrasts/#.YTfIGC1h0Us. November 29, 2020. Accessed September 7, 2021. *See also* https://stateparks.utah.gov/parks/snow-canyon/.

12 Fraleigh, *Dancing Identity*, 171–173.
13 Figure 20.2 is included in Fraleigh's music video, *Sounding Earth* (2016), with Robert Bingham; https://youtu.be/Es2xZIR7aFQ. Figures 20.3 and 20.4 are also included in Fraleigh's original music video, Pulsing Slumber (2016), with Robert Bingham, Megan Brunsvold, Denise Purvis, and Sara Gallo. https://youtu.be/qaI5BjGpHDQ.
14 David W. Johnson, *Watsuji on Nature: Japanese Philosophy in the Wake of Heidegger*, Northwestern University Studies in phenomenology and existential philosophy (Evanston: Northwestern University Press, 2019), 14.
15 Edmund Husserl and Eugen Fink, *Sixth Cartesian Meditation: The Idea of a Transcendental Theory of Method*, trans. Ronald Bruzina (Bloomington: Indiana University Press, 1995), xiv.
16 Edmund Husserl, *The Crisis of European Sciences and Transcendental Phenomenology: An Introduction to Phenomenological Philosophy*, trans. David Carr (Evanston: Northwestern University Press, 1970), 109.
17 Husserl and Fink, *Sixth Cartesian Meditation*, li.
18 Ibid., 21.
19 Ibid., 99.
20 Ibid., 167–189.
21 Kelly Oliver, *Earth and World: Philosophy after the Apollo Missions* (New York: Columbia University Press, 2015), 111.
22 Michael Marder, *Plant-Thinking: A Philosophy of Vegetal Life* (New York: Columbia University Press, 2013), 3.
23 Fred Pearce, *The New Wild: Why Invasive Species will be Nature's Salvation* (Boston: Beacon Press, 2015).
24 Sondra Fraleigh, "Butoh Translations and the Suffering of Nature," *Performance Research: A Journal of the Performing Arts* 21, no. 4 (2016), https://doi.org/10.1080/13528165.2016.1192869.
25 Husserl and Fink, *Sixth Cartesian Meditation*, li.
26 Anna Yeatman, "A Feminist Theory of Social Differentiatioin," in *Feminism/Postmodernism*, ed. Linda J. Nicholson (New York and London: Routledge, 1990), 287.
27 Riley's photograph of herself in performance with peanut shoots is included in Fraleigh's music video, *Decentering the Human: The Story of Us* (2021). www.youtube.com/watch?v=5xi-UApq2bM.
28 *See* www.blackhawkproductions.com/circleville.htm.
29 Patrick Wolfe, "Settler Colonialism and the Elimination of the Native," *Journal of Genocide Research* 8, no. 4 (2006). For contemporary scholarship on settler colonialism and the Mormons in Utah, *see* Gina Colvin and Joanna Brooks, *Decolonizing Mormonism: Approaching a Postcolonial Zion* (Salt Lake City: The University of Utah Press, 2018); P. Jane Hafen and Brenden W. Rensink, *Essays on American Indian and Mormon History* (Salt Lake City: The University of Utah Press, 2019).
30 Most of the histories of the Black Hawk War take the perspective of Mormon settlers. For a discussion of Indigenous perspectives, *see* Phillip B. Gottfredson, *My Journey to Understand ... Black Hawk's Mission of Peace* (Bloomington: Archway Publishing, 2020). *See also* https://historytogo.utah.gov/black-hawk-war/ and www.blackhawkproductions.com/.

Bibliography

Akinleye, Adesola. *Dance, Architecture, and Engineering.* London: Bloomsbury Publishing, 2021.

Bachelard, Gaston. *The Poetics of Space.* New York: Orion Press, 1964.

Casey, Edward S. *Spirit and Soul: Essays in Philosophical Psychology.* 2nd ed. Putnam: Spring Publications, 2004.

———. *Getting Back into Place: Toward a Renewed Understanding of the Place-World.* 2nd ed. Bloomington: Indiana University Press, 2009.

Colvin, Gina, and Joanna Brooks. *Decolonizing Mormonism: Approaching a Postcolonial Zion.* Salt Lake City: The University of Utah Press, 2018.

Ellis, Audrey L. *From Animation to Activation: Improvisational Dance as Invitation and as Interruption.* Ph.D. Doctoral, Stony Brook University, 2021.

Fraleigh, Sondra. *Dancing Identity: Metaphysics in Motion.* Pittsburgh: University of Pittsburgh Press, 2004.

———. *Butoh: Metamorphic Dance and Global Alchemy.* Urbana: University of Illinois Press, 2010.

———. "Butoh Translations and the Suffering of Nature." *Performance Research: A Journal of the Performing Arts* 21, no. 4 (2016): 61–71. https://doi.org/10.1080/13528165.2016.1192869.

———. "Canyon Consciousness." In *Dance and the Quality of Life,* edited by Karen Bond, 23–44. Cham: Springer, 2019.

Fraleigh, Sondra Horton. *Dancing into Darkness: Butoh, Zen, and Japan.* Pittsburgh: University of Pittsburgh Press, 1999.

Fuentes, Agustin. "Naturalcultural Encounters in Bali: Monkeys, Temples, Tourists, and Ethnoprimatology." *Cultural Anthropology* 25 (2010): 600–624. https://doi.org/DOI:10.1111/j.1548-1360.2010.01071.x

Gottfredson, Phillip B. *My Journey to Understand ... Black Hawk's Mission of Peace.* Bloomington: Archway Publishing, 2020.

Hafen, P. Jane, and Brenden W. Rensink. *Essays on American Indian and Mormon History.* Salt Lake City: The University of Utah Press, 2019.

Haraway, Donna. *The Companion Species Manifesto: Dogs, People, and Significant Otherness.* Chicago: Prickly Paradigm Press, 2003.

Husserl, Edmund. *The Crisis of European Sciences and Transcendental Phenomenology: An Introduction to Phenomenological Philosophy.* Translated by David Carr. Evanston: Northwestern University Press, 1970.

Husserl, Edmund, and Eugen Fink. *Sixth Cartesian Meditation: The Idea of a Transcendental Theory of Method.* Translated by Ronald Bruzina. Bloomington: Indiana University Press, 1995.

Johnson, David W. *Watsuji on Nature: Japanese Philosophy in the Wake of Heidegger.* Northwestern University Studies in phenomenology and existential philosophy. Evanston: Northwestern University Press, 2019.

Marder, Michael. *Plant-Thinking: A Philosophy of Vegetal Life.* New York: Columbia University Press, 2013.

Oliver, Kelly. *Earth and World: Philosophy after the Apollo Missions.* New York: Columbia University Press, 2015.

Ovenden, Kathryn, and Nicholas Malone. "Natureculture." In *The International Encyclopedia of Primatology.* John Wiley & Sons, Inc. (2017): 1–2. https://doi.org/10.1002/9781119179313.wbprim0135.

Pearce, Fred. *The New Wild: Why Invasive Species will be Nature's Salvation.* Boston: Beacon Press, 2015.

Tsing, Anna L. Feral Atlas: The More-than-Human Anthropocene. Stanford University Press, 2021. http://feralatlas.org/.

Wolfe, Patrick. "Settler Colonialism and the Elimination of the Native." *Journal of Genocide Research* 8, no. 4 (2006): 387–409.

Yeatman, Anna. "A Feminist Theory of Social Differentiation." In *Feminism/Postmodernism*, edited by Linda J. Nicholson, 287–305. New York; London: Routledge, 1990.

INDEX

For Product Safety Concerns and Information please contact our EU
representative GPSR@taylorandfrancis.com
Taylor & Francis Verlag GmbH, Kaufingerstraße 24, 80331 München, Germany